Advances in Systematic Creativity

Leonid Chechurin • Mikael Collan
Editors

Advances in Systematic Creativity

Creating and Managing Innovations

Editors
Leonid Chechurin
School of Business and Management
Lappeenranta University of Technology
Lappeenranta, Finland

Mikael Collan
School of Business and Management
Lappeenranta University of Technology
Lappeenranta, Finland

ISBN 978-3-319-78074-0   ISBN 978-3-319-78075-7   (eBook)
https://doi.org/10.1007/978-3-319-78075-7

Library of Congress Control Number: 2018950070

This Palgrave Macmillan imprint is published by the registered company Springer Nature Switzerland AG
The registered company address is: Gewerbestrasse 11, 6330 Cham, Switzerland

Preface

The general concept of systematic creativity is based on the simple idea that the creative effort of creating innovations and new designs and ideas can be made in a systematic way, and on the premise that a systematic way of performing the creative process increases the efficiency of the process and enhances the likelihood of ending up with good outcomes. One of the well-known methods for systematic creativity is the "Theory of Inventive Problem Solving," or shortly abbreviated TRIZ after the original name of the method in Russian. Many contributions in this book are connected to TRIZ, but many are general in the sense that they can be used also with other available methods for ideation and systematic creativity. This is especially true for the contributions that describe the management and commercialization of innovations, where the focus is not on any specific ideation or innovation creation method but on the management and commercialization of the created ideas and innovations themselves.

The topics covered are important to the industry from the point of view of the ability to sustain a competitive advantage, because innovation cycles have become shorter and there is constant need for more innovations. The ability to master tools for systematic creativity and efficient management of innovations are pivotal in keeping up with the competition. It can be said that only companies that are able to constantly redefine their products and services are able to stay in business in the highly competitive technical fields we find today in almost every industry.

The authors presenting their work in the book hail from more than ten countries and the industries covered in the examples and illustrations are numerous. The book comprises eighteen chapters divided into three parts.

Part I is dedicated to advances in and applications of the theory of inventive problem solving, better known as TRIZ. These nine chapters present a diversity of novel approaches to extend and to support the use of TRIZ in systematically creating innovations.

With the collapse of the USSR, TRIZ left its cradle, parents, and authorized supervisors and began an "independent life" in the world. In Chap. 1, Abramov and Sobolev discuss the results of an almost thirty-year-long trip, the current stage of TRIZ evolution, and its popularity. While acknowledging the clear and unquestionable achievements and indicators of growth, the authors also highlight the indicators of TRIZ development stagnation and speculate on the fundamental reasons for the stagnation.

One of the critical elements of modern TRIZ-based thinking is the definition of function, function analysis, and the formalism of manipulation with functions (e.g., "trimming"). There are a number of approaches to assist in the construction of a function model for a complex real-world system that can be described by several hierarchical levels. In Chap. 2, Koziolek presents a systematic approach to decompose functions in the built model and for linking them with the system architecture. A case study demonstrates how the presented approach adds functional screening of possible changes in design.

TRIZ is not alone in the list of industrial innovation development tools. In Chap. 3, Livotov et al. compare TRIZ and Process Intensification (PI) principles. It is shown how 155 PI best practices can be clustered around forty inventive principles of TRIZ. Each generic inventive principle is given a subset of specific operations to intensify processes inventively. An illustration is given by the analysis of 150 recent patents in ceramic and pharmaceutical industries.

Can TRIZ make it in a specific engineering subject where the professional knowledge of the background is the prerequisite for any ideation? In Chap. 4, Chechurin et al. enter the field of automation and control and demonstrate how one of the basic TRIZ concepts—Ideal Final Result—can assist the design of nontrivial solutions. The main idea is to modify the design of a plant in a way that control becomes simple, or unnecessary. The latter case is illustrated by three case studies.

It has always been a challenge to describe a real situation that needs improvement by using the language of TRIZ; to model it by TRIZ-related techniques. In Chap. 5, Czinki and Hentschel suggest a TRIZ application protocol called Adaptive Problem Sensing and Solving (APSS). With the help of an example from the automotive industry, they show how the suggested approach assists generating successful solutions in a well-structured and highly efficient manner.

In Chap. 6, Spreafico and Russo suggest a method to improve Failure Mode Effect Analysis (FMEA) by TRIZ and the use of Subversion Analysis. They report that the proposed method delivers a better definition of the failure effects. A case study of a vacuum cleaner design analysis that resulted in a patented solution illustrates the advantages of the approach.

Together with TRIZ, the Lean management approach has been mentioned among the tools of innovation management in terms of process improvement. Chapter 7, by Hammer and Kiesel, adds to the discussion of compatibility between these two concepts. It is shown by a specific industrial example how the combination of TRIZ and Lean can deliver a better process design. A framework for systematic improvement projects is also developed.

In Chap. 8, Efimov-Soini and Elfvengren develop further one of modern TRIZ tools typically used for cost reduction and simplification—trimming. The authors notice that the traditional trimming procedure works on a static functional model, while many real-world systems, and products, work in different modes, situations, and conditions. They extend trimming to variable functional models that reflect changes in system architecture and functions. They show how the proposed approach, called dynamic trimming, can provide results that are different from the traditional static trimming.

Two basic modern TRIZ modelling techniques, Function analysis and Cause Effect Chain analysis (CECA) are often used at the beginning of inventive design projects. Lee develops them both in Chap. 9 and blends them in a new tool named Goal–Direction–Idea–Design. This algorithmic roadmap helps to identify directions for solutions in an almost automated manner. A case study is presented to demonstrate how the proposed methodology works in detail.

Part II concentrates on presenting novel tools and techniques for creating innovations that support and enhance the design, or ideation, process. The chapters in this part also discuss the types of background education needed for the ability to generate new ideas.

In contrast with safe design optimization, TRIZ and other heuristic methods yield conceptually new design ideas. Even if these methods often create disruptive improvement, as far as design specification is considered, they can, and in reality do generate an often unexpected number of secondary problems. In Chap. 10, Livotov et al. present an approach for systematic secondary problem prediction and identification. The method is based on the analysis of inventive patents. An example of secondary problem identification in granulation technology is given as an illustration to the approach.

Chapter 11 by Ohenoja, Paavola, and Leiviskä nicely echoes the two previous contributions on process intensification and on heuristic automation design. It presents a systematic approach for designing control systems for intensified process concepts, generated using TRIZ. For industry practitioners, it is important that the method helps to convert inventive ideas into feasible process designs.

In Chap. 12, Silva and Carvalho discuss the analysis and avoidance of failure situations as a part of the innovation creation process. The chapter is devoted to the comparison of two previously presented failure identification methods, "anticipatory failure determination" and "failure mode and effects analysis," and includes a theoretical and a practical comparison of these two approaches to ex ante avoid failures in designs.

Renev, in Chap. 13, writes about the context of contextual design in construction industry projects and discusses the integration of construction design software with idea-generation techniques. The chapter outlines a three-step process for ideas generation for construction project design and a case that illustrates the discussed issues in practice.

In Chap. 14, Buzuku and Kraslawski discuss how morphological analysis, spiced up with what they call sensitivity analysis, can be used in exploring new feasible solutions to (design) problems. The idea presented is quite straightforward and is based on using known good solutions as a basis for sensitivity analysis, that is, exploring whether changing the morphus (form) of a single design characteristic at a time would yield new, before unseen but satisficing alternatives. The main idea is that by way of sensitivity analysis one does not have to explore the whole universe of possible alternative solutions that may be very large. The new approach is illustrated with an example.

In Chap. 15, Belski, Skiadopoulos, Aranda-Mena, Cascini, and Russo discuss the importance of general knowledge and of domain knowledge in the ability of individuals to create new ideas to solve problems within a given domain. They present the results of an experiment in which they test the hypothesis that more general knowledge is more important than domain knowledge for the ability to create novel solution ideas in engineering and the hypothesis that the use of ideation heuristics makes ideation more efficient. They find evidence to support both claims. Understanding what the favorable circumstances for ideation are is important from the point of view of being able to provide such circumstances—be they in terms of education given, or in terms of the supporting methodologies used.

In Part III, the focus is on managing innovations and the innovation process. The four chapters in this part discuss evaluation of innovations in the different stages of the innovation process and the commercialization of innovations. It is evident from the chapters presented that evaluation is an integral part of a systematic innovation process and hence also of the concept of systematic creativity, even if it is commonly understood as concentrating mainly on the process of innovation creation.

In Chap. 16, Kozlova, Chechurin, and Efimov-Soini discuss how economic evaluation performed already in the early stages of the design process can help make the innovation process more efficient. The idea presented has similarity to "fail fast" piloting: if resources are used in taking an innovation process of a design further, while it is clear from very early on that in any case the design is not feasible from the economic point of view, one is wasting resources. This is why it makes sense to include also economic feasibility measurement in the early stages of the innovation process. The authors propose the use of levelized function cost for this purpose and show how it can be used in weeding out economically unsuitable designs already early on in the innovation process.

Stoklasa, Talásek, and Stoklasová, in Chap. 17, present an interesting essay on how soft and emotional aspects and uncertainty can be incorporated in the evaluation of innovation alternatives. The authors present a new procedure for how this rather difficult task can be accomplished in a multiple-evaluator, multi-criteria environment, building on the semantic differential method and previous work on applications of the semantic differential method for product classification.

In Chap. 18, Collan and Luukka write about selection of innovation designs and concentrate on how scorecards that utilize fuzzy logic to capture estimation imprecision can be used in the task. They concentrate their discussion on how the aggregation of the expert evaluations that are the basis of the scorecard information can and should be done, in order to lose as little as possible of the original information elicited. They illustrate with a numerical example how a recently introduced lossless fuzzy weighted averaging operator can be used in the aggregation of scorecards. The results presented show that the way information about innovation designs is processed may have a significant effect on the decision-making that follows.

Pynnönen, Hallikas, and Immonen discuss in Chap. 19 the commercialization of innovation, and observe that commercialization of innovations is a rather complex issue, as it has to consider not only the issue of developing products from innovations, but also the innovation generation process itself and the product-service system that will surround the developed product. The authors build on the business model innovation (BMI) framework and describe and illustrate a systematic modularized process that has been used in a number of real-world innovation commercialization projects. The presented process is useful in supporting the successful commercialization of innovations.

Finally, the editors thank the anonymous reviewers who performed the "original" reviews for the first versions of the included chapters, and more importantly, we thank all the chapter authors for their contributions. It is our firm belief that this book offers a lot of guidance to the academic reader in terms of showing quite clearly the directions to which research on systematic creativity is evolving, and insight to the industry about how methods of systematic creativity and innovation management support decision-making with regards to innovations setting up innovation systems.

Lappeenranta, Finland
Leonid Chechurin
Mikael Collan

Contents

xiv   Contents

Notes on Contributors

Oleg Abramov is the CTO at Algorithm Ltd., St. Petersburg, Russia—a strategic partner of GEN TRIZ, Boston, USA. He received his undergraduate degree and Ph.D. in Radio Engineering from Saint Petersburg Electrotechnical University (ETU "LETI"). For 15 years, Abramov worked at this university as a researcher and an associate professor until, in 1997, he joined Algorithm Ltd. as head of the department. At Algorithm, he managed over 80 successful innovation consulting projects for companies such as Xerox, FMC, Intel, Honda, and BAT. In 2006, the MaxBeam75 Smart Antenna by Airgain, developed by Abramov's team, received an award from the government of California as the most innovative product of the year. In 2012, he received a TRIZ Master Degree from the International TRIZ Association. He is the author of 38 granted patents and over 50 scientific papers in radio engineering and TRIZ.

Guillermo Aranda-Mena (PhD) is an associate professor, the director of RMIT's Construction Procurement Group, and UNESCO architecture professor at the Politecnico di Milano, Italy. He has worked on research, consulting and education in the built environment disciplines of architecture, planning and procurement for over two decades, in particular, life-cycle design, construction and operations under public-private partnerships. Guillermo is cofounder of MelBIM—the largest building information modeling professional group in Australia—a well-published scholar and regional editor—Australasia—for Facilities by Emerald Scientific Publishers. Guillermo provides independent consulting and expert advice to public and corporate clients at early project stages.

Iouri Belski received his M.Eng. degree in Quantum Electronics in 1981 and a Ph.D. in Physics (Semiconductors and Dielectrics) in 1989 from the Moscow Institute of Physics and Technology, Dolgoprudny, Russia. He is Professor of Engineering Problem Solving in the School of Engineering, the Royal Melbourne Institute of Technology (RMIT). Iouri is the author of a book on systematic thinking and problem solving and over 80 peer-reviewed papers, and has been granted 24 patents. His research interests include engineering creativity and problem solving, as well as novel methods and technologies for education. Iouri is a TRIZ Master (MATRIZ Diploma 75). He is also the recipient of numerous awards including the 2006 Carrick Citation for Outstanding Contribution to Student Learning, the Inaugural Vice-Chancellor's Distinguished Teaching Award (2007), and the Australian Award for Teaching Excellence (2009). In 2016 the Australian government awarded him with the Australian National Senior Teaching Fellowship.

Victor Berdonosov is a professor of Komsomolsk-na- Amure State University. In 1971 he graduated with honors from the Leningrad Institute of Aerospace Instrumentation with specialization in Engineer in Computer Science. He got his Ph.D. on Radiolocation and Radio Navigation. In 1999 he became a project leader of "Application of TRIZ in Higher Education". He has served as the chair of "Design and TRIZ" Department. In 2004 he founded the TRIZ-Amur—a public organization aimed at introducing TRIZ to the public. He also has a certificate of TRIZ Master (number 84). Since 2003 he's been taking part in international scientific conferences, both in Russia and abroad. Berdonosov is the author of several monographs, more than a dozen teaching aids in different languages, and more than a hundred scientific publications.

Shqipe Buzuku is a doctoral researcher in Industrial Engineering and Management, at LUT with a major in Systems Engineering. She received her master's degree in 2010 from University of Pristina, Faculty of Natural Science in Chemistry and Chemical Engineering. In 2014, she received a grant for Ph.D. studies at LUT from the Finnish Cultural Foundation and in 2016, a grant from Foundation for Economic Development in Finland. Her research interest focuses on design-thinking and problem solving, applying systems design methods with their applications including knowledge management, decision-making in systems, and on process engineering. In 2014, she was a visiting researcher at South China University of Technology in Guangzhou, China; in 2015, at University of Washington College of Engineering in Seattle, WA, USA; and in 2018 at Karlsruhe Institute of Technology, Germany. She has authored a number of scientific articles published in several journals and conferences.

Gaetano Cascini is a full professor in the Department of Mechanical Engineering at Italy's Politecnico di Milano. His research interests cover design methods and tools with a focus on the concept generation stages both for product and process innovation. He has coordinated several large European research projects including SPARK: Spatial Augmented Reality as a Key for co-creativity (Horizon 2020—ICT). He is on several journal editorial boards including the *Journal of Integrated Design & Process Science* and the *International Journal of Design Creativity and Innovation*. He is also a member of the advisory board of the Design Society.

Arun Prasad Chandra Sekaran is an academic researcher at Offenburg University of Applied Sciences for European research project entitled "Intensified by Deign", where he is assisting European industrial partners and University researchers for process intensification involving solids handling by application of TRIZ Methodology and Advance Innovation Design Approach. He received his master's degree in Process Engineering at Offenburg University of Applied Sciences, Germany, and Bachelors of Technology in Chemical Engineering from Anna University, India. After completing Bachelor of Chemical Engineering in India, he worked as a quality control trainee in automotive adhesive and sealer manufacturing industry and as a project engineer in the field of Environmental Impact Assessment and Industrial environmental legalization consultation for Industrial, Infrastructure, and mining development projects.

Leonid Chechurin is Full Professor of Industrial Management at Lappeenranta University of Technology (LUT) in Finland. He received his Candidate of Science Degree in 1998 and his D.Sc. degree in 2010 with a thesis on mathematical modeling and analysis of dynamic systems. His research interests focus on the analysis of systems' dynamics, based on mathematical modeling, stability analysis, control, systematic approach for inventive thinking, innovation automation tools, as well as problems of innovative growth of companies, regions and economies. Chechurin has outstanding industrial experience from engineering design groups of some of the leading innovating technology companies in the world. He holds extensive experience also in consulting and training for the industry and business.

Mikael Collan Collan received his D.Sc. degree in information systems in 2004 from Åbo Akademi University, Finland. He works as Professor of Strategic Finance at LUT directing two master's programs, one in business and one in engineering. His research is concentrated on business decision-making in multiple-criteria multiple-expert environments under uncertainty. Mikael is a past-president of the Finnish Operations Research Society and an ordinary member of the Finnish Society of Sciences and Letters, one of the four Finnish academies of science.

Alexander Czinki Prof. Dr.-Ing., Germany, studied Mechanical Engineering at the RWTH Aachen, Germany, and at Michigan State University, USA. After graduation, he started working at the Institute for Fluid Power Drives and Systems (IFAS) at RWTH Aachen, where he did research on anthropomorphic mechanical robot hands. At IFAS he held the position of group leader and vice chief-engineer. After his doctorate, he turned towards the automotive industry, where he worked in the fields of interior design, future product development, and mechatronic systems. Being a professor at the University of Applied Sciences in Aschaffenburg, Germany, his current activities focus on the fields of technology and innovation management, creativity techniques, and mechatronic systems. In parallel, he is working as a trainer, consultant, and speaker.

Renan Favarão da Silva has been working with maintenance and reliability engineering in multinational companies in Brazil. He has a master's degree in Manufacturing Engineering and a postgraduate specialization in Reliability Engineering at Federal University of Technology—Paraná (UTFPR) and in Mechanical Engineering at São Paulo State University (UNESP). His research has focused on improving methods for product and system reliability and on asset management. He supervises Ph.D. students in the Department of Mechanical Engineering at the University of São Paulo (USP) and works as a consultant at Reliasoft.

Marco Aurélio de Carvalho is a mechanical engineer. He holds a master's degree and a doctorate in Product Development. He is an associate professor at the Federal University of Technology—Paraná (UTFPR), in Brazil, where he is active in the areas of systematic innovation, creativity, product development, and project management. He has authored two books and written a number of other articles which have been published in various journals. He is a member of the Brazilian Society of Industrial Engineering, the American Society of Mechanical Engineers, the ETRIA, the Institute of Product Development Management, the Product Development Management Association, the Working Group 5.4—Computer-Aided Innovation at the International Federation for Information Processing, and the World Future Society.

Nikolai Efimov-Soini graduated from Peter the Great St. Petersburg Polytechnic University as a Master of Engineering and Technology in 2007. Now, he is a doctoral student in the School of Business and Management at

LUT. The focus areas of his research are function analysis, TRIZ, inventive methods and Computer Aided-Design in the conceptual design stage. In addition, he works as the lead developer at CompMechLab Inc., in Saint-Petersburg, Russia, that works in the aeronautical industry.

Kalle Elfvengren is an adjunct professor at the School of Engineering Science, LUT, Finland. He has authored over 80 international scientific papers that have been published in various books and journals during his academic career. Currently, he works as a project researcher and is involved in several teaching and research activities at the university. His research interests include process development in the health-care sector, management of technology, and risk management. His interests also cover decision-analysis, creativity tools, such as TRIZ, and the fuzzy front end of the innovation process. Elfvengren is also a board member of the ETRIA.

Jukka Hallikas D.Sc. (Tech), is Professor of Supply Chain Management at the LUT. His research interests focuses on the purchasing and supply chain management, risk management in supply chains, and innovation management. He has authored several scientific articles, books, and book chapters on these topics which have appeared in several publications.

Jens Hammer is director of Digital Business Innovation at Schaeffler AG and Ph.D. student at the Friedrich-Alexander-Universität Erlangen-Nürnberg. He is MA TRIZ Level 3 and has the accreditation for MA TRIZ-Level 1 trainings. He received his diploma in Mechanical Engineering in 2010 and his master's degree in Industrial Engineering in 2013.

Claudia Hentschel Prof. Dr.-Ing., Germany, first studied Economics, switched to Mechanical Engineering, and received an Industrial Engineering Diploma of TU Berlin, Germany, and Ecole Nationale des Ponts et Chaussées ENPC Paris, France. Engineer by profession, she turned to Fraunhofer Gesellschaft/TU Berlin and worked five years as academic assistant, researcher and consultant at the Institute for Assembly Technology and Factory Organization IWF at Production Technology Center PTZ, Berlin, and at the Mechanical Engineering Institute of Technion, Haifa, Israel. After her doctorate in Disassembly in 1996, she started an industrial career at Siemens AG, Information and Communication Mobile, Munich. Since 2000, she is Professor of Innovation, Technology and Production Management at University of Applied Sciences HTW, Berlin, where she holds several functions in administration. Her teaching and research focuses on technical management, systems engineering, product and process design and development, with particular attention to the management of inventions.

Mika Immonen is a post-doctoral researcher at the School of Business and Management at LUT, Finland. He holds a D.Sc. (Tech.) degree from LUT. His main research interests include service innovation and value network management in the intersection of ICT, energy and the healthcare sector. He has worked in various studies focusing on healthcare technology and consumer behavior in health-related services. Via numerous research projects, he has also accumulated experience in the fields of infrastructure service development and smart grid business. His academic work focuses on service systems, emerging business models and customer value creation in multi-stakeholder environments.

Vasilii Kaliteevskii is a junior researcher in the Department of Industrial Engineering and Management, LUT. He received his bachelor's degree in Software Engineering and master's degree in Software and Administration of Information Systems from Saint Petersburg State University. Vasilii works under the Marie Skłodowska-Curie project "Innovative Nanowire Device Design" (INDEED) and studies the conceptual design of products and technologies based on nano-elements. His current research interests include nano-wire production design, software engineering, conceptual design, innovative design, algorithms and automation.

Martin Kiesel works as a system architect in the motion control area of the SIEMENS digital factory division. He is responsible for the architecture of the industrial communication and for the architecture processes of his business unit. Martin has more than 33 years of experience in Systems and Software Engineering in the automation, motion control and drive domain. He has process experience in Lean, Agile, CMMI, usage centered design, and TRIZ (Level 3 since 2011).

Sebastian Koziołek is faculty in the School of Mechanical Engineering at the Wroclaw University of Technology in Poland. He is an associate professor in the Department of Machine Design and Research, working in the area of inventive engineering. His present research is focused on the development of a new method for building a formal design representation space and a new inventive design method, incorporating elements of various heuristic methods, and the principles of big thinkers such as Leonardo da Vinci, Thomas Edison, Genrich Altschuller, and others. Dr. Koziolek is a global scholar of Stanford University, George Mason University, and the University of Sydney in the area of Inventive Engineering as applied to complex mechanical engineering problems.

Mariia Kozlova is a postdoctoral researcher at the School of Business and Management of LUT, Finland. Her current research focuses on decision support under uncertainty, including such fields as investment analysis, design valuation, and multi-criteria decision-making across various industries from renewable energy and mining to e-learning. The approaches she works with include simulation-based techniques, system dynamics modeling, optimization algorithms, and soft computing based systems. She has received a Ph.D. degree from LUT working on Russian renewable energy investment profile. Prior to the current position, she has gained field experience in investment analytics working in a Russian oil company. Being at the beginning of her academic career, Mariia has become an author and coauthor of multiple conference proceedings and journal publications and has been an active reviewer for several academic journals.

Andrzej Kraslawski obtained his Ph.D. from Lodz University of Technology (LoUT), Poland, in 1983. From 1988 to 1990, he worked as visiting professor at ENSIGC Toulouse, France. He has been working at LUT since 1990. He is Professor of Systems Engineering at LUT and Professor of Safety Engineering at LoUT. He is also visiting professor at South China University of Technology, Guangzhou, and Mining Institute, St. Petersburg, Russia. His research interest is focused on the development of methods for knowledge discovery and re-use, sustainability assessment, and process safety analysis. Kraslawski has authored over 130 research papers that have been published in various books and journals and has supervised nine Ph.D. students.

Richard Law is a lecturer in the School of Engineering at Newcastle University, UK, working in the Process Intensification Group. He has worked in the broad field of process intensification since 2010, often with focus on heat transfer enhancement. To date (early 2018), he has co-authored 14 journal articles and one text book and has presented his papers at more than 20 international research conferences.

Min-Gyu Lee is an innovation consultant and director at Uni Innovation Lab and QM&E Innovation Corp. in South Korea. He was the head of innovation department at the AMOLED division of Samsung, a Strategy Consultant at IBM Korea, and an Innovation Strategy Advisor for Korea University. His main expertise is TRIZ, Six Sigma, R&D, and Innovation Strategy & Management. He consulted on hundreds of projects to innovate products, services, processes and businesses for leading Korean companies in various fields. He studied physics at the Seoul National University and Postech and does research in theory, methods and tools for Creative and Systematic Innovation at LUT, Finland.

Kauko Leiviskä received his Ph.D. in Control Engineering from University of Oulu in 1982. He is Professor of Process Engineering (Control Engineering) in the same university, since 1988 and has held several administrative posts: the head of the Department of Process Engineering (1991–2005), vice-dean of Technical Faculty (2000–2005); dean of Technical Faculty (2006–2013), and member of University Board (2006–2009). He has supervised 28 doctoral theses since 1998 and been opponent and/or scientific evaluator of more than 30 theses. He has been the IPC member in numerous international conferences in the fields of intelligent systems, process control and modelling. He is the author of more than 300 scientific papers. He is in Research Gate, LinkedIn and Google Scholar. He is International Federation of Automatic Control (IFAC) Contact Person in Finland since 2001 and the member in four IFAC Technical Committees. He was the Scientific Director in EUNITE European Network of Excellence, (2001–2004).

Pavel Livotov is a professor and the author of more than 80 patented inventions and 70 scientific publications. He has worked with TRIZ methodology since 1980. He received his Ph.D. in St. Petersburg, Russia, in the field of robotics. From 1989 to 1993, he continued his research work at the University of Hanover, Germany, as a senior scientist of Institute for Production Engineering and Machine Tools. From 1993 till 1999, he was the head of R&D department for robotics and material handling at Focke & Co, Germany. In 2000 he co-founded the European TRIZ Association (ETRIA) where he was a president and since October 2017 is a member of the executive board. He is founder and general manager of the TriS Europe Innovation Academy and Professor of Product Development and Engineering Design at the Offenburg University of Applied Sciences, Germany.

Pasi Luukka received his M.Sc. degree in Information Technology in 1999 and D.Sc. degree in Applied Mathematics in 2005 from LUT. He is working as a professor in school of business in LUT. His research interests include fuzzy data analysis, multi-criteria decision-making, soft computing methods and business analytics. He has authored over 50 international journal papers. He is a member of North European Society for Adaptive and Intelligent Systems (NSAIS) where he serves as president.

Mas'udah is an academic researcher at the laboratory for Product and Process Innovation (PPI), Department of Mechanical and Process Engineering, Hochschule Offenburg—University of Applied Sciences (HSO), Germany. She received her master's degree in Process Engineering from HSO and magister in

Environmental Protection and Biotechnology from the University of Warmia and Mazury in Olsztyn, Poland. Her research and publication interests are related to process innovation, process intensification, inventive problem solving, and secondary impact analysis of product innovation. She has been involved in a European Project, entitled "Intensified by Design for the intensification of processes involving solids handling". She co-authored articles about TRIZ-based approach for process intensification in process engineering and secondary problems identification of new technologies by patent analysis.

Markku Ohenoja is a postdoctoral researcher at the Control Engineering Research Group, University of Oulu. He has been working in the area of process engineering in this same group for over ten years and received his doctorate in 2016. His work has focused on process modeling and simulation, control design, and process optimization in number of research project with application areas consisting of chemical processes, water treatment, fuel cells, papermaking, and mineral beneficiation. He has contributed to more than 20 technical and scientific publications.

Marko Paavola received his Ph.D. in 2011 from University of Oulu. Since then, he has been working as a post-doctoral researcher, teacher and project manager in Control Engineering group at the same University. His interests include data analysis, efficient signal processing, modelling and control systems in different applications areas including electronics manufacturing, pulp mills, pharmaceutical manufacturing, minerals processing and wireless sensor networks. He has authored over 30 technical and scientific publications and has additional industrial experience in software testing and manufacturing processes. He participated in European ESNA project, which received ITEA Achievement Award gold medal for "highly successful project with outstanding contributions" from ITEA organization. He is the member of Scientific Advisory Board for Defense, Finland.

Mikko Pynnönen holds a D.Sc. degree in Economics and Business Administration and is Professor of Growth and Internationalization of SMEs at the LUT. He is leading several research projects in the field of services and business development. His main research interests include value creation in business systems. He has written over 60 scientific articles on business models and value creation which have been published in various journals, such as, *Telecommunications Policy, Journal of Business Strategy, Journal of Purchasing and Supply Management* and *International Journal of Production Economics*. He has developed these topics in close connection with a number of firms from a wide variety of industries, including ICT, Healthcare, Forest and Energy.

David Reay is a consulting engineer in energy-related fields and is a visiting professor at Northumbria University. He also is an honorary professor at Nottingham University and works part-time as a senior research associate (RA) at Newcastle University. He graduated in Aeronautical Engineering from Bristol University in 1965. He has co-authored or edited nine energy-related books, including *Heat Pipes* and *Process Intensification*, and was founding editor of the Elsevier Journal *Applied Thermal Engineering*. He is now editor-in-chief of the journal *Thermal Science and Engineering Progress*. For 12 years he advised the European Commission on energy R&D and is a past president of the UK Heat Transfer Society and Honorary Life Member of the Heat Pump Association.

Ivan Renev received the M.Tech. degree (with honors) in Construction from the Saint-Petersburg state polytechnic University, Russia, in 2011. He worked in international pharmaceutical engineering projects as a senior structural engineer with the NNE Pharmaplan in 2011–2014. He was also a chief structural engineer, responsible for the design of pharmaceutical plants with PharmDesign in Saint-Petersburg, Russia, in 2014–2017. He is a lead engineer in the division for design supervision and detailed design, working at the Hanhikivi-1 Nuclear Power Plant project in Northern Finland. Since 2015 he is also a doctoral student at the School of Business and Management at LUT, Finland. His main areas of research are automation of the conceptual design of building structures and inventive problem solving techniques. Ivan is a member of the ETRIA and the Education and research in Computer Aided Architectural Design in Europe (eCAADe) association.

Davide Russo has a master's degree in Mechanical Engineering from the University of Florence in 2003. He is an assistant professor at the Department of Management, Information and Production Engineering (DIGIP), and co-founder and ex-sole director of BIGFLO srl. He is a member of the Center for innovation and Knowledge management at UNIBG, where he teaches "Product and process innovation (TRIZ)" and "Industrial Design". He is the inventor of 13 international patents and the author of over 80 publications in journals, books, and international conference proceedings. His academic activity involves the following topics: methods and tools for systematic innovation, ICT methods and tools for product development, patent search engine, and design ontologies.

Arailym Sarsenova is an academic researcher of the Product and Process Innovation Laboratory at the Offenburg University of Applied Sciences (HSO). She completed her master's degree in Process Engineering at HSO in Germany and her undergraduate studies in Biotechnology at Kazakh Agro Technical University named after S. Seifullin in Kazakhstan. She also has significant experience in patent analysis in the pharmaceutical sector with subsequent identification of secondary problems. She has collaborated in several workshops for European Project entitled "Intensified by Design" for the intensification of processes involving solids handling.

Anne Skiadopoulos is a senior research officer at the Centre for Sport and Social Impact at La Trobe University, Melbourne, Australia. She received her B.Eng. (first class honours) from the RMIT and her Ph.D. in Engineering from Swinburne University of Technology, Melbourne, Australia. Anne has been interested in TRIZ and has co-authored eight papers on application of substance-field analysis for idea generation and failure prevention. Her research interests also include oceanography and public health. In 2012 Anne's work on minimization of plumes that occur during dredging received the Singapore-Netherlands Sustainability Award.

Sergey Sobolev heads a research group in Siemens Corporate Technology department in St. Petersburg, Russia. He received his undergraduate degree as mathematician from St. Petersburg State University and a Ph.D. in Navigation and Radio Systems from ETU "LETI". For seven years, Sobolev worked in various companies as software engineer until, in 2007, he joined Siemens Corporate Technology. At Siemens, he managed various successful innovation R&D projects for Siemens business-units in the areas of energy generation, transmission, and transportation. He is the author or contributor of about 20 invention disclosures, resulting in 8 patent applications.

Christian Spreafico has a master's degree in Mechanical Engineering, received from the University of Bergamo in 2012. He has a Ph.D. in Industrial Engineering from the University of Padova in 2017. He works as a research assistant at the University of Bergamo since 2012, and his research activities focus on methodologies and tools for supporting systematic innovation, eco-design, and problem solving. He is the co-inventor of four patents about innovative devices and systems for renewable energy.

Jan Stoklasa received the MS degree in Applied Mathematics and MS degree in Psychology from Palacký University Olomouc, Czech Republic, in 2009 and

2012, respectively. He received the Ph.D. and D.Sc. degree in Applied Mathematics from Palacký University Olomouc, Czech Republic, and LUT, Finland, in 2014. He is a research fellow at the LUT School of Business and Management and an assistant professor at Palacký University, Olomouc, Czech Republic. His research interests include decision support models, multiple-criteria decision-making and evaluation and linguistic fuzzy models and their practical applications.

Jana Stoklasová Stoklasová received her MS degree in Psychology from Masaryk University, Brno, Czech Republic, in 2013. She is a psychologist and marriage counsellor in the Marital and family counselling centre, Prostějov. Her professional interests include systemic psychotherapy, dyadic developmental psychotherapy, foster care and working with children with attachment disorders.

Tomáš Talášek received the MS degree in applications of mathematics in economy from the Palacký University Olomouc, Czech Republic, in 2012. He has been working toward the Ph.D. degree in Applied Mathematics at Palacký University Olomouc, Czech Republic, and LUT, Finland. His research interests include fuzzy sets, multiple criteria decision-making, linguistic approximation and pattern recognition and their practical applications.

Leonid Yakovis is a professor in the Department of Mechanics and Control Processes of Peter the Great Polytechnic University in Saint Petersburg, Russia. In 1971 he graduated with honours with a degree in Automatic Control Systems from the same university. He specialized in the field of industrial process control. Yakovis is an expert in the design of methods and algorithms for optimization of controlled processes for preparation of multi-component mixtures, and in the field of method development for calculating two-level control systems for a wide class of multidimensional nonlinear control objects. Yakovis is the author of more than 170 publications, among which are three monographs, 13 copyright certificates, and 2 Finnish and United States patents.

List of Figures

**Control Design Tools for Intensified Solids Handling Process
Concepts**

**Anticipatory Failure Determination (AFD) for Product Reliability
Analysis: A Comparison Between AFD and Failure Mode and Effects
Analysis (FMEA) for Identifying Potential Failure Modes**

**Computer-Aided Conceptual Design of Building Systems: Linking
Design Software and Ideas Generation Techniques**

Optimized Morphological Analysis in Decision-Making

Engineering Creativity: The Influence of General Knowledge and Thinking Heuristics

Levelized Function Cost: Economic Consideration for Design Concept Evaluation

Reflecting Emotional Aspects and Uncertainty in Multi-expert Evaluation: One Step Closer to a Soft Design-Alternative Evaluation Methodology

Innovation Commercialisation: Processes, Tools and Implications

List of Tables

Part I

Advances in Theory and Applications of TRIZ

The nine chapters in Part I are dedicated to advances in and applications of the Theory of Inventive Problem Solving, better known as TRIZ. Chapters 1, 2, 3, 4, 5, 6, 7, 8 and 9 present a diversity of novel approaches to extend and to support the use of TRIZ in systematically creating innovations and offer the reader a good overview of the directions in which systematic creativity with TRIZ is evolving.

Current Stage of TRIZ Evolution and Its Popularity

Oleg Abramov and Sergey Sobolev

1 Introduction

Since Genrich Altshuller introduced the Theory of Inventive Problem Solving (TRIZ) at the end of the 1940s, it has been greatly developed and refined both by Altshuller and by his numerous colleagues and followers.

Over time, TRIZ has demonstrated great efficacy in solving difficult technical problems, many books on TRIZ have been issued, and thousands of people have been taught TRIZ and become certified TRIZ specialists.

O. Abramov (✉)
Algorithm Ltd., Saint Petersburg, Russia
e-mail: Oleg.Abramov@algo-spb.com

S. Sobolev
Siemens LLC, Saint Petersburg, Russia

© The Author(s) 2019
L. Chechurin, M. Collan (eds.), *Advances in Systematic Creativity*,
https://doi.org/10.1007/978-3-319-78075-7_1

3

TRIZ has not, however, become a standalone best industry practice for developing new products, technologies and services. In fact, very few innovations have been developed using TRIZ.

Moreover, even after years of intensive development, TRIZ still has not manifested itself as a serious science. For example, as shown in a recent review by Chechurin (2016), only 1200 publications with the word "TRIZ" were indexed in Scopus (the largest database of peer-reviewed literature from scientific journals, books and conference proceedings) by July 2014; another paper by Chechurin et al. (2015) indicates 1333 publications indexed by mid-2015. Considering that Scopus indexes about 21,000 scientific journals and contains about 50 million records, this number is quite small.

The goal of this work is to clarify the current status of TRIZ and its acceptance in the world, and to identify why TRIZ does not play the important role it deserves.

Research on these topics was recently done by Abramov (2016). In this chapter, the authors present further elaboration on the matter.

2   Method

The current status of TRIZ was determined by studying the following parameters:

- How far TRIZ has spread around the world;
- How much world interest in TRIZ there is;
- How intensively TRIZ is used in industry and what its recognized area of application is; and
- How aware the world is of TRIZ compared to other innovation methodologies.

The first three items were evaluated by analyzing available reports and research papers, while the last parameter was assessed by analyzing the number of web pages relating to TRIZ and other popular innovation methodologies revealed by advanced Google search.

3   Results of the Research

3.1   Worldwide Propagation of TRIZ Is Decelerating

At first glance, TRIZ has circulated around the world fairly successfully: as pointed out by Goldense (2016), the number of certified TRIZ experts worldwide has grown steadily, reaching the impressive number of 18,000 in 2015. Based on the International TRIZ Association (MATRIZ) data, in 2017 the number of certified TRIZ experts exceeded 24,000 (see Fig. 1).

This number, however, is distributed across countries very unevenly (Goldense 2016):

- Of certified TRIZ specialists, 65% are now located in South Korea, where the government has actively supported the propagation of TRIZ;
- Most of the remaining 35% are in China, Germany and Russia; and
- A few other countries have a miniscule share of TRIZ specialists.

Using Goldense's data, Abramov (2016) has shown that, after peaking in 2014, the number of specialists certified annually has begun decreasing

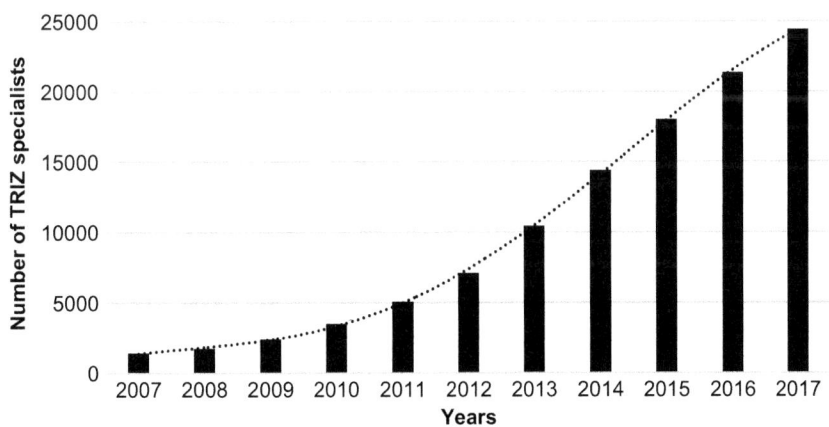

Fig. 1   The total number of certified TRIZ specialists: growth over years

(see Fig. 2), which likely reflects the fact that the popularity of TRIZ in South Korea, a major contributor to the number of TRIZ specialists, started to decrease at that time.

From Figs. 1 and 2, it can be concluded that the popularity of TRIZ reached its peak in about 2014; that is, TRIZ in its current/classical form is either at the third (maturity) stage of its evolution or in the beginning of the fourth (stagnation) stage.

3.2   World Interest in TRIZ Is Declining

Research conducted by Patrishkoff (2012) revealed that world interest in TRIZ is currently diminishing. The research is based on Google statistics of web searches, which shows that since 2004 worldwide interest in "TRIZ" has steadily decreased, and in 2011 it was down 55% while worldwide interest in "Innovation" decreased only ~25% by 2007 compared to 2004, and after 2007 it remains stable (Patrishkoff 2012b).

In contrast, worldwide interest in "Lean Six Sigma" has steadily increased and in 2011 it was up 110% relative to 2004; worldwide interest in "Lean" demonstrated only a slight down trend during 2004–2011 (Patrishkoff 2012a).

The current decline of world interest in TRIZ is confirmed indirectly by the dramatic reduction in the amount of web pages containing the

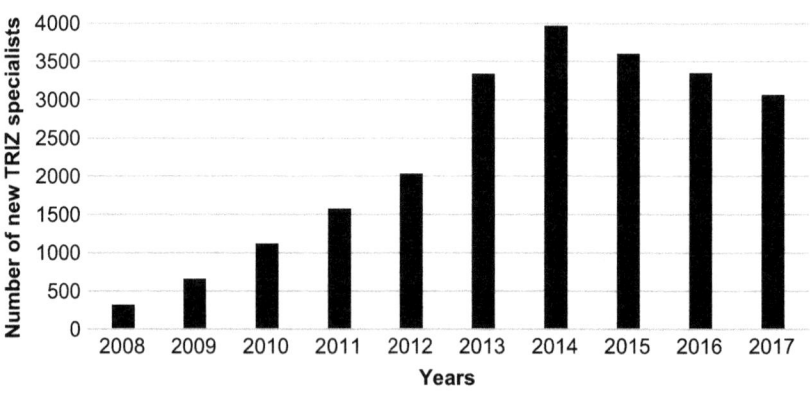

Fig. 2 Number of new TRIZ specialists certified annually

word "TRIZ," which has been observed in the last few years (Abramov 2016).

Based on this data, we can conclude that worldwide popularity of TRIZ has already passed its peak and is now declining, despite the fact that world interest in innovation remains stable.

This most probably means that competing methods for innovation, such as Lean Six Sigma, have become more widely adopted than TRIZ.

3.3   World Awareness of TRIZ Is Low

In order to identify how well TRIZ-related information is presented in the public domain, the authors have conducted a brief study. This involved a Google web search for a few popular competing methods and processes for solving technical and business problems, and developing new products (NPD). Besides TRIZ, these methods included Lean methodology, Theory of Constraints (TOC), Six Sigma, crowdsourcing and Design Thinking.

The following keywords were used to perform the search: Lean Method; Theory of Constraints (TOC); "Six Sigma"; "Design Thinking"; crowdsourcing; TRIZ. Only exact matches were searched.

The authors performed the search twice: in July 2016 and in January 2018 (see Fig. 3).

As seen from Fig. 3, there is far less TRIZ-related information on the Internet than information on other problem-solving and NPD methods. For example, the number of web pages related to Lean, Six Sigma and TOC are about two orders of magnitude larger than that of TRIZ-related pages.

This seems to be an accurate representation of how little the world knows about TRIZ compared to the other methodologies for innovating considered in the current research.

Moreover, Fig. 3 shows that TRIZ-related information on the Internet further decreased ~10% since July 2016, while the information on almost all other methods considered in this research noticeably increased over the same period. This may indicate that the downward trend in world interest in TRIZ, identified by Patrishkoff (2012), continues.

Keywords googled:

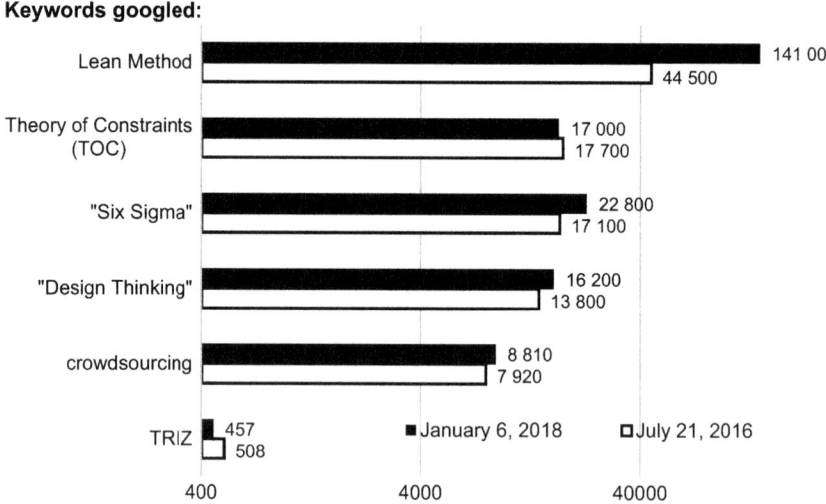

Fig. 3   Results of Google search for innovation methods

3.4   Recognized Area of TRIZ Application Is Narrow

Despite the fact that TRIZ is not very well known to the world and that whatever world interest does exist is falling, it must be admitted that TRIZ has been recognized and even adopted by popular best industry practices for NPD, such as Design for Six Sigma (DFSS).

Unfortunately, as shown by Kim et al. (2012), TRIZ is used in the DFSS process only at the concept development stage—when, or if, it is necessary to solve difficult technical problems.

It is clear from literature on Six Sigma, including the DFSS handbook by Yang and El-Haik (2009), that TRIZ tools employed in the DFSS process include only basic problem-solving tools from older "classical TRIZ" such as the Contradiction Matrix and 40 Inventive Principles, S-curve analysis, Trimming and so on.

The survey of TRIZ industrial case studies performed by Spreafico and Russo (2016) also identifies the Contradiction Matrix and 40 Inventive Principles as being the most frequently utilized tools.

TRIZ tools reduce technical risks associated with an NPD process. This is why the idea of integrating TRIZ into best industry practices has been popular among TRIZ developers since early 2000–2001. However, all publications on this matter so far have included only basic TRIZ tools—see, for example papers by Domb (2001), Sibalija and Majstorovic (2009), and Ilevbare et al. (2011).

The more advanced tools developed in modern TRIZ, for example, Function Oriented Search (FOS) (Litvin 2005), Main Parameters of Value (MPV) analysis (Malinin 2010; Litvin 2011) and Voice of the Product (VOP) (Abramov 2015a), are not yet recognized by the world, and, therefore, not used in existing best industry practices.

The authors' conclusions, which correlate with those found in Chechurin's review (Chechurin 2016), are:

- TRIZ has been adopted for use in some popular NPD methods alongside other (non-TRIZ) tools;
- The area for applying TRIZ, as currently recognized by the world, is too narrow because it is limited to the design of products/processes; and
- The TRIZ tools that are recognized and most frequently used are the simpler, basic tools from old, classical TRIZ.

3.5   Practical Application of TRIZ in Industry Is Not Huge

In their survey of TRIZ industrial case studies, Spreafico and Russo (2016) said: "the spread of TRIZ has never reached the level of capillarity expected."

Moreover, based on the time distribution of the published industrial case studies that they considered (see Fig. 4), it may be concluded that the practical application of TRIZ in industry has been rapidly declining since 2011.

The decrease in the application of TRIZ in industry can be illustrated by the following example: Adunka (2008) reported that by 2008 about 41 engineers had passed a five-day TRIZ training course at Siemens,

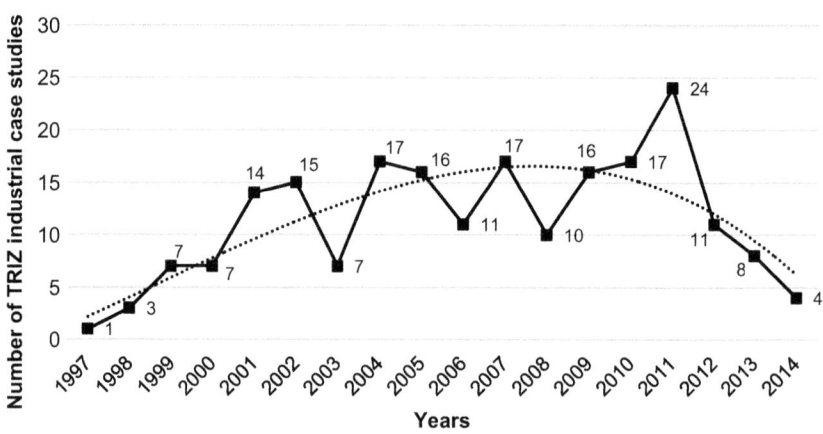

Fig. 4 Time distribution of TRIZ industrial case studies

while by mid-2016 there had been 104 participants of MATRIZ level 1 training courses—which seems to represent the total number of engineers trained in TRIZ at Siemens. This means that only about 60 people had passed TRIZ training in the eight years since 2008, which is not a very impressive number for a company with almost half a million employees (in those years). This may reflect a general decline of industry's interest in using TRIZ.

4   Discussion

As mentioned, classical TRIZ seems to be reaching the maturity, or even the stagnation, stage of its evolution just as world interest in TRIZ is declining.

According to a TRIZ S-curve analysis, it is fair to expect a more advanced innovation methodology to spark a new S-curve in the near future.

This new innovation methodology may be a modern, next-generation TRIZ—providing that it overcomes the main flaws in classical TRIZ.

One of these flaws, mentioned by Chechurin and Renev (2016), is a lack of specific tools for individual industries. In an industry-specific environment, universal TRIZ tools can be too cumbersome for practical use.

Some other researchers also consider classical TRIZ as not being particularly practical for industry. For example, Howard et al. (2009) said: "Creative stimuli in the form of the TRIZ inventive principles have shown much potential, however the industrial uptake of such stimuli is limited due to the practicalities of using this TRIZ approach."

Another—and more serious—TRIZ flaw is the neglect of business and market needs.

In their report, Ilevbare et al. (2011) clearly describe the strength and flaw of TRIZ in current use: "TRIZ has its major strength in its ability to solve difficult innovation problems in a systematic and logical manner. However, it appears to pay little attention to linking the inventive problems and their solutions to market needs and drivers. Therefore there exists the unpleasant possibility of TRIZ providing a solution to a problem which has little or no profitability or commercial benefit to an organization."

Modern TRIZ, however, has tools such as MPV analysis (Malinin 2010; Litvin 2011) and the VOP approach (Abramov 2015a), which are aimed specifically at addressing business/market needs. These tools may eliminate the main drawback of classical TRIZ and allow for more comprehensive integration of TRIZ into best industry practices.

This integration should involve using modern TRIZ tools at all stages of the NPD process, as suggested by one of the authors in an earlier paper (Abramov 2014).

Such comprehensive integration of TRIZ into the product development process dramatically reduces technical and business risks, which may be especially beneficial for businesses related to technological start-ups, specifically those that implement the Lean Startup methodology (Ries 2011). In lean startups rapid prototyping is the key, and utilization of TRIZ tools can make this process more efficient.

Moreover, before starting the development of any technical solutions, the Lean Startup methodology assumes performing so-called customer development. This involves validating assumptions about customer needs and checking the correctness of a customer portrait, which is further used as an input for developing new technical solutions—possibly using TRIZ. The TRIZ VOP approach (Abramov 2015a) can make the customer development process more reliable and objective.

An example of another opportunity to link TRIZ with business needs involves enhancing such popular systems-management methodology as the Theory of Constraints (TOC) with TRIZ tools as suggested by Domb and Dettmer (1999).

In the TOC, TRIZ tools can be beneficial for finding root causes using, for example, TRIZ-based Cause and Effect Chain Analysis (Abramov 2015b), and for efficiently solving the root causes.

5   Conclusions

Based on the results of this research, the following conclusions can be made.

World interest in TRIZ as well as the practical application of TRIZ in industry is declining, and classical TRIZ seems to be reaching the maturity, or even stagnation, stage in terms of its propagation and popularity.

The world-recognized application of TRIZ is currently limited to solving difficult technical problems at the concept generation stage.

Only basic, classical TRIZ tools have been adopted for this purpose by best industry practices, for example, by DFSS methodology.

Further development of TRIZ should focus on (but not be limited to) addressing business and market needs, which may include:

- Developing business/market-oriented tools that are missing in classical TRIZ. Examples of such tools are VOP and MPV analysis;
- Integrating TRIZ more fully with the most popular best industry NPD practices, such as Six Sigma, DFSS, TOC, and so on; and
- Incorporating TRIZ tools into the most popular business approaches, for example, into the Lean Startup method.

Addressing business and market needs may initiate a new S-curve of TRIZ popularity and result in much wider adoption of TRIZ.

Acknowledgements The authors would like to thank Deborah Abramov (Saint Petersburg, Russia) for her helpful comments and for editing this paper; and Alex Zakharov (Boston, USA) for providing the statistical data on the amount of web pages containing the word "TRIZ."

References

Abramov, O. (2014). TRIZ-assisted stage-gate process for developing new products. *Journal of Finance and Economics*, *2*(5), 178–184. http://pubs.sciepub.com/jfe/2/5/8. Accessed 7 Jan 2018.

Abramov, O. (2015a, September 10–12). 'Voice of the product' to supplement 'voice of the customer'. *Proceedings of TRIZFest-2015 conference*, Seoul, South Korea, pp. 309–317. http://matriz.org/wp-content/uploads/2012/07/TRIZfest-2015-conference-Proceedings.pdf. Accessed 7 Jan 2018.

Abramov, O. (2015b, September 10–12). TRIZ-based cause and effect chains analysis vs root cause analysis. *Proceedings of TRIZFest-2015 conference*, Seoul, South Korea, pp. 283–291. http://matriz.org/wp-content/uploads/2012/07/TRIZfest-2015-conference-Proceedings.pdf. Accessed 8 Jan 2018.

Abramov, O. (2016). Warning call: TRIZ is losing popularity. TRIZ in evolution/collection of scientific papers. *Library of TRIZ Developers Summit*, *8*, 240–248. http://triz-summit.ru/file.php/id/f300871-file-original.pdf. Accessed 7 Jan 2018.

Adunka, R. (2008). Teaching TRIZ within Siemens. *Proceedings of the TRIZ-future conference*, The Netherlands, The European TRIZ association, pp. 91–93. https://www.brainguide.de/upload/event/00/ou9r/ec1772c4dd-3ba053471d75ce7a75b3db_1311536006.pdf. Accessed 7 Jan 2018.

Chechurin, L. (2016). TRIZ in science. Reviewing indexed publications. *Procedia CIRP*, *39*, 156–165. http://www.sciencedirect.com/science/article/pii/S2212827116001979. Accessed 7 Jan 2018.

Chechurin, L., & Renev, I. (2016). Application of TRIZ in building industry: Study of current situation. *Procedia CIRP*, *39*, 209–215. http://www.sciencedirect.com/science/article/pii/S2212827116002055. Accessed 7 Jan 2018.

Chechurin, L., Elfvengren, K., & Lohtander, M. (2015, June 23–26). TRIZ integration into product design roadmap. *Proceedings of the 25th international conference on flexible automation and intelligent manufacturing (FAIM)*, Wolverhampton, vol. 2, pp. 198–205. https://www.researchgate.net/publication/281235320_TRIZ_Integration_into_Product_Design_Roadmap. Accessed 7 Jan 2018.

Domb, E. (2001, April). Using TRIZ in a Six Sigma Environment. *Proceedings of TRIZCON2001: The third annual Altshuller institute for TRIZ studies conference*. https://ru.scribd.com/document/236878361/DOMBai2001sixsigma. Accessed 7 Jan 2018.

Domb, E., & Dettmer, W. (1999, May 15). Breakthrough innovation in conflict resolution. *The TRIZ Journal*. https://triz-journal.com/breakthrough-innovation-conflict-resolution-marrying-triz-thinking-process/. Accessed 7 Jan 2018.

Goldense, B. (2016, March 21). TRIZ is now practiced in 50 countries. *Machine Design.* http://machinedesign.com/contributing-technical-experts/triz-now-practiced-50-countries. Accessed 7 Jan 2018.

Howard, T., Culley, S., & Dekoninck, E. (2009, August 24–27). Stimulating creativity: A more practical alternative to TRIZ. *Proceedings of ICED'09, the 17th international conference on engineering design,* Palo Alto, CA, vol. 5, pp. 205–216. https://www.designsociety.org/publication/28693/stimulating_creativity_a_more_practical_alternative_to_triz. Accessed 8 Jan 2018.

Ilevbare, I., Phaal, R., Probert, D., & Padilla, A. (2011). Integration of TRIZ and roadmapping for innovation, strategy, and problem solving: Phase 1 – TRIZ, roadmapping and proposed integrations. *Report on a collaborative research initiative between the centre for technology management,* University of Cambridge and Dux Diligens. http://www.ifm.eng.cam.ac.uk/uploads/Research/CTM/Roadmapping/triz_dux_trt_phase1_report.pdf. Accessed 7 Jan 2018.

Kim, J. H., Kim, I. S., Lee, H. W., & Park, B. O. (2012). A study on the role of TRIZ in DFSS. *SAE International Journal of Passengers Cars – Mechanical Systems, 5*(1), 22–29. https://doi.org/10.4271/2012-01-0068. Accessed 7 Jan 2018.

Litvin, S. (2005). New TRIZ-based tool Function-Oriented Search (FOS). *The TRIZ Journal,* August 13. http://www.triz-journal.com/new-triz-based-tool-function-oriented-search-fos/. Accessed 7 Jan 2018.

Litvin, S. (2011, September 9). *Main parameters of value: TRIZ-based tool connecting business challenges to technical problems in product/process innovation.* 7th Japan TRIZ symposium, Yokohama, Japan. http://www.triz-japan.org/PRESENTATION/sympo2011/Pres-Overseas/EI01eS-Litvin_(Keynote)-110817.pdf. Accessed 7 Jan 2018.

Malinin, L. (2010). The method for transforming a business goal into a set of engineering problems. *International Journal of Business Innovation and Research, 4*(4), 321–337. http://www.inderscienceonline.com/doi/abs/10.1504/IJBIR.2010.03335. Accessed 7 Jan 2018.

Patrishkoff, D. (2012a, January 4). The most popular business initiatives in 2011…per Google data. *David Patrishkoff's Blog.* http://dave-patrishkoff.blogspot.ru/2012/01/most-popular-business-initiatives-in.html. Accessed 7 Jan 2018.

Patrishkoff, D. (2012b, January 16). The worldwide popularity of TRIZ & other innovation-related search terms is dropping…per Google data. *David Patrishkoff's Blog.* http://dave-patrishkoff.blogspot.ru/2012/01/worldwide-popularity-of-triz-other.html. Accessed 7 Jan 2018.

Ries, E. (2011). *The lean startup: How todays entrepreneurs use continuous innova tion to create radically successful businesses.* New York: Crown Publishing Group.

Sibalija, T., & Majstorovic, V. (2009). Six Sigma – TRIZ. *International Journal "Total Quality Management & Excellence"*, *37*(1–2), 375–380. https://www. researchgate.net/publication/236855204_Six_Sigma_-_TRIZ. Accessed 7 Jan 2018.

Spreafico, C., & Russo, D. (2016). TRIZ industrial case studies: A critical sur vey. *Procedia CIRP*, *39*, 51–56. http://www.sciencedirect.com/science/arti cle/pii/S2212827116001803. Accessed 7 Jan 2018.

Yang, K., & El-Haik, B. (Eds.). (2009). *Design for Six Sigma: A roadmap for product development.* New York: The McGraw-Hill Companies, Inc.

Design for Change: Disaggregation of Functions in System Architecture by TRIZ-Based Design

Sebastian Koziołek

1   Introduction

The innovativeness of products is characterized by improved performance at a decreased cost, thereby improving value. Increasingly diverse and ever-changing customer demands provide constant stimuli for industry to adapt their products to remain innovative. As a result, products are systematically re-designed from one product generation to the next in order to follow changing market requirements and maintain a degree of performance enhancement that is attractive to customers. Hence, when starting the development process for a new product generation, it is already anticipated that the product will be modified again in the future by adapting, adding, or removing certain functions and/or components.

The ability to adapt the product functionality depends on the design of the product and its architecture. Depending on the specific structural

S. Koziołek (✉)
Wroclaw University of Science and Technology, Wrocław, Poland
e-mail: sebastian.koziolek@pwr.edu.pl

L. Chechurin, M. Collan (eds.), *Advances in Systematic Creativity*,
https://doi.org/10.1007/978-3-319-78075-7_2

17

aggregation, adapting the functionality of the product may thus require considerable effort and substantial changes to multiple components. This frequently requires changes to associated components, thus increasing required efforts further.

2   Background

Design innovation depends on performance and expenses of invented products (Martinsuo and Poskela 2011). Usually, design solutions are the reflection of the present or future customer needs and the entire design process is focused on satisfying these needs. The market requirements are changeable in the entire Product Life Cycle (PLC) and the same features of product that are attractive at the beginning become ordinary at the end of the PLC. Therefore, the ability to change product systematically and easily has a significant impact on competitiveness of companies. Thus, the TRIZ-based Re-design Methodology is intended for use by research and design (R&D) engineers in order to support decision makers and improve long-term development strategies of a company.

2.1   Function Modeling

To understand the multiple relations between function, behaviour and performance, the system is modeled as a multilayer bipartite network (see Fig. 2). In the model, the primary nodes belong to function (f) as intended purpose of the system. Next, type of nodes presents system behavior (b), which is a method or technique describing how function is achieved. The last layer of nodes represents performance (p) as a nominal range of function output. It is strongly recommended to create the multilayer bipartite network based on the energy-material-signal (EMS) model (Pahl et al. 2007).

The ability of the system to add or erase a function depends mostly on number of connections between the nodes. For example, function $(f1)$ has 20 connections and is aggregated with function $(f2)$ in the range of eight connections. The aggregation is counted by number of connections between functions and intersected behavior nodes related to those functions. This modeling technique shows the system complexity and ability to group components in system architecture. If the number of connections is relatively high, the system should be re-designed on the functional level first. Moreover, if the model of multilayer bipartite network presents separated functions with a low number of function aggregation, the system is qualified for re-designing on the architectural level (see Fig. 1).

2.2   System Architecture Modeling

When the function model is disaggregated, the system is prepared for architecture modeling. In the proposed methodology, architectural description of the system is based on the TRIZ System Operator (Altshuller 1990). In this stage, relations of functions and system components are identified and structured in a Function–Architecture model (see Fig. 2).

Behaviors *(b1–b6)* are delivered in the system by sub-systems. In the process of product development, when the change of system behavior is required, all the related sub-systems must be re-designed. Therefore, in

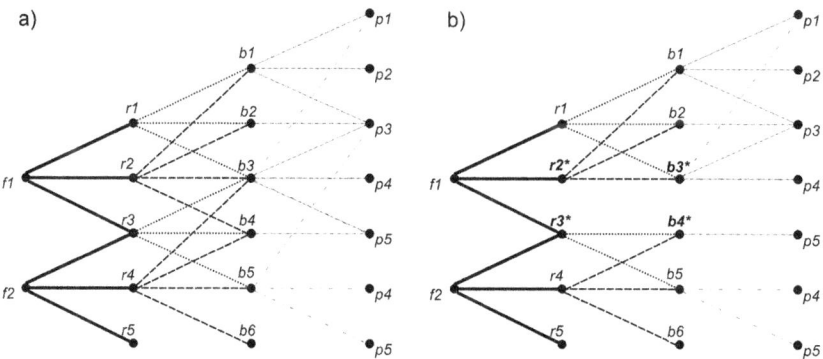

Fig. 1 Model of multilayer bipartite network of a system: (a) model of system with aggregated functions; (b) model of system with disaggregated functions, where: *(r2*)* and *(r2*)- replacement of resources; (b3*)* and *(b4*) – represents re-design methods*

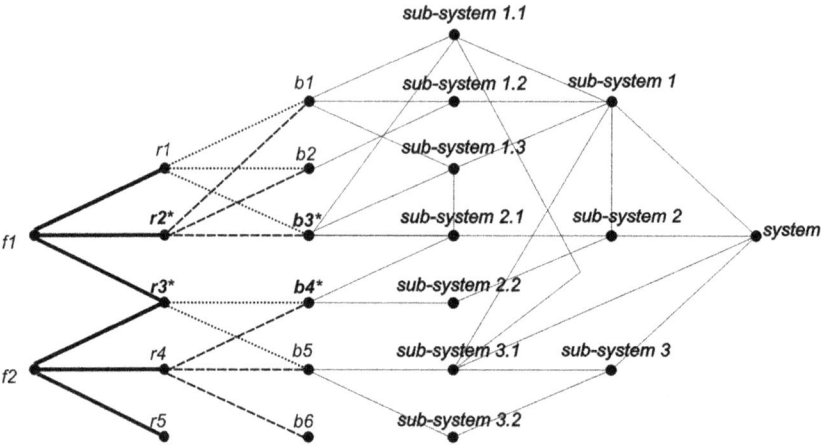

Fig. 2 Function–architecture model of system

this case, the number of connections between the nodes of sub-systems and system behavior has also a significant impact on design process.

In the aspect of design for change, the ideal model has only three connections in the Function–Architecture model: *f1–r1*, *r1–b1*, *b1–sub-system 1.1*. Therefore, the number of node connections should be as few as possible.

2.3   Grouping in System Architecture

Modularization is a known and appreciated strategy in R&D of many manufacturing companies (Gu and Sosale 1999; Nepal et al. 2008; Seol et al. 2007). Mostly, the purpose of modularization is to simplify both manufacturing process and standardization (Sered and Reich 2006). In the proposed TRIZ-based Re-design Methodology, grouping in system architecture is a method for systematical product change in PLC following the changeable market requirements.

The first step of grouping is identifying components with the highest number of node connections (single element of the sub-system). These components are Special Connectors (SCs) qualified for standardized and unchangeable elements of the system architecture in single run of

PLC. The sub-systems are dedicated for systematic change using the stan dardized SC elements. Nested sub-systems are also changeable, even more frequently than sub-systems.

In the final stage, the plan of product re-design is prepared. There are three principles of design for change based on the grouping model. First, the most frequent changes are dedicated to the nested sub-systems. Second, sub-systems may be changed a few times in the PLC. Third, SCs are intended for change only as a next-generation product, when a new run of PLC is needed (Aurich et al. 2006; Gu and Sosale 1999). All the changes have to be planned according to the principles following the market requirements (Hansen and Sun 2011; Nepal et al. 2008; Sered and Reich 2006; Steve et al. 2013).

3   Proposed Solution

Producing electricity from biogas is a well-known solution of waste disposal (Berglund and Börjesson 2006; Weiland 2010; Wellinger et al. 2013; Wiśniewski et al. 2015). Nevertheless, one of the main disadvantages of this technology is the low efficiency of the energy used in cases where the heat from biogas combustion is not fully consumed. Biogas is produced mainly in waste treatment systems or in agricultural biogas plants, which are usually located outside the urbanized zones. The highest efficiency of biogas utilization is achieved when the heat and electricity are consumed by nearby industrial companies (Herout et al. 2011). However, location of the biogas plants outside the urbanized zones limits industry's ability to use the waste gas efficiently. The presented solution is the mobile biogas station, which makes biogas available for every industrial company with a high need of electrical and thermal energy consumption. The essential problem to resolve is simple customization of the station in order to maximize the market share of the new product. The main functions of the systems are storage and distribution of biogas in a safe and economically justified way. The new mobile compressed biogas filling station was designed using the TRIZ-based Re-design Methodology. The main function of the device is to provide compressed biogas for machinery and/or equipment adapted for biogas or natural gas. This system was designed to deliver biogas with the specifications listed in Table 1. The design process was supported by technology forecasting and market research.

Based on market research and technology forecasting (Cascini et al. n.d), the biogas delivery system was described with the use of an IDEF0 model (Kim et al. 2003). First, this detailed description of the process was used for business modeling in order to confirm that the concept of mobile biogas is potentially expected on the market. The business model was approved and then the system was analyzed in the context of useful, harmful functions and available resources (inc. expenses). Finally, the function model was integrated with system architecture in order to identify the number of node connections in the Function–Architecture model (see Fig. 3).

In the function model of the mobile biogas station, the system has 17 node connections. Space and Biogas are identified as intersected resources.

Table 1 Biogas specification

Compound	Molecular formula	Percentage
Methane	CH_4	50–85
Carbon dioxide	CO_2	25–55
Nitrogen	N_2	0–5
Oxygen	O_2	0–0.34
Hydrogen sulphide	H_2S	0–1
Hydrogen	H_2	0–1

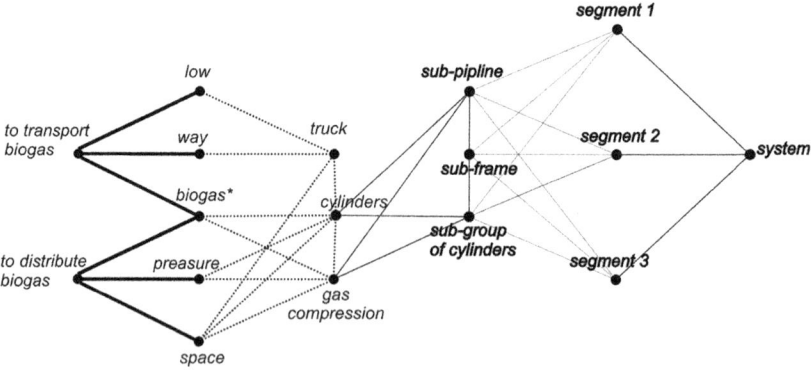

Fig. 3 Mobile biogas station – function–architecture model

The functions aggregation of <to transport biogas> and <to distribute biogas> is high, because there is a strong contradiction between resource of *<space>* and behavior *<truck>*. Therefore, this problem had to be solved in order to disaggregate the functions. As a result, the aggregation range was reduced from eight to six connections. It is satisfactory because all the connections are related to the main resource *<biogas>*.

In the next step of the system architecture modeling, the SCs were identified. In the presented case, there are two SC elements: *connecting pipeline* and *frame* (see Fig. 4).

The entire system is comprised of three segments. Each of the segments is connected with the other segments by the SC elements only.

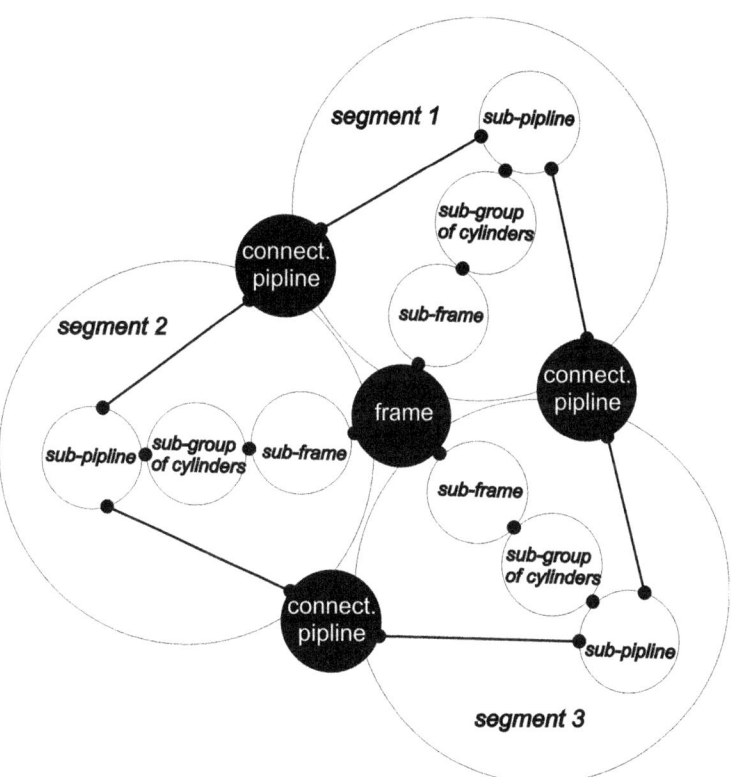

Fig. 4 Mobile biogas station – grouping model

Internally, each of the segments is equipped with a group of 12 cylinders connected by sub-pipelines and supported by sub-frames. Each of the system elements may be easily re-designed; even the function <to transport biogas> will be changed to <to transport methane> or <oxygen> if required.

4   Discussion and Recommendations

In the presented approach of the case study, the project leader used problem description as <aggregation of the system is too high>, which is measurable by number of connections between functions and intersected behavior nodes related to those functions. After problem identification, the project leader decided to build the problem-solving team. There was no rule for selecting team members; he intuitively collected participants who were the most experienced and well known to him. Unlike heuristic design, in a systematic design approach the mobile biogas station was carefully described in order to identify the problem in the perspective of its harmful result. Finally, the problem of high-function aggregation was described parametrically with the use of a Function–Architecture Model (see Fig. 3). Based on these properties, the system complexity was defined as unacceptable. The parametrical specification limit of function aggregation was also determined experimentally by manual prototyping with the use of design thinking methodology. Therefore, the mobile biogas station was modeled and design as a modularized system with benefits of less manufacturing complexity and high ability to re-design in a short time according to changeable market requirements.

Acknowledgements This chapter demonstrates the design concept for a mobile biogas station and presents results of experimental research of the proposed energy solution. The presented technology was developed as part of Project LIDER/034/645/L-4/12/NCBR/2013 *"Mobile gas supply station for treated and compressed biogas"* and funded by the National Centre for Research and Development. The functional modeling of the new product was conducted at the University of Sydney in range of the Go8 European Fellowship.

References

Altshuller, G. S. (1990). On the theory of solving inventive problems. *Design Methods and Theories, 24*, 1216–1222.

Aurich, J. C., Fuchs, C., & Wagenknecht, C. (2006). Life cycle oriented design of technical product-service systems. *Journal of Cleaner Production, 14*(17), 1480–1494. https://doi.org/10.1016/j.jclepro.2006.01.019.

Berglund, M., & Börjesson, P. (2006). Assessment of energy performance in the life-cycle of biogas production. *Biomass and Bioenergy, 30*(3), 254–266. https://doi.org/10.1016/j.biombioe.2005.11.011.

Cascini, G., Ramadurai, B., Slupiński, M., Becattini, N., Kaikov, I., Kucharavy, D., Nikulin, C., & Sebastian Koziolek, E. F. (n.d.). *FORMAT – The handbook: Knowing the future is possible.* Retrieved from https://www.amazon.com/dp/B06XTJWB87

Gu, P., & Sosale, S. (1999). Product modularization for life cycle engineering. *Robotics and Computer-Integrated Manufacturing, 15*(5), 387–401. https://doi.org/10.1016/S0736-5845(99)00049-6.

Hansen, P. K., & Sun, H. (2011). Complexity in managing modularization. In *Proceedings – 2011 4th international conference on information management, innovation management and industrial engineering, ICIII 2011* (Vol. 3, pp. 537–540). Shenzhen, China: IEEE. https://doi.org/10.1109/ICIII.2011.410

Herout, M., Malaták, J., Kucera, L., & Dlabaja, T. (2011). Biogas composition depending on the type of plant biomass used. *Research in Agricultural Engineering, 57*(4), 137–143.

Kim, C. H., Weston, R. H., Hodgson, A., & Lee, K. H. (2003). The complementary use of IDEF and UML modelling approaches. *Computers in Industry, 50*(1), 35–56. https://doi.org/10.1016/S0166-3615(02)00145-8.

Martinsuo, M., & Poskela, J. (2011). Use of evaluation criteria and innovation performance in the front end of innovation. *Journal of Product Innovation Management, 28*(6), 896–914. https://doi.org/10.1111/j.1540-5885.2011.00844.x.

Nepal, B., Monplaisir, L., Singh, N., & Yaprak, A. (2008). Product modularization considering cost and manufacturability of modules. *International Journal of Industrial Engineering: Theory Applications and Practice, 15*(2), 132–142.

Pahl, G., Beitz, W., Feldhusen, J., & Grote, K.-H. (2007). Engineering design: A systematic approach. *Engineering Design: A Systematic Approach.* https://doi.org/10.1007/978-1-84628-319-2.

Seol, H., Kim, C., Lee, C., & Park, Y. (2007). Design process modularization: Concept and algorithm. *Concurrent Engineering Research and Applications, 15*(2), 175–186. https://doi.org/10.1177/1063293X07079321.

Sered, Y., & Reich, Y. (2006). Standardization and modularization driven by minimizing overall process effort. *CAD Computer Aided Design, 38*(5), 405–416. https://doi.org/10.1016/j.cad.2005.11.005.

Steve, J., Heilemann, M., Culley, S. J., Schluter, M., & Haase, H. J. (2013, August). *Examination of modularization metrics in industry* (pp. 1–10). International Conference On Engineering Design, ICED 13.

Weiland, P. (2010). Biogas production: Current state and perspectives. *Applied Microbiology and Biotechnology, 85*, 849. https://doi.org/10.1007/s00253-009-2246-7.

Wellinger, A., Murphy, J., & Baxter, D. (2013). The biogas handbook: Science, production and applications. *The Biogas Handbook: Science, Production and Applications.* https://doi.org/10.1533/9780857097415.

Wiśniewski, D., Gołaszewski, J., & Białowiec, A. (2015). The pyrolysis and gasification of digestate from agricultural biogas plant/Piroliza i gazyfikacja pofermentu z biogazowni rolniczych. *Archives of Environmental Protection, 41*(3). https://doi.org/10.1515/aep-2015-0032.

Systematic Innovation in Process Engineering: Linking TRIZ and Process Intensification

Pavel Livotov, Arun Prasad Chandra Sekaran, Richard Law, Mas'udah, and David Reay

1   Introduction: Literature Review and Objectives

Process Intensification (PI) as a part of knowledge-based engineering (KBE) can be defined as any significant technological development leading to more efficient and safer processes in chemical, petrochemical, and pharmaceutical industries. The PI databases of new technologies and equipment allow one to quickly achieve the typical goals of innovation, such as reduced energy and raw material consumption, increased process

---

P. Livotov (✉) • A. P. Chandra Sekaran • Mas'udah
Faculty of Mechanical and Process Engineering, Laboratory for Product and Process Innovation, Offenburg University of Applied Sciences, Offenburg, Germany
e-mail: pavel.livotov@hs-offenburg.de

R. Law
Newcastle University, Newcastle upon Tyne, UK

D. Reay
David Reay & Associates, Newcastle upon Tyne, UK

© The Author(s) 2019   **27**
L. Chechurin, M. Collan (eds.), *Advances in Systematic Creativity*,
https://doi.org/10.1007/978-3-319-78075-7_3

flexibility, safety and quality, and better environmental performance. However, some of these objectives are often contradictory in their realization. In order to accelerate the implementation of PI technologies and solutions, the identified engineering contradictions can be eliminated with the help of the Theory of Inventive Problem Solving (TRIZ), which is today considered as one of the most comprehensive invention methodologies. Both approaches—PI and TRIZ—were developed and are currently used independent of each other. Therefore, an attempt has been made to analyse how the various methods and technologies of PI can be linked to the components of TRIZ.

The concept of PI dates back to the research of Prof. Ramshaw and his colleagues (Cross and Ramshaw 1986; Reay et al. 2013) and subsequently became more diverse in its implementation and practice. Today it can be generally defined as a knowledge-based methodology for "development of innovative apparatus and techniques that offer drastic improvements in chemical manufacturing and processing, substantially decreasing equipment volume, energy consumption, or waste formation, and ultimately leading to cheaper, safer, sustainable technologies" (Stankiewicz and Moulijn 2000). The modern interpretation of PI also includes benefits related to business, process, and environmental aspects of process engineering (Boodhoo and Harvey 2013). The PI technological databases are continuously evolving and currently cover more than 150 components, representing two distinct categories—equipment and processing methods for liquid/liquid and liquid/gas reactions, separations, absorption, solids handling, and so on (Reay et al. 2013; Stankiewicz and Moulijn 2000; Boodhoo and Harvey 2013; Wang et al. 2008), as shown in Table 1.

The existing PI databases enable engineers to identify and implement appropriate process-intensifying solutions faster in accordance with the objectives and constraints of their development tasks. However, some of the PI objectives can be often contradictory in their specific realization (Benali and Kudra 2008; Kardashev 1990). For example, decreasing equipment volume may cause product quality deviations, make the process control more difficult, or lead to other negative side effects or limitations. The analysis of 150 recent patent documents in the field of solid handling demonstrates that all inventions promise to solve numerous problems but also generate so-called secondary problems (Casner and

Table 1 Process intensification equipment and methods with examples according to Stankiewicz and Moulijn (2000), Boodhoo and Harvey (2013)

Equipment		Processing methods			
Operations involving chemical reactions	Operations not involving chemical reactions	Multifunctional reactors	Hybrid separations	Alternative energy sources	Other methods
Spinning disk reactor, Static mixer reactor, Static mixing catalysts, Monolithic reactors, Microreactors, Heat exchange reactors, Supersonic gas/liquid reactor, Jet-Impingement reactor, Rotating packed-bed reactor, others	Static mixers, Compact heat exchangers, Microchannel heat exchangers, Rotor/stator mixers, Rotating packed beds, Centrifugal absorber, others	Reverse-flow reactors, Reactive distillation, Reactive extraction, Reactive crystallization, Chromatographic reactors, Periodic separating reactors, Membrane reactors, Reactive extrusion, Reactive comminution, Fuel cells, others	Membrane absorption, Membrane distillation, Adsorptive distillation, others	Centrifugal fields, Ultrasound, Solar energy, Microwaves, Electric and electromagnetic fields, Laser and plasma technologies, others	Supercritical fluids, Dynamic (periodic) reactor operation, others

Livotov 2017). Moreover, a growing demand for faster transformation from research to market requires new engineering and inventive approaches for a) rapid optimization and adaptation of the existing PI solutions, and b) early development of entirely new PI equipment or processing technologies.

These challenges can be met by a dramatic enhancement of engineers' inventive skills and technological competences with the methods and tools of TRIZ for identification and elimination of the technical contradictions. Since it was established by Altshuller and co-workers (Altshuller 1984), modern TRIZ is considered the most comprehensive, systematically organized invention and creative thinking methodology for knowledge-based innovation (KBI) (Cavallucci et al. 2015; VDI 2016). One of the main advantages of TRIZ is that it allows for finding new, inventive solutions for a given problem in a systematic way by using the entire potential of science and engineering, even outside of the field of the originally formulated problem (Altshuller 1984, Livotov and Petrov 2013).

The application of TRIZ in process engineering started relatively recently and progressed in three directions: (1) direct application of existing TRIZ methods and tools, (2) adapting TRIZ for the domain of process engineering, including development of new approaches, and (3) extending or combining specific process engineering solutions with TRIZ solution principles.

In addition to the successful TRIZ applications in process engineering by the TRIZ experts, there belong new developments of chemical or biochemical products and technologies (Abramov et al. 2015), problem solving with inventive principles and standard solutions (e.g. Rahim et al. 2015; Ferrer et al. 2012; Kim et al. 2009; Kraslawski et al. 2000; Srinivasan and Kraslawski 2006), and TRIZ evolutionary forecast of equipment (Berdonosov et al. 2015) and technologies (Cascini et al. 2009).

Numerous researches outline the necessity to adapt TRIZ for the domain of process engineering, such as for an environment-oriented eco-innovation approach for the chemical industry with a reduced abstraction level of TRIZ (Ferrer et al. 2012), TRIZ modifications in context with safety issues of chemical reactors (Kim et al. 2009) and processes (Srinivasan and Kraslawski 2006). Process engineering interpretations and examples for 40 TRIZ inventive principles are presented in Grierson et al. (2003) and Hipple (2005). Two contradiction matrix versions adapted for problem solving in process engineering are condensing the number of engineering parameters to six categories—process disturbance, design, mechanics, human operator, natural hazard, and materials (Kim et al. 2009)—and to 14 general characteristics (Pokhrel et al. 2015) such as complexity, concentration, conversion, economics, and so on. Based on the analysis of research articles, eight solution principles for process engineering are proposed in Pokhrel et al. (2015): change equipment type or design, change operation sequence, change process chemistry or conditions, convert harmful effects into benefits, generate material just on time and on place, make operation simpler, and use simple design. In Yakovis and Chechurin (2015), the standard and creative design approaches are integrated in a new process control design method illustrated with examples related to cement manufacturing.

Combining specific food-processing standard solutions and technologies with some TRIZ solution principles is proposed in the "agro-food equipment design" method (Totobesola-Barbier et al. 2002). Another approach merges the Case-Based Reasoning (CBR) in chemical engineering with TRIZ (Robles et al. 2009). The CBR is applied to solve problems using existing chemical solutions, which can be further enhanced by accessing other engineering domains with TRIZ. However, more recent investigations have shown that directly merging two innovation approaches, CBR and TRIZ, may weaken each approach if the execution of one is dominated by another (Houssin et al. 2014).

Based on the presented literature review and the practical experience of the authors, the approach for linking PI and TRIZ methodologies in process engineering should ensure their complementary and mutual reinforcement, and involve the following features:

(a) Mutual adaptation of methodologies, making them understandable and reliably applicable by non-experts in TRIZ and PI.
(b) Completeness and repeatability of TRIZ and PI methods and principles, and avoidance of simplifying generalizations.
(c) Universality, flexibility, and adaptability to varying requirements or limitations in practice.

2   Research Approach

The fundamentals and objectives of PI (Boodhoo and Harvey 2013; Wang et al. 2008; Gerven and Stankiewicz 2009) are highly consistent with the postulates and evolution laws of technical systems in the TRIZ methodology (VDI 2016; Altshuller 1984; Livotov and Petrov 2013). At the same time, numerous problem-solving tools and methods of TRIZ correspond to only one PI database of equipment and processing methods, classified into thermodynamic, functional, spatial, or temporal domains (Gerven and Stankiewicz 2009), as presented in Table 2.

Practically, all these TRIZ tools can be applied in combination with PI in accordance with the basic algorithm (Casner and Livotov 2017) shown in Fig. 1. However, using one universal TRIZ tool seems to be more

Table 2 Comparing fundamentals and methods of PI and TRIZ

	Process Intensification	TRIZ Methodology
Fundamentals	Decreasing energy and raw material consumption, waste, costs (Stankiewicz and Moulijn 2000)	Law of increasing the degree of ideality of technical systems
	Transition from the macro- to meso- and molecular scale (Boodhoo and Harvey 2013)	Transition from macro to micro level (evolution pattern)
	Enhancement of the force fields: mechanical-acoustic-electromagnetic-light energy	Increasing the controllability of fields (evolution pattern)
		Other evolution patterns of technical systems
Main problem-solving tools and methods	Equipment: reactors, mixing, heat- or mass-transfer devices, etc.	40 inventive principles
		76 standard solutions
	Methods: extraction, separation, absorption, techniques using alternative energy sources and new process-control methods, etc.	Inventive algorithm ARIZ
		Separation principles
		Database of physical, chemical, biological, and geometrical effects

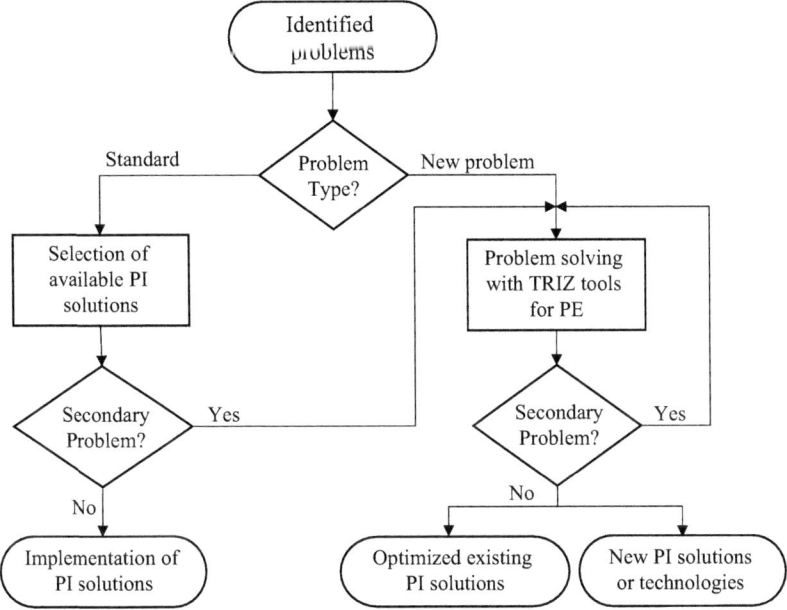

Fig. 1 Possible TRIZ and PI combination in process engineering PE according to Casner and Livotov (2017)

convenient for both process engineers and researchers. Selection of only one TRIZ tool is also favourable for faster revealing of concrete opportunities for synergies between TRIZ and PI in process engineering. The choice was made for 40 inventive principles (Altshuller 1984), which over decades have remained the most popular and usable TRIZ components in industrial TRIZ practice (Livotov and Petrov 2013; Grierson et al. 2003; Hipple 2005). The reason for this decision is that inventive principles are good for newcomers to TRIZ; they are simple to use or modify for a specific technical domain and can be easily integrated in brainstorming sessions or an engineer's daily work. Another established part of industrial practice is the composition of the specific groups of inventive principles for solving different kinds of problems, for example, statistically most often used principles (Nos. 35, 10, 1, 28, etc.), principles for solving design problems, principle sets for cost reduction

(Livotov and Petrov 2013), other customized sets, or even the principles selection with the contradiction matrix (Altshuller 1984).

Each TRIZ inventive principle can be characterized by its ID or number, title, and detailed explanations containing in the classical version (Altshuller 1984) one to five sub-principles. These sub-principles act as inventive operators, defining directions for system transformation and ideation. The TRIZ application and development of recent decades has demonstrated that many sub-principles were extended, updated, or adapted for specific technical domains by numerous authors, for example, Livotov and Petrov (2013), Grierson et al. (2003), and Hipple (2005). Under these circumstances, the following research plan was proposed to analyse the relationship of 155 PI technologies to TRIZ inventive principles, sub-principles, some TRIZ inventive standards, and evolution patterns:

1. Extension of the principle titles and sub-principles with the inventive operators relevant to the process engineering (based on the analysis of research and practitioner literature, and the practical experience of the authors).
2. Removal of some repetitions of similar sub-principles and minor reassignment of the sub-principles to the 40 inventive principles.
3. Renaming of terms used in sub-principles in case they coincide with the specific terms in the process engineering, such as separation, extraction, and so on.
4. Introduction and assignment of the new sub-principles corresponding to practically all relevant evolution patterns and some TRIZ standard solutions.
5. Introduction of the new sub-principles, identified as core inventive operations in the PI technologies (equipment and methods).
6. Limitation of the maximum number of sub-principles to five for each principle and to 200 in total for 40 invention principles.
7. Identification and statistical analysis of 40 inventive principles and sub-principles used in the 155 PI technologies.

8. Identification of 40 inventive principles and sub-principles used in 150 patent documents published between 2008 and 2016 in the field of solid handling in ceramic (100 documents) and pharmaceutical (50 documents) industries.

Table 3 shows exemplarily the new version of principle 31. Porous materials, which includes a typical pattern of evolvement and utilization of porous structures from introducing cavities or porosity (31a) to utilization of structured porosity and capillaries in combination with filler and external fields (31c, d, e).

The identification of the inventive principles in PI is illustrated in Table 4.

For each of 155 PI technologies, a systematic analysis of the processing methods and design solutions has resulted in extracting of corresponding TRIZ 40 inventive principles with their sub-principles (inventive operators), as shown in the example of the Spiral Flash Dryer (SFD). Only the

Table 3 TRIZ inventive principle "porous materials" with updated sub-principles

Inventive principle	Updated sub-principles (inventive operators)	Classical sub-principles (Altshuller 1984)
31. Porous materials	(a) Make an object or its surface porous, or add porous elements (inserts, coatings, etc.). Utilize objects with hollow spaces or cavities	(a) Make an object porous or add porous elements (inserts, covers, etc.)
	(b) If an object is already porous, fill the pores with a useful substance	(b) If an object is already porous, fill the pores with a useful substance
	(c) Utilize capillary and micro-capillary effects in porous materials	
	(d) Use the filler in combination with physical effects, (e.g. ultrasound, electromagnetic field, temperature differences, osmosis, etc.)	
	(e) Use structured porosity, like pipes, canals, or capillaries on the molecular level	

Table 4 Identification of TRIZ inventive principles for the Spiral Flash Dryer SFD

N	Key characteristics of the PI technology	TRIZ Inventive Principles	Corresponding sub-principles (inventive operators)
1	The product is fluidized by the drying air or gas without mechanical moving parts	29 Pneumatic or hydraulic constructions	(a) Gas as a working element (d) Fluidization of powders or granulates
2	Due to the cone, the gas flow is toroidal and highly turbulent	14 Spheroidality and Rotation	(b) Use sphere. cylinders, cones (d) Swirling motion
3	Toroidal shape with the highest surface area to the gas stream	17 Shift to another dimension	(e) Increase contact area between objects or substances
4	High gas impact minimizes the insulating gas layer around particles, increasing heat transfer	21 Skipping/Rushing through	(b) Boost the process that may result in new useful properties

inventive operators significant for the distinguishing characteristics of the PI technologies were taken into consideration.

3   Discussion of Results

The performed analysis of the PI technologies and other research and patent literature has resulted in the creation of the advanced set of 160 sub-principles, assigned to the 40 TRIZ inventive principles, which are presented in the appendix. Compared with the original number of sub-principles varying between 88 and 90 (Livotov and Petrov 2013; Grierson et al. 2003; Hipple 2005), at least 70 additional inventive operators relevant for process engineering have been introduced. Many of them were identified or refined due to the PI database, such as, for example, 14(d) Swirling motion, 29(d) Fluidization of powders, and 30(e) Membrane operations and processing.

The top 10 TRIZ inventive principles, most frequently used in the PI technologies, are presented in Fig. 2 Nearly all of them are related to five major classes of process operations, such as heat and mass transfer processes, fluid flow processes, and thermodynamic and mechanical processes.

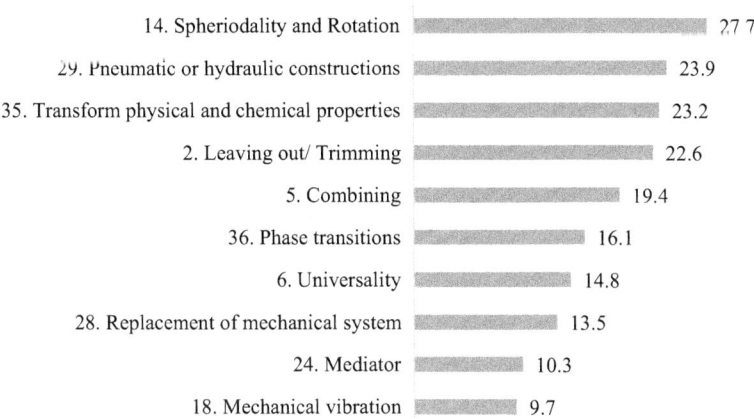

Fig. 2 Top 10 TRIZ inventive principles most frequently encountered in the 155 analysed PI technologies, in [%]

These top 10 inventive principles with corresponding sub-principles can be generally recommended for the new development or optimization of PI equipment or methods. The suggested application of the selected sub-principles instead of the principles may help to reduce a number of ideation efforts in the early stage of the innovation process. For example, the often applied principle 35—Transformation of physical and chemical properties—with the total frequency of 23.2%, is represented only by three sub-principles: 35(d) Change temperature, 8.4%; 35(b) Change concentration, 8.4%; followed by 35(a) Change aggregate state, 6.4%. Another benefit of using the sub-principles is a possibility to build a specific set of those inventive operators that are most appropriate for definite PI objectives: inventive operators for reduction of energy consumption, for waste reduction, for cost cutting, as well as specific groups of operators for solid handling or other process engineering domains or industries. For example, the comparison of the TRIZ inventive sub-principles extracted from the patent literature in the field of solid handling is illustrated in Fig. 3.

The sets of inventive sub-principles frequently used in patents for analogous ceramic and pharmaceutical processing operations show clear differences. The main reason is that the ceramic processing operations are mostly focused on controlling the mechanical properties, and the

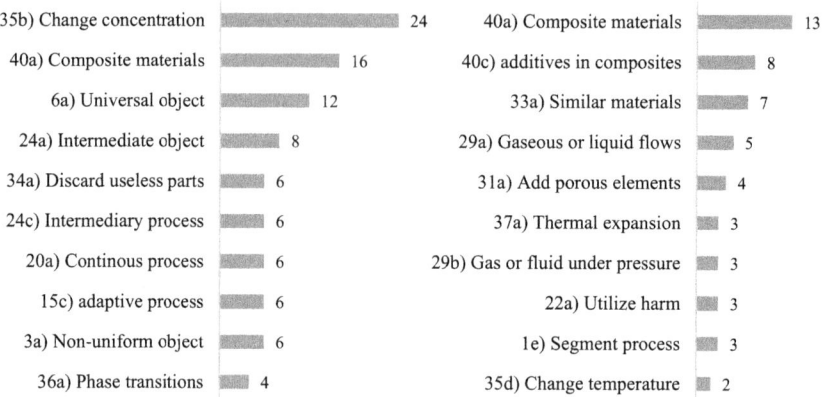

Fig. 3 Top 10 innovation sub-principles most frequently encountered in 150 patent documents in [%]: *left* – pharmaceutical; *right* – ceramic operations

pharmaceutical operations with solids deal with the physical, chemical, and biological properties. As a consequence, even in the case of similar process intensification demands and operations, engineers of different industrial sectors require specific sets of inventive principles for problem solving. It is to be anticipated that different recommendable sets of inventive principles exist for liquid/liquid, liquid/gas reactions, and for solids handling.

Another notable finding is that many statistically strong inventive principles, such as Nos. 2, 5, 14, 18, and 28 (see Fig. 3), frequently used in the PI technologies, practically don't appear in the analysed patents or patent applications.

4   Concluding Remarks and Outlook

The objective of the described approach for linking PI and TRIZ was to reveal synergies and mutual benefits between both methodologies. The performed analysis of large numbers of PI equipment and methods, and of the related patents and research literature resulted in the comprehensive listing of 160 inventive operators in the framework of 40 TRIZ inventive principles. On one hand, the specific PI knowledge has been transferred to the abstract knowledge domain of TRIZ. On the other hand, the extended inventive principles were refined for their intuitive application in process engineering and especially for the optimization of existing and the creation of novel PI techniques. In addition to the classical usage of the whole TRIZ inventive principles, the suggested application of the specific sub-principles groups seems to be a promising and more precise technique, adaptable for a large variety of potential problem situations like

- mobilization of resources of the existing processes to reach the maximum efficiency with minimum expenditure,
- limitation of the side effects of new PI techniques to enable their smooth loss-free implementation,
- inventive solving of the bottle-neck problems in process engineering, and
- prediction of the technological evolution for processes and equipment, and others.

Acknowledgments The authors wish to thank the European Commission for supporting their work as part of the research project "Intensified by Design® platform for the intensification of processes involving solids handling" within international consortium under H2020 SPIRE programme.

Appendix

Advanced TRIZ Inventive Principles with 160 sup-principles for Process Engineering (without description and examples).

1 Segmentation	**21 Skipping/Rushing through**
1(a) Segment object	21(a) Skip hazardous operations
1(b) Dismountable design	21(b) Boost the process
1(c) Segment to microlevel	**22 Converting harm into benefit**
1(d) Segment function	22(a) Utilize harm
1(e) Segment process	22(b) Remove harm with harm
2 Leaving out/*Trimming*	22(c) Amplify harm to avoid it
2(a) Take out disturbing parts	**23 Feedback and automation**
2(b) Trim components	23(a) Introduce feedback

2(c) Trim functions
2(d) Trim process steps
2(e) Extract useful element
3 Local quality
3(a) Non-uniform object
3(b) Non-uniform environment
3(c) Different functions
3(d) Optimal conditions
3(e) Opposite properties
4 Asymmetry
4(a) Asymmetry
4(b) Enhance asymmetry
4(c) Back to symmetry
5 Combining
5(a) Combine similar objects
5(b) Combine functions
5(c) Combine different properties
5(d) Combine complementary properties
5(e) Combine opposing properties
6 Universality
6(a) Universal object
6(b) Universal process
7 Nesting/Integration

7(a) Nested objects
7(b) Passing through cavities
7(c) Telescopic systems
8 Anti-weight
8(a) Use counterweight
8(b) Buoyancy)
8(c) Aero- or hydrodynamics
8(d) Use gravitation
9 Prior Counteraction of harm
9(a) Counter harm in advance
9(b) Anti-stress
9(c) Cooling in advance
9(d) Rigid construction
10 Prior useful action
10(a) Prior useful function
10(b) Pre-arrange objects
10(c) Prior process step
11 Preventive measure/Cushion in advance
11(a) Safety cushion

23(b) Enhance feedback
23(c) Automation
23(d) Data processing
24 Mediator
24(a) Intermediate object
24(b) Temporary mediator
24(c) Intermediary process
25 Self service
25(a) Object serves itself
25(b) Utilize waste resources
25(c) Use environmental resources
26 Copying
26(a) Simple copies
26(b) Optical copies
26(c) Invisible copies
26(d) Digital models
26(e) Virtual reality
27 Disposability/cheap short-living objects
27(a) Short-living objects
27(b) Multiple cheap objects
27(c) One-way objects
27(d) Create objects from resources
28 Replace mechanical working principle
28(a) Use electromagnetics
28(b) Optical systems
28(c) Acoustic system
28(d) Chemical and biosystems
28(e) Magnetic particles and fluids
29 Pneumatic or hydraulic constructions
29(a) Gaseous or liquid flows
29(b) Gas or liquid under pressure
29(c) Use vacuum
29(d) Fluidization
29(e) Heat transfer and exchange
30 Flexible shells or thin films
30(a) Flexible shells or films
30(b) Flexible isolation
30(c) Piezoelectric foils
30(d) Use rushes
30(e) Use membranes
31 Porous material
31(a) Add porous elements

11(b) Preventive measures
12 Equipotentiality
12(a) Keep altitude
12(b) Equipotentiality
12(c) Avoid fluctuations
13 Inversion
13(a) Inversed action
13(b) Make fixed parts to movable
13(c) Upside down
13(d) Reversed sequence
13(e) Invert environment
14 Spheroidality and Rotation
14(a) Ball-shaped forms
14(b) Spheres and cylinders
14(c) Rotary motion
14(d) Swirling motion
14(e) Centrifugal forces
15 Dynamism and adaptability

15(a) Optimal performance
15(b) Adaptive object
15(c) Adaptive process
15(d) Flexible elements
15(e) Change statics to dynamics
16 Partial or excessive action
16(a) One step back from ideal
16(b) Optimal substance amount
16(c) Optimal action
17 Shift to another dimension
17(a) Multi-dimensional form
17(b) Miniaturization
17(c) Multi-layered structure
17(d) Tilt object
17(e) 3D interaction
18 Mechanical vibration
18(a) Oscillate object
18(b) Ultrasound
18(c) Resonance
18(d) Piezo-electric vibrators
18(e) Ultrasound with other fields
19 Periodic action
19(a) Periodic action
19(b) Change frequency
19(c) Use pauses
19(d) Match frequencies

31(b) Fill pores with substance
31(c) Use capillary effects
31(d) Physical effects and porosity
31(e) Structured porosity
32 Change colour
32(a) Change colour
32(b) Change transparency
32(c) Coloured additives
32(d) Use tracer
33 Homogeneity
33(a) Similar materials
33(b) Similar properties
33(c) Uniform properties
34 Rejecting and regenerating parts
34(a) Discard useless parts
34(b) Restore parts
34 (c) Create parts on time and on site
35 Transform physical and chemical properties
35(a) Change aggregate state
35(b) Change concentration
35(c) Change physical properties
35(d) Change temperature
35(e) Change chemical properties
36 Phase transitions
36(a) Phase transitions
36(b) 2nd order phase transitions
37 Thermal expansion
37(a) Thermal expansion
37(b) Bi-metals
37(c) Heat shrinking
37(d) Shape memory
38 Strong Oxidants
38(a) Oxygen-enriched air
38(b) Use pure oxygen
38(c) Use ionized oxygen
38(d) Use ozone
38(e) Strong oxidants
39 Inert environment
39(a) Inert environment
39(b) Inert atmosphere process
39(c) Process in vacuum
39(d) Inert coatings or additives
39(e) Use foams
40 Composite materials

19(e) Separate in time	40(a) Composite materials
20 Continuity of useful action	40(b) Use anisotropic properties
20(a) Continuous process	40(c) Additives in composites
20(b) Operate at full load	40(d) Composite microstructure
20(c) Eliminate idle work	40(e) Combine different aggregate states

References

Abramov, O., Kogan, S., Mitnik-Gankin, L., et al. (2015). TRIZ-based approach for accelerating innovation in chemical engineering. *Chemical Engineering Research and Design, 103*, 25–31.

Altshuller, G. S. (1984). *Creativity as an exact science. The theory of the solution of inventive problems.* Gordon & Breach Science Publishers, issn 0275–5807.

Benali, M., & Kudra, T. (2008). *Drying process intensification: Application to food processing.* Retrieved September 11, 2016, from https://www.researchgate.net/publication/266211018

Berdonosov, V. D., Kozlita, A. N., & Zhivotova, A. A. (2015). TRIZ evolution of black oil Coker units. *Chemical Engineering Research and Design, 103*, 61–73.

Boodhoo, K. V. K., & Harvey, A. (2013). Process intensification: An overview of principles and practice. In V. Kamelia, K. Boodhoo, & A. Harvey (Eds.), *Process intensification for green chemistry: Engineering solutions for sustainable chemical processing* (pp. 1–31). Wiley.

Cascini, G., Rotini, F., & Russo, D. (2009). Functional modeling for TRIZ-based evolutionary analyses. DS 58–5: Proceedings of ICED 09, the 17th International Conference on engineering design, vol. 5, *Design methods and tools* (pt. 1), Palo Alto, CA, 24.-27.08.2009.

Casner, D., & Livotov, P. (2017). Advanced innovation design approach for process engineering proceedings of the 21st international conference on engineering design (ICED 17) vol. 4, *Design methods and tools*, Vancouver, 21–25.08.2017, pp. 653–662, isbn 978–1–904670-92-6.

Cavallucci, D., Cascini, G., Duflou, J., Livotov, P., & Vaneker, T. (2015). TRIZ and knowledge-based innovation in science and industry. *Procedia Engineering, 131*, 1–2.

Cross, W. T., & Ramshaw, C. (1986). Process intensification: Laminar flow heat transfer. *Chemical Engineering Research and Design, 64*, 293–301.

Ferrer, J. B., Negny, S., Robles, G. C., & Le Lann, J. M. (2012). Eco-innovative design method for process engineering. *Computers & Chemical Engineering, 45*, 137–151.

Gerven, T. V., & Stankiewicz, A. (2009). Structure, energy, synergy, time – The fundamentals of process intensification. *Industrial and Engineering Chemistry Research, 48*, 2465–2474.

Grierson, B., Fraser, I., Morrison, A. et al. (2003). 40 Principles – Chemical illustrations. *The Triz Journal.* Retrieved September 11, 2016, from triz-journal.com/40-principles-chemical-illustrations

Hipple, J. (2005). 40 inventive principles for chemical engineering. *The Triz Journal.* Retrieved September 11, 2016, from triz-journal.com/40-inventive-principles-for-chemical-engineering

Houssin, R., Renaud, J., Coulibaly, A., et al. (2014). TRIZ theory and case based reasoning: Synergies and oppositions. *The International Journal on Interactive Design and Manufacturing, 9*, 177–183.

Kardashev, G. A. (1990). *Physical methods of process intensification in chemical technology.* Moscow, Khimia. 208 p. (in Russian).

Kim, J., Kim, J., Lee, Y., Lim, W., & Moon, I. I. (2009). Application of TRIZ creativity intensification approach to chemical process safety. *Journal of Loss Prevention in the Process Industries, 22*, 1039–1043.

Kraslawski, A., Rong, B. G., & Nyström, L. (2000). Creative design of distillation flowsheets based on theory of solving inventive problems, European symposium on Computer Aided Process Engineering, Elsevier 10, pp. 625–630.

Livotov, P., & Petrov, V. (2013). TRIZ innovation technology. Product development and inventive problem Solving. *Handbook.* 284 pages, Innovator (06) 01/2013, issn 1866–4180.

Pokhrel, C., Cruz, C., Ramirez, Y., & Kraslawski, A. (2015). Adaptation of TRIZ contradiction matrix for solving problems in process engineering. *Chemical Engineering Research and Design, 103*, 3–10.

Rahim, Z. A., Sheng, I. L. S., & Nooh, A. B. (2015). TRIZ methodology for applied chemical engineering: A case study of new product development. *Chemical Engineering Research and Design, 103*, 11–24.

Reay, D., Ramshaw, C., & Harvey, A. (2013). *Process intensification: Engineering for efficiency, sustainability, and flexibility* (2nd ed.). Oxford: Butterworth-Heinemann.

Robles, G. C., Negny, S., & Le Lann, J. M. (2009). Case-based reasoning and TRIZ: A coupling for innovative conception in chemical engineering. *Chemical Engineering and Processing: Process Intensification, 48*, 239–249.

Srinivasan, R., & Kraslawski, A. (2006). Application of the TRIZ creativity enhancement approach to design of inherently safer chemical processes. *Chemical Engineering and Processing: Process Intensification, 45*, 507–514.

Stankiewicz, A., & Moulijn, J. A. (2000). Process intensification: Transforming chemical engineering. *Chemical Engineering Progress, 96*, 22–34.

Totobesola-Barbier, M., Marouz, C., & Giroux, F. (2002). A TRIZ-based creativity tool for food processing equipment design. *The Triz Journal.* Retrieved July 10, 2017, from https://triz-journal.com/triz-based-creativity-tool-food-processing-equipment-design

VDI. (2016). VDI Standard 4521. Inventive problem solving with TRIZ. *Fundamentals, terms and definitions.* Berlin.

Wang, H., Mustaffar, A., Phan, A. N., Zivkovic, V., Reay, D., Law, R., & Boodhoo, K. (2008). A review of process intensification applied to solids handling. *Chemical Engineering and Processing: Process Intensification, 118,* 78–107.

Yakovis, L., & Chechurin, L. (2015). Creativity and heuristics in process control engineering. *Chemical Engineering Research and Design, Inventive Design and Systematic Engineering Creativity, 103,* 40–49.

Heuristic Problems in Automation and Control Design: What Can Be Learnt from TRIZ?

Leonid Chechurin, Victor Berdonosov, Leonid Yakovis, and Vasilii Kaliteevskii

1   Introduction

There are two distinct periods in the evolution of the research and application field called automation and control theory. The era of ancient inventions left us the descriptions and drawings of machines and mechanisms that empowered human beings, and some that even completely

L. Chechurin (✉)
School of Business and Management, Lappeenranta University of Technology, Lappeenranta, Finland
e-mail: leonid.chechurin@lut.fi

V. Berdonosov
Komsomolsk-on-Amur State Technical University, Komsomolsk-on-Amur, Russia

L. Yakovis
St. Petersburg State Polytechnical University, Saint Petersburg, Russia

V. Kaliteevskii
Department of Industrial Engineering and Management, Lappeenranta University of Technology, Lappeenranta, Finland

© The Author(s) 2019
L. Chechurin, M. Collan (eds.), *Advances in Systematic Creativity*,
https://doi.org/10.1007/978-3-319-78075-7_4

45

replaced the need for human intervention by making some things happen automatically, as in nature. The gate-opening mechanism by Heron of Alexandria (10–70 AD) could be an example of the first drive design. The rise of needs and engineering ambitions evolved into the problem of the automatic governing of a device output, referred to in modern terminology as tracking or error stabilization. There are two known remarkable inventions of the early Age of Discovery that illustrate a typical approach to self-governing. One is found in the drawings of Leonardo da Vinci and depicts a self-rotating roasting jack. Its rotation rate follows the fire intensity thanks to the propeller in the chimney. Another seems to be the first thermostat by Cornelis Drebbel (1624), where the incubator's vent openings follow the temperature in the incubator thanks to a mercury piston. An unknown ingenious Dutch mind developed a mechanism that controlled the gap of a windmill's running stone in respect to the speed of wind. More than likely, James Watt simply adopted the same idea to the steam turbine rotational rate stabilization in his famous patent of 1788.

The scaling of steam machines revealed cases of instable rotation of some turbines with Watt's governor. The invention yielded obviously new phenomena and these phenomena had to be scientifically explained. J. Maxwell and A. Vyshnegradsky independently modeled the closed-loop steam turbine behavior and provided the safe governor's parameters set. This analysis opened a new era in control system design: the pure inventive concept of the self-governing device became supported by mathematical performance analysis and optimization. One of the brightest examples is the famous negative feedback amplifier invented by Harold Stephen Black (1932) with the help of Harry Nyquist's stability analysis. These two mathematical treatments of feedback system stability are typically referred to as marking the beginning of control theory. Any theory is to turn inventing into systematic routine sooner or later, and it basically happened by the middle of the twentieth century when the synonym of control and automation became the mathematical model-based feedback design and optimization. The automatic control design arrived at the following general algorithm:

1. Choose the model for the Object;
2. Identify its parameters;
3. Define the control goals, model them;

4. Design a Controller;
5. Optimize its parameters; and
6. Implement in hardware.

Indeed, that is a great achievement of control theory. This formalization almost excluded the heuristic (and therefore unpredictable) component from the automation design process. And at the same time, nothing comes without drawbacks and the standardization is not an exclusion. We are going to highlight various difficulties of the "classic automation approach."

1. Control and automation are expensive. They require measurements (sensors), controllers and drives plus automation engineer work.
2. The required controller is not always feasible, it may be feasible in theory only, or it is feasible at the expense of big power losses.
3. The complexity of the closed-loop system is equal to the complexity of the plant plus the complexity of the controller. More elements in general would mean higher failure probability.
4. Automation/control engineer starts her/his project when the object of control has already been designed. It is assumed that the object of control cannot be changed. It is the starting point for formal control design, although not so often the case in reality.

And, at a more generic level, we want to change the nature, the existing way of doing things, to design a useful machine (not just understand and explain the nature, like in most physics) and/but we want the designed device to work itself, like in nature.

The main idea of the study is to provide a strategy of automation that might add inventive ideas to the standard model-based control designs. The inventive part of the design is inspired by TRIZ (Ideal Final Result [IFR], resource analysis).

If we use the concept of IFR at the macro-level of automation system design, we have to get the situation when there is no need for control at all; the object operates itself the way we need. What if we modify the object to be controlled in such a way that the control either is not needed or becomes much simpler?

We illustrate this approach with three case studies. One presents the analysis of hydraulic booster control system design, where the inventive redesign dismissed the necessity to introduce the feedback system. The inventive part of redesign is inspired by IFR and contradiction elimination models. The second example is the treatment of a classical sway stabilization problem. The systematic generating of conceptual ideas is based on a mathematical model of the object. The third example is the analysis of a technology process control design problem. Standard automation design and inventive object redesign give an idea for a hybrid approach and its mathematical optimization.

2   Case Study 1: Hydraulic Power Steering System

Let us consider a hydraulic power assisted steering (HPAS) mechanism as an example. Power steering is designed to make driving safer through assisting the driver in guiding the car in normal situations (parking, highway driving, etc.) and in emergency situations (front tire rupture) (Marcus 2007; Karim 2016). The primary power steering function is to reduce forces exerted on the steering wheel and at the same time reduce the steering ratio. Other functions are:

1. To reduce driver fatigue;
2. To improve car maneuverability; and
3. To provide better "road feeling" for the driver.

2.1   Rack-and-Pinion Steering Without Hydraulic Power

Steering without hydraulic power is a rack-and-pinion mechanism where the steering wheel rotates through a cardan system, a pinion that moves the rack connected through the rods with steering wheels.

Let us note that in such system, a reduction in force on the steering wheel is achieved either by increasing wheel diameter or by increasing the transmission ratio by means of reducing the diameter of the pinion.

The main disadvantage of this steering control is a strong dependence of the force on the steering wheel and the rotation angle, with corresponding forces and rotation angles of the car wheels. Usually, to reduce the steering wheel force, you reduce the diameter of the pinion, and, of course, in this case the steering wheel angle increases. To reduce the turning radius of the car, the steering wheel needs to be turned several times (Kyosuke et al. 1992).

The contradiction (CD_1) in this case is: "When the force on the car steering wheel is reduced by increasing the 'pinion:rack' ratio, the steering wheel angle unacceptably increases."

We begin with IFR formulation: "The steering system ITSELF reduces the force on the steering wheel while maintaining the 'pinion:rack' ratio." IFR was achieved (and contradiction resolved) obviously by means of feedback introduction (the principle of "feedback").

2.2   Steering Control with Servo Drive

Let us consider one of the first hydraulic power steering systems—a system where a regulator and an executive mechanism are separated (Fig. 1a). Such a system consists of an executive mechanism (steering wheel), comparator (slide valve), actuating mechanism (working cylinder), external power source (hydraulic pump) and a feedback system (rods and hinges system) (Kloos and Pfeffer 2017).

This hydraulic power steering works as follows: when turning the steering wheel, the turned upper part of the slide valve directs the working fluid to the required side of the hydraulic cylinder piston, and as a result, the steering wheel turns. At the same time, the hydraulic cylinder piston rod through the rack mechanism rotates the lower part of the slide valve to align the rotation angles of the upper and lower parts of it (slide valve), and by this way, feedback realizes. In such a system, a small steering effort to move the slide valve is converted in the hydraulic cylinder into a significant effort to turn the car steering wheel (determined by the fluid pressure) (Susumu and others 1997).

The design has some weaknesses, however (Kazumasa et al. 1989; Takeshi and Noguchi 1984; Kyosuke et al. 1992). The working

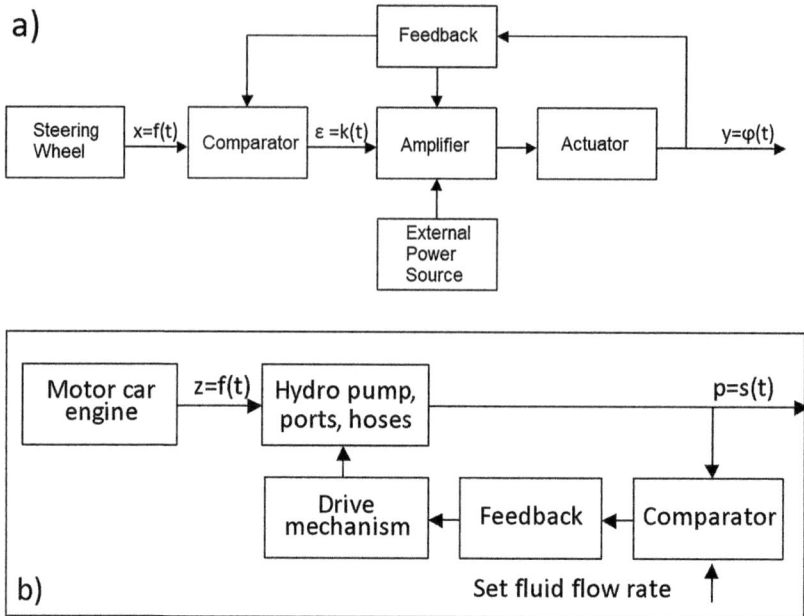

Fig. 1 (a) Schematic design of steering with servo drive, x – steering wheel rotation angle; ε – error; y – steering wheels angle. (b) Schematic design of flow rate stabilizing servo with steering wheel constant angle, where z is the engine rpm, p is the outlet flow rate

fluid flow rate depends on the engine rpm and therefore the engine rpm causes the steering wheel to feel either light or heavy. And the higher the car velocity, the harder the steering wheel rotation and the higher the driver's fatigue. Let us address these weaknesses one by one.

Hydro Pump Capacity Irrespective of Engine RPM

The power steering pump is driven by the car engine via a drive belt. The pumped fluid flow rate is proportional to the pump speed. As a result, different engine rpm will change the force exerted on the steering wheel, which is unacceptable. In typical servo design, the hydro pump schematic diagram is as shown in Fig. 1b.

Let us approach the situation with inventive design tools. We begin with IFR formulation: "The pump (or inlet and outlet ports) shall maintain the constant outlet flow rate ITSELF regardless of the engine rpm." Let us formulate a contradiction (CD$_2$): "Better flow stabilization makes the hydro pump unacceptably complicated."

What resources do we have? A material resource is the liquid that returns to the pump; the excess fluid from the pump can be added. In addition, the inlet and outlet ports are the resources; the drain valve may be fitted in between (Fig. 2) to drain the fluid excess. We can apply the "local quality" and "continuity of useful action" inventive principles: to incorporate a flow control valve in the hydro pump.

With low rpm and constant wheel angle (Fig. 2b), the fluid flows from the discharge line directly to the steering gear through a small opening. As the rpm increases, the fluid flow rate and the pressure in chamber A increases. This allows the flow control valve to overcome the spring force. The flow control valve starts moving to the left (Fig. 2c) thus enabling the fluid to escape through to the suction pipe while excessive pumped fluid is drained (reduced). Thus, the outlet flow rate is stabilized.

On the other hand, the steering wheel turn causes problems. When the driver turns the steering wheel, pressure in chambers A and B in the flow control valve equalizes and the flow control valve shifts to the left. Obviously, more fluid will move from the discharge to the suction channel and the outlet flow rate will decrease. In other words, we face a secondary problem: to stabilize the flow rate against the pressure in chamber B. For this purpose, the servo schematic diagram should be redesigned (Fig. 3a).

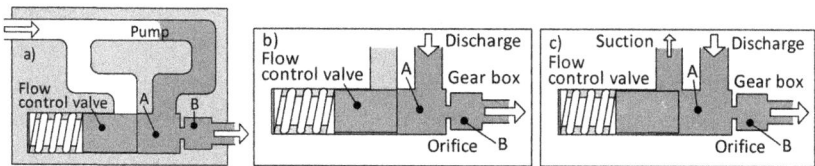

Fig. 2 Flow control valve: (**a**) normal flow; (**b**) flow slightly increased; (**c**) flow considerably increased

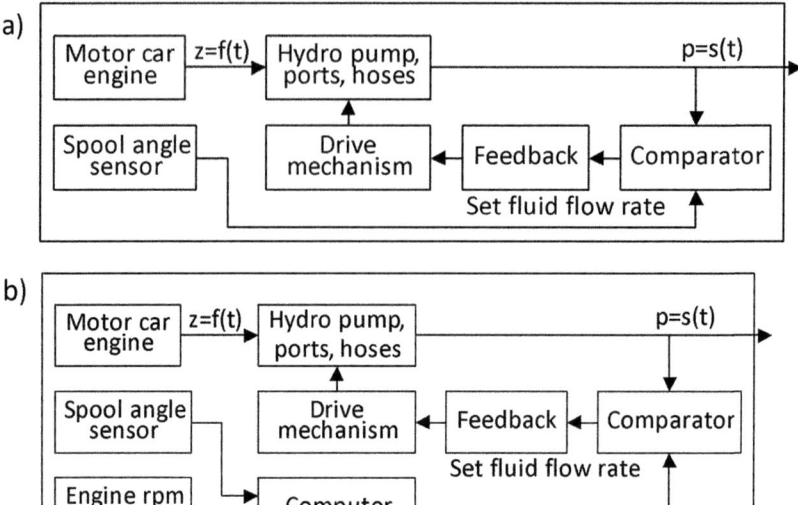

Fig. 3 (a) Schematic design of the flow rate stabilizing servo with varying steering wheel angle. (b) The servo schematic diagram of Required Flow Rate and RPM Relationship

The position-monitoring loop of control spool directly linked to the steering wheel is added to the schematic design diagram. The control spool position governs the flow rate.

We again approach the situation with the inventive technique. The formulation of IFR yields: "The flow control valve shall increase the flow rate ITSELF once the control spool position changes." Let us formulate the contradiction (CD_3): "The improved hydro pump performance (flow rate stabilizing with both stable and unstable steering wheel) makes the hydro pump more complicated."

What resources do we have? The power source can be in the fluid pressure. Let us ideate around the "feedback" inventive principle. On the left side of the flow control valve, communicate the fluid under the same pressure as in chamber B. To do this, we introduce the bypass chambers B and C on the left side of the flow control valve (Fig. 4). The spring will of course remain in place.

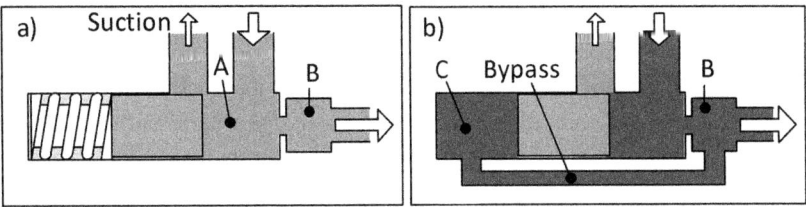

Fig. 4 High flow rate stabilizing against steering wheel turn: (a) without pressure feedback, (b) with pressure feedback provided

Hydro Pump Capacity Depending on Engine RPM

Let us turn to another weakness of the basic design, that is, the wheel does not feel heavy at high speed (Kazumasa et al. 1989; Brunner and Harrer 2017). To avoid this, the fluid flow to the power steering shall decrease as the car speed increases. At high and very high speed, the engine rpm variation range is minor (speed mainly depends on the gear ratio). Keeping this in mind, it would be enough to just maintain the required flow rate and rpm relationship with no regard to the gear ratio.

In typical servo design approach, the block diagram (Fig. 3a) should incorporate an engine rpm meter and computer. The resultant schematic diagram is shown in Fig. 3b.

Proceed to solving the problem by an alternative technique. Set IFR as: "The hydro pump ITSELF shall reduce the flow rate (steering wheel is unstable), against the engine rpm increase." Let us formulate contradiction (CD$_4$): "Better hydro pump performance (required flow rate and rpm relationship is ensured with both stable and unstable steering wheel) makes the hydro pump unacceptably complicated."

What resources do we have? Space: A, B and C chambers. Power: Fluid pressure. Use the "feedback" principle to solve the problem: fit the control spool in chamber A that reduces the orifice area against the chamber pressure rise (Fig. 4).

The control spool is fitted between the flow control valve and orifice. The control spool reduces the flow rate by reducing the orifice area. With low rpm, hydraulic pressure in chamber A is not enough to overcome the spring force and the control spool remains in its position. Therefore, there is no flow rate decrease. As rpm increases, hydraulic pressure in chamber A rises and overcomes the spring force and the control spool shifts to the right and partly covers over the orifice. Pressure in chambers B and C drops. This results in big differential pressure in chambers A and C and the flow control valve shifts to the left, hence the outlet flow rate decreases. The higher the rpm, the more orifice area is covered by the control spool. This makes the flow control valve shift to the left and the outlet flow rate stabilizes. The whole process with low and high rpm depicted in Fig. 5.

As a result, two springs, flow control valve and control spool were able to substitute complex servo design (Brunner et al. 2017) as shown in the schematic in Fig. 3b.

3   Case Study 2: Sway Stabilization System Conceptual Design

Let us illustrate the application of the ideality concept for more model-based control design. We consider the general problem of sway stabilization, which can be found in many engineering fields, such as in crane load stabilization, gondola sway stabilization, free vibration damping in mechanisms, and so on. Let us assume we observe substantial sway of a

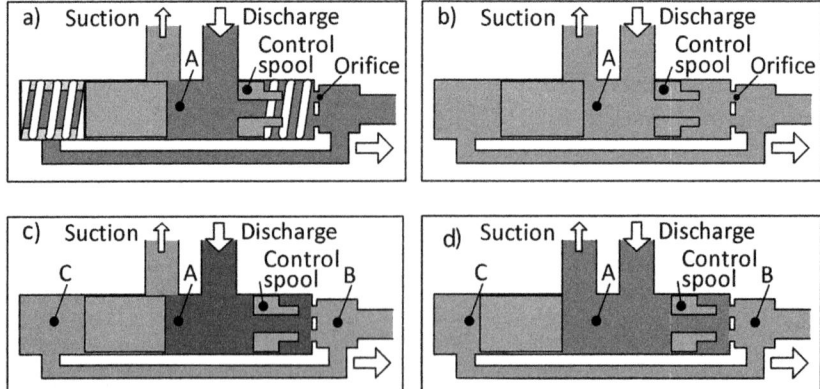

Fig. 5 Decrease in fluid flow with increasing rpm: **(a)** design; **(b)** low rpm. Flow rate decrease against RPM rise: **(c)** high rpm, **(d)** decreasing flow rate

crane load and need to generate conceptual ideas for reducing it. We would like to apply standard active damping strategy architecture first and add more concepts that engage the resources of the object itself. We would like to stress that any TRIZ modeling tool (function model, contradictions, subfields, etc.) would not develop any further the general strategy of "the gondola is to stabilize itself." The concepts are systematically coming out of the mathematical model of the problem. The model "contains" those physical phenomena that can be used for self-stabilizing.

We depart from one of the basic models for oscillating body description

$$a_2\ddot{x} + a_1\dot{x} + a_0x = 0, \tag{1}$$

where $x(t)$ is the oscillation variable, for example, a pendulum's angle, and a_i are the oscillator's parameters. For simplicity reasons we may see a_2 as inertia parameter, a_1 as damping parameter and a_0 as elasticity parameter. We assume small damping in the system ($a_1 \ll 1$), otherwise we would have not been faced with the oscillation problem at all. Given the initial conditions are non-zero or an external impulse force appears (a sharp wind gust), the oscillations of natural frequency f_0 would asymptotically approach its equilibrium point. However, it takes too long.

3.1 Standard Feedback Control Framework

We would speculate that the standard feedback stabilization approach means applying an external force F. The function $F(x,t)$ is to be chosen in such a manner that the oscillations in the system

$$a_2\ddot{x} + a_1\dot{x} + a_0x = F(x,t)$$

vanish faster. If such a controller $F(x,t)$ is found, the closed-loop system is stable itself, that is, the stabilization is automatic.

We can then generate various implementation models within this standard paradigm. For example, the proportional feedback controller $F=kx$ for the simplicity, or PD (proportional-derivative) controller for faster stabilization, or PID (PI+integrative) controller for better accuracy and elimination of static error, and so on. The physical embodiment of this formal model(s) with external stabilizing force is not easy, but it is possible. For example, in Vieira et al. (2010), the force is generated by moving of mass inside the cabin. In Kokoev's (1999) patent, the force is generated by the air jet. If the longitudinal motion of the suspension point is controllable, the change of its position is equal to applying an external torque to the load. In this case, the load displacement angle can be fed back to the position of the suspension load, see, for example the patent of Olli and Ahvo (1994). Any of the controller types discussed above can be applied. However, all these standard feedback control concepts would require sensor(s), controller and drive servomotor but do not require any change in the object.

3.2   Inventive Design, Based on Mathematical Model

Now, let us return to the model of the object (1). Let us assume now that we can change object design that can be followed by the changes in the governing equations. We are interested in the implementable changes of object design that reduce the settling time or make the object less sensitive to external disturbances. In other words, we are looking for ideal feedback concepts, in which the object stabilizes itself and no feedback is needed. What phenomena can stabilize the sways?

Anti-Resonance Absorber

One of the conceptual frameworks could be passive feedback, in which an absorbing oscillator is added to the design. In the case of a pendulum-like object, the new design would mean two oscillators as shown in Fig. 6a. In this case, (1) is replaced by

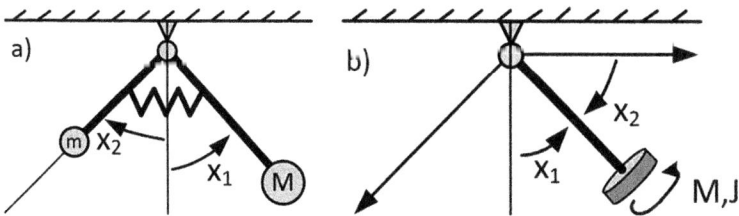

Fig. 6 (a) Pendulum stabilization: passive feedback case. (b) Pendulum stabilization: gyroscopic effect in use (here M stays for the external torque)

$$a_{12}\ddot{x}_1 + a_{11}\dot{x}_1 + a_{10}x_1 - a(x_2 - x_1) = 0$$

$$a_{22}\ddot{x}_2 + a_{21}\dot{x}_2 + a_{20}x_2 - a(x_1 - x_2) = 0,$$

where x_i are the oscillator displacements and a_{ij} can be seen as the parameters of the oscillators reflecting inertia, damping factor and gravity constant; a reflects the spring elasticity. Having carefully chosen a_{2i} and a, we can show that the magnitude of oscillations at frequency f_0 can be reduced to zero. Thus, the passive damper can effectively attenuate the periodic disturbing force with the most dangerous for (1) dominating frequency f_0.

Self-Stabilization by Gyroscopic Effect

Another idea for self-stabilization of (1) is the introduction of the gyroscopic effect. More precisely, the gyroscopic effect provides less sensitivity of an object to external disturbances. The rotation of the load (or a part of the load) can provide the gyroscopic effect for the pendulum. We have to consider a new mathematical model instead of (1):

$$a_{12}\ddot{x}_1 + a_{11}\dot{x}_1 - a_G\dot{x}_2 - a_{10}x_1 = 0$$

$$a_{22}\ddot{x}_2 + a_G\dot{x}_1 - a_{21}\dot{x}_2 - a_{20}x_2 = 0,$$

where x_i are the orthogonal deviation angles of the load, a_{ij} represent the parameters of oscillations in these angles and a_G is the parameter of gyroscopic effect (function of the load rotational inertia and rotational rate). The idea made its way to the patents, for example, in Behbudov and Goldobina (2001).

Variable Length Pendulum Stabilization

Even less obvious phenomena is that the variation of the length of the load suspension can also be used to control its oscillation. Indeed, the mathematical model of the pendulum of variable length can be simplified to the form

$$a_2\, \ddot{x} + a_1 \dot{x} + a(t) x = 0,$$

where $a(t)$ explicitly denotes the variation of the length with time. The model belongs to the class of linear time variant systems. The periodic changes of certain profile (frequency and phase) of the parameter can lead to a new form of instability called parametric resonance. It is natural to expect that the same periodic changes of different profile (phase) can lead to sway stabilization. The idea is filed as the patent in Chechurin et al. (2009).

In all these cases, we stayed within the simplest linear mathematical models of the object and its modification. Still, we were able to mobilize system resources by various linear phenomena (anti-resonance, gyroscopic stabilization, parametric resonance). We can use more careful mathematical modeling that requires nonlinear dynamics or we can enlarge the class of possible object modification by those, described by nonlinear differential equations. In this case, the analysis of the oscillation becomes much more complicated, but the pallet of physical phenomena becomes much wider. For example, the problem of synchronizing two oscillators can be analyzed within a simpler linear model framework. However, the synchronizing would require direct mechanical linking of the feedback control ("non-ideal design"). Having modeled the situation by nonlinear equations, we can reveal the phenomenon of self-synchronizing ("ideal design"). In fact, any nonlinear differential equation can hide many known and unknown phenomena. They can be mined by the mathematical analysis. In this sense, any "database of physical effects" can never be complete.

4   Case Study 3: Technology Process Control Design

In the previous examples relating to individual devices, it has been shown that the same goals can be more easily achievable by changing the design of a technical object than by creating external feedback control systems. In this section, we consider similar problems for technological complexes, consisting of a number of units that perform certain technological operations. At the same time, we try to generalize the concept presented in the previous sections.

The generalization is that the final aim (IFR in terms of TRIZ) is to overcome some certain technical or technological problem at low economic costs in the framework of integrated solutions in the field of technology and control. Moreover, the most rational projects may be complex and costly in technology but simple in the implementation of control systems. An opposite situation may arise when the use of modern tools and control algorithms make it possible to simplify and reduce the cost of technological solutions to the exclusion of the project—the number of technological operations and units. Finally, as mostly happens, the most economical and so the most rational solution is not on the edges, but somewhere in the middle, that is, it combines the most sensible technology and control solutions. Due to the complexity of the tasks arising from the presence of this "Golden mean," there are great prospects for the integrated application of scientific and inventive approaches.

As a typical example to illustrate the given thesis, let us consider the technological complex of the production of raw mix in cement production (Duda 1984). The instability of the chemical composition of the raw mix is due to the chemical heterogeneity produced in the quarries of mixed materials. We discuss different ways to ensure the required stability parameters of the raw material mix in terms of the random variations of its components' composition.

4.1   Maintaining a Constant Raw Mix Formulation

To simplify the real technological process, consider a two-component raw mix of limestone and clay, characterized by the only chemical composition indicator—the percentage of calcium oxide CaO. Let $\beta_1(t)$ and $\beta_2(t)$ indicators of chemical composition of two mixed materials in the current time t, $u_1(t)$ and $u_2(t)$ are their mass fractions in the mixture. Then from relations of material balance follows a model of mixing operation

$$\beta(t) = \beta_1(t)u_1(t) + \beta_2(t)u_2(t), \tag{2}$$

that determines the dependence of the mixture composition, $\beta(t)$, on the composition and proportions of its components.

Geological exploration of deposits of mineral raw materials allows for determining the average mining area characteristics of a chemical composition $\bar{\beta}_1$ and $\bar{\beta}_2$. These data provide the ability to calculate the mass fraction of the mixed materials $\bar{u}_1$ and $\bar{u}_2$, which ensure the equality of the average chemical composition of the prepared mixture and specified technological regulations to the value $\bar{\beta}$. The maintenance of constant proportions of mixed materials can be performed with simple control systems with which modern batchers are equipped. A block diagram of the process for preparing the mixture with the maintenance of a constant formulation of the raw mix is shown in Fig. 7a.

As chemical heterogeneity of mineral raw materials exists, as well as due to errors of dosing, the indicators of the raw mix composition is different from the calculated values by the value $\Delta\beta(t)$ representing the total perturbations.

Since the perturbations are of a random nature, the theory of random processes should be used to analyze the mixture-preparation scheme. In particular, the magnitude of the perturbations should be assessed according to standard deviation (SD) $\sigma_{\Delta\beta}$. Without going into the mathematics,

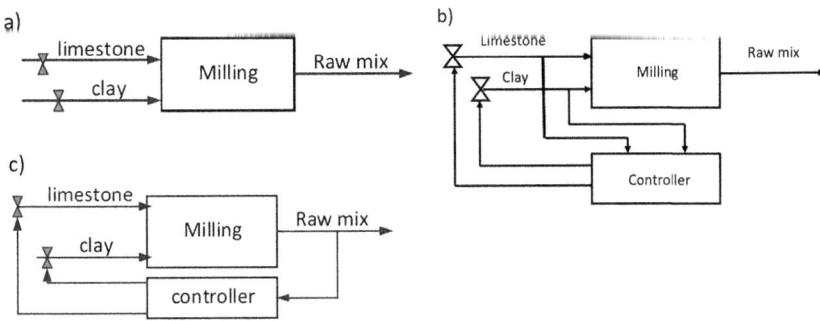

Fig. 7 Mixing process for cement manufacturing. (**a**) Without control. (**b**) Disturbance control. (**c**) Feedback control design

we present the results of calculations for specific values of parameters typical for cement manufacture. For the average chemical composition, we take the values: $\bar{\beta}_1 = 50\%$, $\bar{\beta}_2 = 5\%$, $\bar{\beta} = 42\%$, and SD of indicators of the composition of mixed materials $\sigma_{\Delta\beta_1} = 2.5\%$, $\sigma_{\Delta\beta_2} = 2\%$. Then, as a result of the calculations, we get: $\bar{u}_1 = 0.82$, $\bar{u}_2 = 0.18$, $\sigma_{\Delta\beta} = 2.1\%$. Compared with the maximum acceptable value, $\sigma_{max} = 0.5\%$ indicates that the SD of variations of mixture composition significantly exceeds the technological standard, and therefore the scheme shown in Fig. 7a does not provide the required stability of the raw mix chemical composition.

4.2 Perturbations Control

It is clear that controlling of $\beta_1(t)$ and $\beta_2(t)$ make it possible to maintain the composition of the mixture at the required level $\bar{\beta}$, if synchronously with the changes in the composition of the mixed materials, their proportions are appropriately changed. Thus, it would seem that the task of stabilizing the mixture composition at a required level can be simply solved by equipping the technological process control sensors of mixed material current composition, and also a control system, which, using obtained information, synchronously changes the intensity of material flows. The block diagram of the system control is shown in Fig. 7b.

It appears, however, that in real conditions the proposed solution is feasible, then, at least not optimally. The main reason is the difficulty of accurately controlling the chemical composition of unmilled materials, and in errors of dosing. Overcoming the difficulties of preliminary control of the mixed materials would require the development of special costly installations for the sampling, preparation and continuous analysis of the chemical composition. In addition, to eliminate dosing errors, it would have to use expensive batchers of raw materials.

4.3   Control with Feedback According to the Mixture Composition Data

The above disadvantages of the control by perturbations monitoring can be eliminated if the idea of feedback control is used according to the current monitoring of the mixture composition at the outlet of a milling unit. The block diagram of such a control system is shown in Fig. 7c.

The advantages of feedback control are that not crushed, but finely milled material is under chemical composition control. With the help of X-ray analyzers, both high accuracy and high speed of determination of the chemical composition of the milled mixture can be achieved. In addition, a single point of control allows for determining reaction mixture composition for both types of perturbations, that is, to variations in the chemical composition of raw materials and to the errors of their dosage. From an economic standpoint, it is important also that unlike the previous control scheme, where each raw component should be analyzed for its appliance, only one analyzer of chemical composition is needed when using feedback.

In order to understand the satisfaction of such a system for assigned task—stabilization of the mixture composition—it is needed to evaluate the SD of the output variable, that is, the milled mixture chemical composition $\beta_m(t)$. If simply take to a milling unit the model of transport delay, and for the correlation function of given perturbation $\Delta\beta(t)$ use a frequently used approximation $R(\theta) = \sigma_{\Delta\beta}^2 \exp(-\alpha|\theta|)$ (where α characterizes the smoothness of the perturbation of the milling output), then it is possible to obtain the required estimate achievable level of stability in the control system with feedback (Yakovis and Chechurin 2015a). We assume that the total delay in the system resulting from the transport delay in technological process and the delay of chemical analyses is one hour.

Adding to the previous example the numerical values for the newly introduced parameter $\alpha - 0.3 \ h^{-1}$, we get the estimated value SD of the mixture at the milling output $\sigma_{\Delta\beta_m}$ = 1.4%, which significantly exceeds the allowed maximum σ_{max} = 0.5%. Thus, despite the advantages of the system for controlling the proportions of the materials being mixed, which actively uses the current information about the output characteristics of the process, it does not solve the task, which is to achieve the required degree of stability of the chemical composition of the raw mix. The main reason for this lies in a significant total delay in the control loop, which does not allow the control system to suppress high-frequency components of perturbations.

4.4   Homogenization Systems Application

Moving on to the IFR, let us turn to the analysis of the perturbations effective smoothing possibilities by purely technological methods. To do this, we supplement the process with the homogenization of the milled raw mix (Fig. 8a) (Duda 1984).

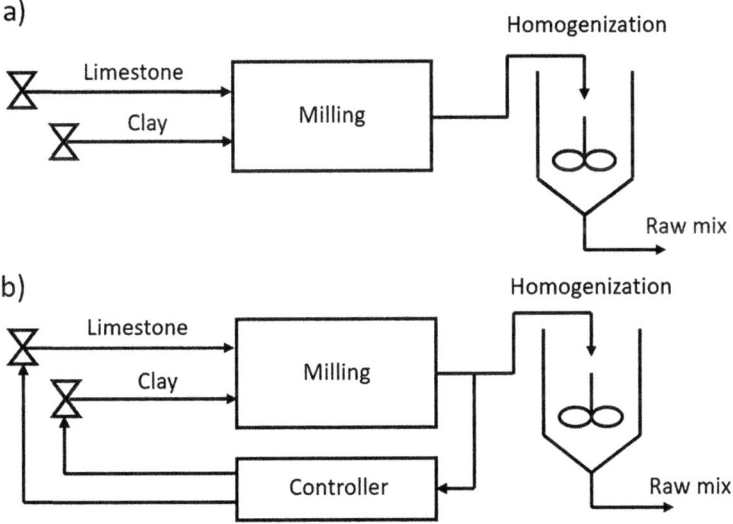

Fig. 8 Scheme of process with mixture homogenization. (**a**) Without control. (**b**) Control system in combination with homogenization system

Homogenizers used in industry represent a container with forced intensive mixing of entering material (Duda 1984). The energy cost of homogenization nonlinearly grows with increasing capacity, so for economic reasons the volume of the homogenizer needs to be minimized. In order to calculate the required volume of the homogenizer, a mathematical model is necessary. The simplest mathematical description of homogenization in averaging capacity gives the model of an ideal mixer (Fitzgerald 1974). It shows how the chemical composition at the output of the flow homogenizer $\beta_a(t)$ varies depending on changes in the chemical composition of the material flow at its input $\beta_m(t)$. The model has a single parameter, which is time of filling of the averaging capacity.

Using the ratios of the statistical dynamics, we can calculate the SD of the output variable $\beta_a(t)$ for various values of T with respect to the scheme shown in Fig. 8a with the homogenizer, which supports constant proportions of mixed materials. As a result, we can find the smallest needed homogenizer filling time, T_{min}, to guarantee the required blending ability. For the considered parameter values T_{min} equals to 56 h. With a representative for cement production lines productivity, $Q = 100 t/h$, the minimum capacity of the averaging capacity should be 5600 tons. Such a significant amount, entailing both large capital costs for the construction of a homogenizer and the serious operating costs of forced homogenization, motivates us to search for better methods of solving the original problem.

4.5   The Combined Use of Feedback Control and Systems of Homogenization

It is known that the averaging systems well suppress high-frequency perturbations, but do not cope well with low frequency. The feedback control systems, on the contrary, effectively compensate low-frequency components of the disturbances, but because of the significant delay they can't handle high-frequency perturbations. Taking into account both circumstances, it is hoped that the significant technological benefits can be achieved by combined use of mixed materials proportions control discussed in Sect. 3 and averaging the milled mixture in containers equipped with a homogenization system discussed in Sect. 4. Figure 8b shows the block diagram of such a system.

Calculations show that the minimum value of the homogenizer filling time is $T_{min} = 3.55\ h$ (Yakovis and Chechurin 2015a). With productivity equal to $Q = 100t/h$, the minimum capacity of the averaging capacity is 355 tons, which is 15 times less than in a scheme that does not use feedback control. Without giving here specific cost values for implementation of both compared schemes of mixture production, it can be safely asserted that the scheme obtained by the serial connection of mixing controlled part with working in pass-through mode, the low-capacity homogenizer is much better in economic terms than the uncontrolled scheme with a large homogenizer. At the same time, it appears that the economic effect can be increased further if we consider that the aim of the control scheme with the homogenizer should be achievement of the minimum SD of the mixture composition not at the milling output, but at the homogenization output. The point is that the feedback control algorithm has to compensate the perturbations predicted for the time delay in the control loop. In the scheme with forced averaging, compensation of perturbations is needed at the output of the homogenizer. As perturbations at the output of the homogenizer contain only relatively low-frequency components, their prediction for the delay time is much more efficient than at the milling output. Hence, the additional effect of reducing the SD of the output variable $\beta_a(t)$. It can be quantitatively estimated using statistical dynamics of control systems (Yakovis and Chechurin 2015a). The minimum time of homogenizer filling $T_{min} = 1.25\ h$, when productivity $Q = 100t/h$ and the minimum capacity of averaging tank is 125 tons, which is almost three times less than in the scheme where controlled mixture production part and homogenizing part are considered separately.

So, by combining heuristic considerations of inventive nature with the exact calculations performed with the use of statistical dynamics of control systems (Aström 1970), it makes it possible to significantly move toward IRF. However, the end point in this movement has not been achieved (and, most likely, will never be achieved), since there is still a problem of reducing the energy costs for forced homogenization. The solution can be sought in the direction of a fundamental redesign of the homogenizer. The system of forced homogenization should be replaced by the sequence of technological operations of separation of the input flow for a number of "subflows" with waiting for each "subflow" in the buffer tank for some time and then the subsequent merging of the "subflows". The correct choice of the intensities of the "subflows" which should be calculated with statistical dynamics, should lead this system to the desired averaging effect without any significant energy costs.

Since this inventive idea is still in the research stage, we will not go into details. Let's just say that here we need a simple control system that will have to maintain the required values of the intensities of "subflows" at the calculated values.

We demonstrated in the specific example that the optimal solution may not be to abandon the control process in favor of purely technological methods. In addition, as shown, the purely controlling methods as another extreme does not always lead to the achievement of the IFR. The optimal solution was found in combination of the technological object and controlling it with automation as a combination of two modified parts of a single complex (Sharifzadeh 2013). In this case, it is possible to take into account and use the beneficial properties of both components in the most reasonable manner.

The simplified nature of the considered example made it possible to perform analysis in an analytical way. In a real situation, when:

- The mixture does not consist of two, but of a larger number of components;
- The chemical composition is characterized by not one, but several parameters;
- Along with mixture homogenization, prior mixed materials averaging can be used;
- The more or less accurate batchers can be chosen;
- There are alternatives in control methods of chemical composition;
- For milling, homogenizing and prior averaging, units of different type and size can be used,

behavior analysis of the "Object–Controller" system under random perturbations is seriously complicated (Gorenko et al. 1987). Mostly, such studies are carried out using computer simulation and automated comparative analysis of various options according to their technological efficiency and value characteristics (Doroganitch et al. 1989). However, the general sense of the problem, optimization of economic indicators with technological requirements, remains. Moreover, the nature of the man-machine decision is compounded because the preliminary selection, which requires the experience of design-technologists and design-managers, is difficult due to the huge number of options and the complexity of formalizing. Eventually, fruitful results can be expected only in the connection of experience based on inventions and scientific approaches that allow for quantitative evaluation of the effectiveness and suitability of various alternatives (Yakovis and Chechurin 2015b).

5   Conclusions

The study provides an idea for an alternative (inventive) approach to design automation. It shows how the TRIZ concepts of IFR and contradiction analysis assist in simplifying the controller design through inventive changes in the plant. The use of a mathematical model of the plant can enrich the design ideas by providing an additional domain of resource analysis.

Three examples demonstrate the approach to inventive automation design. In the first two examples, the object design is changed in such a way that the control feedback is either not required or it becomes much simpler. The third example, which is related to technological complexes, demonstrates the generalizing idea. The optimal economic plan combines the capabilities of technology and management and should be formed on the basis of experience, invention and science.

Future research will focus on the systematization of heuristic methods in controller design and the development of concurrent plant and control design in the framework of control theory.

Acknowledgements L. Chechurin would like to acknowledge the support of TEKES, Finnish agency for innovation support, and its Finnish distinguished professor (FiDiPro) program.

The authors would like to acknowledge the EU Marie Curie program INDEED project for its support.

References

Aström, K. J. (1970). *Introduction to stochastic control theory.* New York: Academic.

Behbudov, M. B., & Goldobina, L. A. (2001). *Mechanism for load swing damping hanged by crane.* Patent RU 2224708.

Brunner, S., & Harrer, M. (2017). Steering requirements: Overview. In M. Harrer & P. Pfeffer (Eds.), *Steering handbook.* Cham: Springer. https://doi.org/10.1007/978-3-319-05449-0_3, Print ISBN978–3–319-05448-3, Online ISBN978–3–319-05449-0.

Brunner, S., Harrer, M., Höll, M., & Lunkeit, D. (2017). Layout of steering systems. In M. Harrer and P. Pfeffer (Eds.), *Steering handbook.* Cham: Springer. https://doi.org/10.1007/978-3-319-05449-0_3, Print ISBN978–3-319-05448-3, Online ISBN978–3–319-05449-0.

Chechurin, S., Chechurin, L., & Mandrik, A. (2009). *A method of stabilizing of output signal of oscillating system.* Patent RU 2393520.

Doroganitch, S., Edvabnik, J., Shtengel, E., & Yakovis, L. (1989). *Optimization of parameters of automated process complex for raw mix preparation.* Presented at the the second NCB international seminar on cement and building materials, New Delhi, pp. 56–63.

Duda, W. H. (1984). *Cement-data-book, volume 2, Electrical engineering.* Automation. Storage. Transportation. Dispatch. Wiesbaden/Berlin: Bauferlag GmbH.

Fitzgerald, T. J. (1974). Theory of blending in single inlet flow systems. *Chemical Engineering Science, 29*(4), 1019–1024.

Gorenko, I., Doroganitch, S., & Yakovis, L. (1987). *Multilevel process control in multi-component mixtures blending.* Presented at the the 10th world congress on automatic control, Munich, vol. 2, pp. 218–222.

Karim, N. *How car steering works,* last visited 14 July 2016. http://auto.howstuffworks.com/steering4.htm

Kazumasa, K., et al. (1989). *Control apparatus for power-assisted steering system.* Patent EP 0430285 A1.

Kloos T., & Pfeffer P. E. (2017). Steering system models – An efficient approach for parameter identification and steering system optimization. In P. Pfeffer (Ed.), *Proceedings of the 8th international Munich chassis symposium 2017.* Wiesbaden: Springer Vieweg. https://doi.org/10.1007/978-3-658-18459-9_36, Print ISBN978-3-658-18458-2, Online ISBN978-3-658-18459-9.

Kokoev, M. N. (1999). *Mechanism for swing damping and leveling of the load.* Patent RU 2141926.

Kyosuke, H., et al. (1992). *Power steering apparatus.* Patent EP 0562426 A1.

Marcus, R. (2007). Hydraulic power steering system design in road vehicles analysis, testing and enhanced functionality Linkoping. ISBN 978-91-85643-00-4.

Olli, M., & Ahvo, R. (1994). *Method for damping the load swing of a crane.* Patent US 5799805.

Sharifzadeh, M. (2013). Integration of process design and control: A review. *Chemical Engineering Research and Design, 91*(12), 2515–2549.

Susumu, H., et al. (1997). *Flow control device of power steering apparatus.* Patent US 6041807 A.

Takeshi, F., & Noguchi, M. (1984). *Fluid pressure control device in power steering apparatus.* Patent US 4619339 A.

Vieira, D. M., Ibrahim, R. C., & Torikai, D. (2010). *Active control system to stabilize suspended moving vehicles in cables.* ABCM symposium series in mechatronics, Rio de Janeiro, Brasil, vol. 4, pp. 120–126.

Yakovis, L., & Chechurin, L. (2015a). Systematic design of automated processing complexes » 25th international conference on flexible automation and intelligent manufacturing – FAIM, international conference on flexible automation and intelligent manufacturing, vol. 1, pp. 438–445

Yakovis, L., & Chechurin, L. (2015b). Creativity and heuristics in process control engineering. *Chemical Engineering Research and Design, 103*, 40–49.

The Adaptive Problem Sensing and Solving (APSS) Model and Its Use for Efficient TRIZ Tool Selection

Alexander Czinki and Claudia Hentschel

1   Introduction

In the scenario of ever more interconnected and interdependent tasks, problem-solving capabilities become a progressively important capability (Funke et al. 2018). The focus of this chapter is the very early step in problem solving, which—as it will be suggested—should include "sensing" a problem situation. The step of "sensing the problem" is especially crucial, since it (at least it should) significantly influences the problem formulation and the design of the subsequent problem-solving process.

The Cynefin framework (Snowden and Boone 2007) offers a (relatively) new and interesting approach for sensing problems that allows assignation of certain attributes (such as complex, complicated, chaotic, etc.) to given problem situations. The authors of the chapter have started

A. Czinki (✉)
University of Applied Sciences Aschaffenburg, Aschaffenburg, Germany
e-mail: Alexander.Czinki@h-ab.de

C. Hentschel
University of Applied Sciences HTW Berlin, Berlin, Germany
e-mail: claudia.hentschel@HTW-Berlin.de

© The Author(s) 2019
L. Chechurin, M. Collan (eds.), *Advances in Systematic Creativity*,
https://doi.org/10.1007/978-3-319-78075-7_5

to link the topics "complexity," "problem solving" and "Cynefin framework" and generated an adaptive problem-solving process that aims for both a more systematic and a more efficient approach for problem solving.

This chapter explains the approach of the Adaptive Problem Sensing and Solving (APSS) Model in combination with selected TRIZ tools. The approach is depicted by a problem taken from the automotive industry.

2   Background: The Cynefin Framework in a Nutshell

In the real world, whenever facing a problem, one tends to think that all there is to do is to find options, select one and put the chosen solution into practice. This assumes that effect follows cause and that there is a good reason why one option is chosen and the others are not. If this works out, all right. But if it doesn't? It usually gets very tricky if the problem solver realizes that the chosen option does not show the desired effect. It is usually just these situations, when one restarts all over—in the hope of a better choice—for better results during the next iteration. Again, the next iteration often starts without really knowing why the selected tool failed before and what is "reasonable" to do next.

Here, a very interesting, relatively new and highly acknowledged model comes into play: the Cynefin model, a framework for *sensing* causal differences in any kind of problem that seeks for more adequate decision making in problem situations (Snowden and Boone 2007) than classical problem-solving strategies can provide. Cynefin originated from knowledge management as a means for distinguishing between formal and informal behavior (Kurtz and Snowden 2003), but has emerged into all kinds of sciences and applications, from management, politics and branding to leadership and supply chain management, to name only a few. The Cynefin framework most importantly distinguishes between ordered and unordered problems, meaning that for the latter, one cannot determine cause and effect (Brougham 2015; Waldrop 1992).

If the cause–effect relationship is not obvious, but the causes can be detected by analysis, the model speaks of a complicated problem. These situations are not self-evident, but can be resolved by experts that have established good practice. These two domains—Simple Domain and Complicated Domain—belong to the "Ordered space" (Snowden and Boone 2007).

For situations where causality cannot be determined, and the directions in which the system evolves cannot be easily predicted, the model calls it a complex problem. For all unstable situations that behave (or seem to behave) randomly, the model provides the term "chaotic" (Snowden and Boone 2007; Waldrop 1992). Chaotic systems are highly turbulent, show no evidence of any constraint and old certainties and rules (in case there have been such before) cannot be extended to the present or the future. Therefore, responding to it is limited to spontaneous action.

Cynefin is a sense-making model, where data acquisition precedes framework application. A person has to assess a problem according to the available knowledge to solve it and at the same time assess the situation according to its behavior. There is no consensus about the classification of problems; it is heavily bound to knowledge (Funke et al. 2018). Only after a decision depending on the available knowledge, a sensing about "which domain it is in" can be undertaken—always starting with a decision about order or unorder. If the situation is assessed as an unordered problem, there only remains the complex or chaos field. This is very much an upside-down approach, as any classic categorization models, for example, portfolio management, rely on two variables, where a system is classified according to its degree of fulfilment in terms of the chosen variables. Those categorization models thus more or less reinforce the desire and assumption of order—which fails in cases where the situation has unordered characteristics. So more traditional approaches try to reduce any problem to a set of rational actions and choices and might misleadingly define it as an ordered problem situation.

Cynefin, on the contrary, acknowledges that there might be situations where we cannot predict the outcomes. With the addition of two unordered domains into the model—the situation or problem to be solved can be sensed as complex or even chaotic—the model acknowledges that the outcome might be unpredictable.

Subsequently, a new, general problem-solving model is introduced that uses problem domains, which are inspired by the problem spaces of the Cynefin framework. However, the new general problem-solving model does not only provide a model for the domains but also offers a process of how to "navigate" between the domains and rules on how to select appropriate problem domains for a given situation and constraints.

3   Proposed Solution: Problem Solving Based on the APSS Model

3.1   General Findings While Working with the Cynefin Framework

The Cynefin framework inspired the authors to deepen their understanding in dealing with general problem-solving strategies. The general history and findings on solving problems and the different types of problems is depicted in numerous publications (e.g., text and references in Fischer et al. 2012) and should therefore not be repeated in this work. In addition, the authors have shared their views on complex problems in three recent publications (Czinki and Hentschel 2016a, b; Hentschel and Czinki 2016). During applying the Cynefin framework, the authors observed some characteristics that triggered the creation of the APSS problem-solving model. These key findings are listed below and will be explained in more detail:

1. *Results generally improved when a given problem was processed in consideration with the Cynefin framework*
   It was interesting to learn that results generally improved when being processed in consideration of the Cynefin framework. This likely underlines the potential of the problem domain approach suggested by the Cynefin framework.
2. *One and the same problem can be processed in different domains*
   This insight was especially important since it showed that there is a possibility to process the same problem in different domains. Obviously, a problem does not have an inherent belonging to one of the problem

domains. Rather, it seems that every problem—at least as soon as we consider a sufficiently high number of corresponding aspects of it—is of a chaotic nature. This is presumably true, since the human mind is limited concerning understanding its environment and therefore needs to reduce reality in order to keep it manageable (Sargut and McGrath 2011). Although the insight of our limited perception and processing abilities and—as a consequence—the permanent reduction of reality by our mind might sound frightening, in reality it is not: in many cases, problem solvers process problems in other domains than the chaotic domain and they do so with very much success. This is true, since problem solvers usually do not seek to solve a problem in an all-encompassing way, but rather look for efficient solutions, thus regularly accepting a suboptimal solution for the benefit of a low consumption of resources.

3. *The character of the results strongly depends on the domain the problems were sensed in*

Another important aspect is the observation that the character of the results was strongly influenced by the domain a problem has been processed in. Basically, different domains generated different types of solutions. Some general qualities of solutions—as they were observed—depending on the domain they were processed in, are listed in Table 1.

Table 1 Observed solution characteristics as a function of the solution space (text form)

Solution space	Main characteristics/likely character of solution
Chaotic	deductive, generates little system understanding, solutions cannot (easily) be transferred to other situations, relatively low probability of success
Complex	Strengthen the intuitive understanding of the users, makes use of own and other peoples (unconscious) experiences, can often be formulated as abstract rules, have an increased likelihood of success (compared to the "chaotic" domain)
Complicated	Generate well-directed, structural changes, generate a high level of understanding concerning the interdependencies within a system, in case of a proper system modelling: high probability of success
Simple	Generate targeted effect in terms of enhancement, conservation or depreciation of a given cause-effect relation, effect of changes can usually be well predicted in advance, with correct knowledge of effects: very high probability of success

Obviously, it is possible to influence—to some degree—the character of the results in advance by selecting a certain problem domain.

4. *The selection of domains was very much done unconsciously and was very much based on personal preferences rather than on systematic processes, knowledge or a clear strategy*
   This finding is striking, since it raises the question in how far problem solvers are making full use of the domains and their specific advantages. It also raises the question of whether a general process can be defined that helps problem solvers to identify adequate domains for a given problem.
5. *Prevalent problem-solving tools performed very differently, depending on which domain they were applied to*
   The finding was that typically the very same tool that performed well in a certain domain performed insufficiently—or even failed—in other domains. Obviously, there is a need to assign tools to domains in order to allow problem solvers to pick appropriate tools for solving a given problem.

3.2   The APSS Model

The results of the preceding chapter suggest that people tend to solve a given problem directly, without reflecting on which domains are suitable to solve the given problem. This can lead to a situation where a problem is processed in an ordered domain, although the prerequisites in terms of system and functional understanding are not fulfilled. In this case, usually neither the domain nor the tools applied will be adequate to process the given problem successfully.

In order to avoid these disadvantages, the authors developed the Adaptive Problem Sensing and Solving (APSS) Model.

The objective of the APSS Model is to provide problem solvers a tool that allows them to:

1. gain a deeper problem understanding,
2. identify appropriate domains for their problem, and
3. effectively select and apply adequate tools to their problem.

As a result, the APSS Model helps users to achieve successful solutions in a well-structured and highly efficient manner. The flowchart (Fig. 1) demonstrates the general idea of the model together with its major process steps.

In general, the chart depicts a sequence of steps (with numbers from 0 to 5). Depending on the satisfaction with a solution found, the model allows getting back to former steps in order to iteratively sharpen the problem understanding and to explore different domains and therefore to explore different types of solutions.

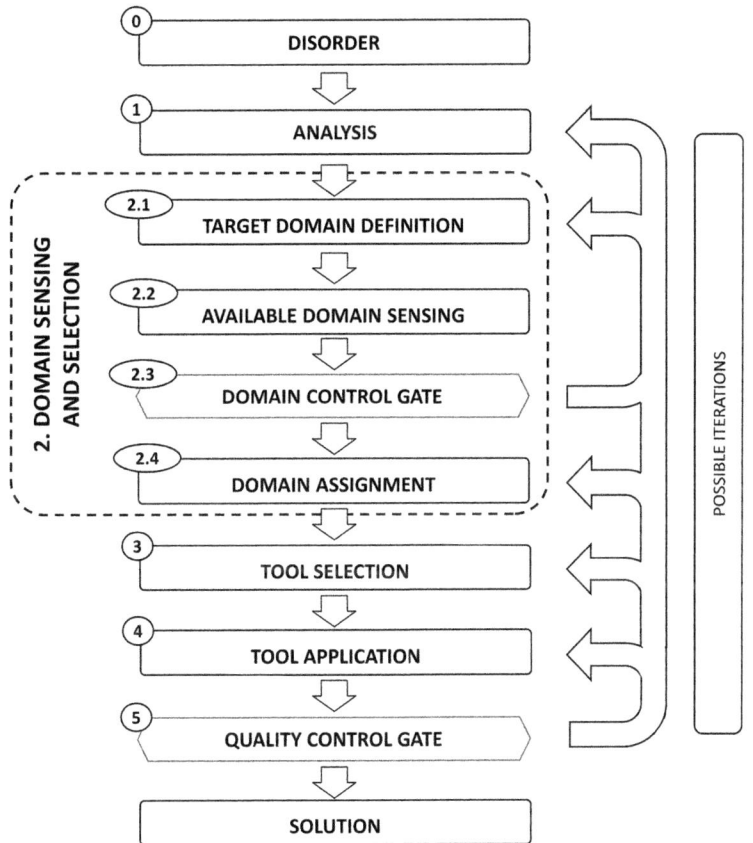

Fig. 1 Adaptive Problem Sensing and Solving Model (APSS Model): General process

Problem solving usually starts with a disordered state: the start is characterized by a fuzzy problem awareness, basically being a deviation between a desirable state—usually insufficiently known—and an actual state—prevalently also insufficiently known. The first step in the APSS Model therefore is a system analysis step. It will provide the user with a better understanding of the actual and the desired state of the system. Additionally, it will tell the user how well he/she understands:

(a) the structure of the system and
(b) the functions and effects within the system structure.

This assessment is an important prerequisite of the succeeding steps. During "Target Domain Definition" (step 2.1), the user will define the desirable character of the solution (and might make use of Table 1 to do so). This can happen before any major solution attempt is initiated.

However, usually not all domains will be available to the user for solving his problem. The availability of domains strongly depends on the level of structural and functional understanding of the user. A high level of structural understanding is characterized by a deep knowledge of the main contributors to the problem and by a good understanding of the amount of impact that the various contributors have on the problem.

Basically, the chaotic domain and—to some extent—the complex domain are available in all cases, since they come along with (almost) no system—and (almost) no function/effect—understanding. If both the structure and the functions are well understood, the simple and complicated domains become available. The more a deeper understanding is lacking, the more the user will be limited to the complex or even to the chaotic domain.

The availability of domains is evaluated in the "Available Domain Sensing" step (step 2.2). As soon as the user has identified both the target and the available domains, the user can assess the fit of both (step 2.3). The main question at this point is: "Is the targeted domain within the set of available domains?" A non-fit will require the start of a new iteration. In many cases, there is more than one domain available, which forces the problem solver to take a decision concerning which domain to use or which domain to start with, respectively (step 2.4).

The knowledge of the domain in use will help the problem solver to identify problem-solving tools that are especially powerful in combination with the selected domain. Problem tool selection will therefore become much more systematic than it usually is during classical problem-solving processes. Consequently, the application of problem tools (step 4) will be more efficient.

In step 5 ("Quality Gate: Verification"), the user will assess the solutions found based on the desired qualities defined earlier. Based on the deviation between the qualities of the solutions being found and the targeted qualities, the user will decide to either start a new iteration or stay with the solution already found.

4   Recommended Assignment of TRIZ Tools to the APSS Model

Now, after having an abstract problem-solving model, the question is how to assign the TRIZ tools to the problem domains. Actually, this turned out to be the tricky part of it. What finally helped was the consideration of the question: "What happens if a given problem is solved within a certain domain?"

Solving a problem in the "simple" domain should result in the application of a known effect or a known effect chain. If we solve a problem in the complicated domain, the solution will mainly have the character of a structural change of the system. If we solve a problem in the complex domain, the solution will mainly have the character of an observation-based assumption. Finally, solutions in the chaotic domain will carry characteristics of "trial and error-approaches."

This phenomenological description helped to assign TRIZ tools to the different problem domains. However, not all TRIZ tools could clearly be assigned to one particular domain. The reasons for this are easy to understand, since some TRIZ tools

- are not meant to solve a specific problem, but rather analyze it,
- are more processes that walk through different domains, rather than belonging to a single one, or
- cover aspects of more than only one domain.

However, in case of many tools, an assignment to one or several problem areas can be suggested (Table 2).

Interestingly, no TRIZ tool could be identified that particularly meets the demands of the chaotic domain. This might suggest that there is a need within TRIZ to fill this gap and to provide a tool (or a strategy) that allows for dealing with systems that are of a chaotic nature.

5 Applying the APSS Model to an Exemplary Problem in Automotive Industry

The main idea of the APSS Model shall be depicted by a short example taken from automotive engineering. The example illustrates that success often cannot be achieved by staying within a certain domain, but rather by identifying a (more) suitable domain and then processing the problem within this domain anew.

Table 2 Current suggestion for the assignment of TRIZ tools to the different domains

Tool name	Problem analysis	Problem solving (domains)			
		Simple	Complicated	Complex	Chaotic
Innovation Situation Questionnaire	++				
Root-Conflict-Analysis	++				
Cause-Effect Chain Analysis	++				
Function Analysis	++				
Catalogue of Effects		++			
Inventive Principles		++	+	++	
Trimming			++		
Substance-Field-Analysis		+	++		
Physical Contradictions			++		
Laws of Engineering System Evolution				++	
Ideality				++	
Smart Little People Model		+	+	++	
System-Operator (Nine Windows)				++	

'++' good fit, '+' fit; Status: December 2017

The example briefly described here is from the automotive industry, where the inflation control of an airbag is as long a problem as airbags have been around (Rokosch 2011). The problem refers to the so-called Out of Position (OoP) situation, where a driver is seated too close to the steering wheel. This situation usually refers to small, lightweight persons having shorter legs but finding the same seats and gas pedal position as larger persons. The main danger is when—during an accident—the inflating airbag creeps under the chin of the driver before being fully blown up, leading to an abrupt extension of the neck by the air-filled cushion (Fig. 2).

In those cases, the measured neck momentum of the driver is exceeding biomechanical limits, which was the reason for the development engineers to find a solution. The company then tried to solve the problem with so-called Low-Risk Deployment features. In the first instance, they did so by changing parameters such as layout, pre-cut and surface of the airbag tissue to prevent it from fitting under the driver's chin. Ideas were a flatter design and a more stable front area. Although this showed some remarkable results—with a worsening parameter of size of the airbag, though—the problem could not be eliminated by such a change of design.

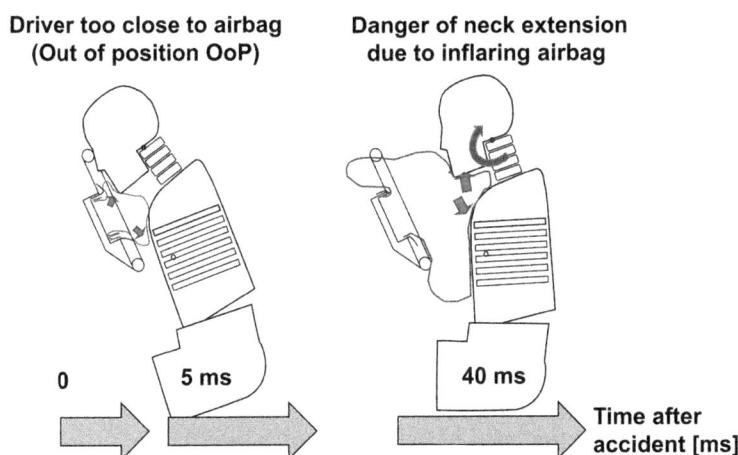

Fig. 2   Out-of-Position (OoP) problem: Driver sitting too close to steering wheel

This is where the APSS Model was used to sense the problem differently and to apply a different TRIZ tool than the previously chosen Catalogue of Effects. According to the authors' approach, the first problem-solving process of the company referred to an understanding of the problem as a complicated problem, where the cause-effect chain was understood as follows: A smaller person sits too close to the steering wheel, due to shorter legs and a need to reach the gas pedal, which leads to the danger of neck extension during an accident due to underinflated airbag material creeping under the chin of the driver. Tests revealed a good functioning of the idea, but a deeper analysis showed that there were cases that could not be explained with the given cause-effect chain.

A deeper analysis of the problem revealed that any person, even a tall person, can suffer from critical neck extension when sitting too close to the steering wheel. The domain sensing was such that there are uncontrollable parameters, as the seat position in a car is still and only influenced by the driver. As it became even clearer, the functioning of the entire restraint system "airbag" mainly depends on the room for inflation and therefore on the distance between the person and the steering wheel. The seat position of the driver is a variable that cannot be influenced by the airbag supplier. Rather, it is a parameter that is deliberately chosen by the person driving, and very often, even tall drivers are not aware of the fact that sitting as far from the steering wheel as possible is best for the functioning of the airbag in a dangerous situation.

Rather than changing the form, material thickness and pre-cut of the airbag, which would have negatively influenced size, volume, stability and longevity of the tissue, it became clearer that restraint systems must become more adaptive to the driver and/or the driver's behavior—and to the driver's faults and misbehavior, too. Sensing the problem in such a complex way, where the position of the seat cannot yet be influenced by the airbag supplier but is depending on the driver's seat position, the solution space opened.

This was done by the previous selection of the TRIZ tool—System Operator (Nine Windows)— that revealed that the problem could be influenced before the problem occurs. A measuring of the distance between driver and steering wheel could lead to a warning sound or light to the driver that the driver is sitting too close, and to a directive to move the seat back, which is easily possible for taller persons. Based on this understanding, the problem was then modeled by those models in TRIZ that were earlier assigned to the complex problem domain. This led to completely different solutions that were not available in the first place.

Such active solutions do not exist yet, but these solutions were only thought of when the problem domain was sensed as complex and modeled with a TRIZ tool for complex problem solving. The influence of the driver's behavior was taken as a parameter that is not yet to be controlled by the airbag supplier but has massive impact on the functioning of the entire airbag system. Now that the problem was sensed differently, TRIZ tools were selected that were not thought of before; the Method of Smart Little People (MSLP) in TRIZ led to solutions that it would be desirable to move the seat in a backward position in case of the accident, or move the steering wheel to the front end of the car, or both, in order to free the person from the steering wheel and gain more space for inflation. The solutions generated herewith completely differed from the solutions generated in the first round of problem solving, as it became clear that the elements surrounding the airbag that are not yet an active part of the restraint system will become more so in the future.

6   Conclusions

This chapter describes the development and application of a new problem-solving model called the Adaptive Problem Sensing and Solving (APSS) Model and its application to TRIZ. The APSS Model is inspired by the problem domain approach introduced by the Cynefin framework. However, in contrast to the Cynefin framework, the APSS Model is based on the assumption that all problems may belong to the "chaotic" domain and that the other domains only represent a reduced and therefore a usually "easier-to-manage" perspective on a problem.

A key finding—that triggered the development of the APSS Model—was the observation that depending on the number and types of domains selected, the quantity and the characteristics of the solutions usually varies. Therefore, a problem solver has both the possibility and the responsibility to decide in which domain(s) he or she is processing a given problem. The APSS Model is seeking to provide problem solvers a robust process to do so.

Thus, the APSS Model not only delivers a process to solve problems in different domains, but it also provides a set of criteria that helps the user to assess which problem domains he or she should/could use. As a result, the APSS Model helps users to generate successful solutions in a well-structured and highly efficient manner.

In addition to presenting the APSS Model itself, this chapter also describes the use of the APSS Model in combination with tools from systematic innovation (TRIZ tools). The use of TRIZ tools in combination with the APSS Model turned out to be especially advantageous. It was shown with a problem from industry that was initially assigned to the complicated domain, which led to an unsatisfactory solution that focused on changing system components. When the complex domain was sensed and the adequate TRIZ tools were selected, the problem understanding widened and led to solutions that changed the overall system, and provided promising new solutions that were not thought of before.

References

Brougham, G. (2015). *The Cynefin mini-book – An introduction to complexity and the Cynefin framework*. Poland: C4Media.

Czinki, A., & Hentschel, C. (2016a). *Adaptive Problem Sensing and Solving Model (APSS-model) inspired by the cynefin framework and its application to TRIZ*. Proceedings of the TRIZ future conference TFC 2016, Wrocław, 6 pages.

Czinki, A., & Hentschel, C. (2016b). Solving complex problems and TRIZ. In I. Belski (Ed.), *Structured Innovation with TRIZ in Science and Industry – Creating Value for Customers and Society, S* (pp. 27–32). Amsterdam: Elsevier B.V.

Fischer, A., Greiff, S., & Funke, J. (2012). The process of solving complex problems. *The Journal of Problem Solving, 4*(1), 19–42.

Funke, J., Fischer, J., & Holt, D. V. (2018). Competencies for complexity: Problem solving in the twenty-first century. In E. Care et al. (Eds.), *Assessment and teaching of 21st century skills, educational assessment in an information age* (pp. 41–53). Cham: Springer International Publishing.

Hentschel, C., & Czinki, A. (2016). Taming complex problems by systematic innovation. In L. Chechurin (Ed.), *Research and practice on the theory of inventive problem solving (TRIZ) – Linking creativity, engineering and innovation* (pp. 77–93). Berlin: Springer International Publishing.

Kurtz, C. F., & Snowden, D. J. (2003). The new dynamics of strategy: Sense-making in a complex and complicated world. *IBM Systems Journal, 42*(3), 462.

Rokosch, U. (2011, Auflage 2). *Airbag und Gurtstraffer.* Würzburg: Vogel Industrie Medien.

Sargut, G., & McGrath, R. (2011). Learning to live with complexity. *Harvard Business Review, 89*(9), 68–76.

Snowden, D. J., & Boone, M. E. (2007, November). A leader's framework for decision making. *Harvard Business Review, 85*(11), 68–76, 9 pages.

Waldrop, M. M. (1992). *Complexity – The emerging science at the edge of order and chaos.* New York: Simon & Schuster Paperbacks.

Case: Can TRIZ Functional Analysis Improve FMEA?

Christian Spreafico and Davide Russo

1   Introduction

Nowadays, even if there were multiple efforts in standardization, a unique reference structure for Failure Mode and Effect Analysis (FMEA) does not exist, unlike in other sectors such as Life Cycle Assessment and quality management. Some efforts at standardization come from the US Department of Defense, which developed and revised the MIL-STD-1629A guidelines during the 1970s, and from Daimler Chrysler, Ford and General Motors who jointly developed an international standard named SAE J1739–2006 documentation for FMEA. Other guidelines include FMEA from the Automotive Industry Action Group (AIAG), ARP5580 from the Society of Automotive Engineers (SAE) for non-automotive applications, and EIA JEP 131 from the electronic industry.

Despite of the multitude of FMEA improvements, some problems remain unanswered, such as the request for a more practical and less

C. Spreafico • D. Russo (✉)
Department of Management, Information and Production Engineering,
University of Bergamo, Bergamo, Italy
e-mail: davide.russo@unibg.it

© The Author(s) 2019
L. Chechurin, M. Collan (eds.), *Advances in Systematic Creativity*,
https://doi.org/10.1007/978-3-319-78075-7_6

time-consuming way for determining the Failures (Failure Effects and Failure Modes). In particular, this problem continues to be repurposed in the literature, especially by industrial papers.

The objective of this chapter is to propose a new approach for simplifying FMEA by determining the Failure Effects and the Failure Causes in a more practical way and by better involving the problem solving in a more pro-active and creative approach.

The proposed approach is based on FMEA analysis and involves some TRIZ tools (Functional Analysis, Contradictions), along with other tools (Film Maker) and the logic of Subversion Analysis.

2   Background

Over the years, several authors criticized different aspects of FMEA methodology.

Some of them expressed their doubts and uncertainties about the application of the methodology. Denson et al. (2014) criticized the method as tedious and time-consuming, while Mader et al. (2013) criticized the non-influence of FMEA in product development because it is generally performed too late and it cannot affect the design process and decision making.

Other authors criticized the representation of the cause-and-effect chain, that is, the inefficiency of FMEA to represent the combinations between multiple, simultaneous effects (Augustine et al. 2011).

Still others criticized the phase of risk analysis by explaining the inconsistencies of Risk Priority Number (RPN) in evaluating risks (Liu et al. 2013).

Lastly, some authors criticized FMEA for its ineffectiveness in decision making and problem solving (Xiao et al. 2011).

Many FMEA improvements responding to these limitations can be found in the literature.

Some authors propose improvements for ameliorating the applicability of the method (Price and Taylor 2002) by suggesting reducing or excluding human intervention in performing the analysis. Other authors (Yang et al. 2010) improved the cause-and-effect chain representation. Still others proposed methodologies for improving the risk analyses, by involving qualitative methodologies (i.e., statistical evaluations; Kara-Zaitri et al. 1991, and cost-based approaches; Xu et al. 2002).

Finally, some authors try to improve the decision making by studying better modalities of representations of FMEA outputs (Llu et al. 2002), and the problem-solving phase by suggesting more suitable approaches for reducing and eliminating the possible faults (Annamalai and Muthukarapan 2010).

Among the few authors that try to answer to this problem, the following trend has been identified in scientific literature and in patent databases. Kmenta and Ishii (2000) worked on the behaviour of the technical system and perturbed it in order to find possible failure modes. In Regazzoni and Russo (2011), the authors determine only the most significant failure modes after the determination of the failure effects, in turn determined through a reasoning at a functional level.

In particular, some attempts to merge TRIZ tools with Cause and Effect Chain Analysis (CECA) and FMEA can be found in the literature (Spreafico and Russo 2016).

CECA-TRIZ (Koltze and Souchkov 2011 and Dobrusskin 2016) uses TRIZ function analysis in addition to expert knowledge to determine the cause-and-effect chain and it performs the analysis at a deeper level of detail to investigate the root causes. The technical contradictions are instead used to eliminate those in conflict with the project requirements.

Subversion Analysis, developed by Mann (2002) and implemented in the Anticipatory Failure Determination (AFD) by Kaplan et al. (1999), helps the designer in finding all the ways to destroy the product that he is designing in order to find the failure modes and eliminate or correct them before they occur. Different from FMEA, with this logic, the designer is involved in a creative and pro-active approach in finding the failures: he uses TRIZ not to invent a new system but to provoke a damage by using TRIZ tools, suggestions and resources. Once the failure has been determined, the designer re-uses TRIZ to fix it.

The method proposed by Regazzoni and Russo (2011) is a new paradigm for enhancing risk management through the integration of FMEA, TRIZ tools and the logic of Subversion Analysis based on the following steps: (1) Identification of the primary function, (2) Definition of elements and effects, (3) Effects modelling with ENV model, (4) Assessment of risk via RPN, (5) Subversion Analysis to determine the most critical failure causes, and (6) Application of TRIZ 76 Standard Solutions for solving the identified problems.

This method, already tested in real case studies with medium and large companies (Russo and Birolini 2011; Russo et al. 2016), presents multiple advantages for simplifying the FMEA approach, however, some details of the procedure remain to be defined.

3   Proposed Solution

Compared to previous approaches, the proposed one aims to clarify the effects description of CECA-TRIZ by dividing them into Failure Modes and Failure Effects, through a deeper use of the function analysis by perturbing it according to the logic of Subversion Analysis and the deepening of the temporal description of the effects occurrence.

In the following, we explain the proposed approach in detail.

3.1   Step 1. Main Useful Function Determination

The main useful function represents the reason for which the technical system has been created.

3.2   Step 2. Main Elements Identification

As in traditional FMEA, in the proposed approach, we consider only the main groups of elements instead of the entire bill of material and we detail them only when required. The suggestion provided is to consider only the main elements by maintaining an abstract level of detail in order to not complicate excessively the analysis described below. In this way, all the elements in the functional analysis are generally assemblies and no single components.

3.3   Step 3. System Map Through Functional Analysis

Different from traditional FMEA, which suggests modifying only useful design functions in order to identify main failure effects, the proposed approach suggests mapping them by using the classical TRIZ functional analysis. In this way, more actions (useful and harmful, non-design and secondary functions) are taken into account, enlarging the network of relationships shared between the elements. Starting from the more abstract level allows including at the first step also the user and the environment.

3.4   Step 4. Determination of the "Perturbed" Functional Analyses (PFA)

Once the functional analysis is set, we can systematically produce a list of failure effects by analysing the current functional map, and then perturbing it with a specific technique conceived for modifying functional links among elements. Perturbation can be created simulating new off-design conditions, new environmental conditions, and so on.

Through this approach, the problem solver modifies one element at time by hypothesizing off-design configurations due to possible anomalies during manufacturing and product use (i.e., an anomalous user manipulation, an unconsidered variation in the environment condition, etc.). In this way, the suggestion is to describe the assemblies as black boxes and their off-design conditions as the variations in their input and output parameters. In addition, since the aim of this procedure is to simplify the analysis of the system, the presence of the assemblies in the analysis should be more common than that of the single components.

For identifying in what manner the elements of the functional analysis can change, the designer can use the list of noise factors (environmental variation during the use of the product, manufacturing variation and component deterioration) from robust design theory, by hypothesizing the variations for both components and assemblies.

Table 1 summarizes an example of noise factors adapted from Byrne and Taguchi (1987).

Through the variation of the elements in functional analysis, the shared actions between the considered element and the others, and the relations between the other elements, can change in different ways: a sufficient action can turn, for example, into an insufficient one, or a new negative action could appear, and so on. The task of the problem solver is to redefine the shared actions for each identified configuration.

Table 1 Examples of noise factors. (Adapted from Byrne and Taguchi 1987)

Product design	Process design
Consumer usage conditions	Ambient temperature
Low temperature	Humidity
High temperature	Seasons
Temperature change	Incoming material variation
Shock	Operators
Vibration	Voltage change
Humidity	Batch to batch variation
Deterioration of parts	Machinery aging
Deterioration of material	Tool wear
Oxidation (rust)	Deterioration
Piece to piece variation where they are supposed to be the same (e.g., Young's modulus, shear modulus, allowable stress)	Process to process variation where they are supposed to be the same (e.g., variation in feed rate)
All design parameters (e.g., dimension, material selection)	All process design parameters
	All process setting parameters

Let's consider, for example, a painted iron gate exposed to the weather. Considering noise factors as humidity, temperature change, consumer usage and deterioration of parts is easier to understand how these factors, alone or combined, can influence a modification of the functional triads of Paint–Adheres–Iron (positive action) and/or Paint–Blocks–Condensate. In this case, we can produce very easily many different failure modes for identifying a same final failure effect, that is, the corrosion of the iron gate. This technique can be useful both for quickly creating a list of the most critical failure effects and preparing for the next stage of failure modes identification.

3.5   Step 5. Failure Effects Determination

To better describe the Failure Effects, the Film Maker tool (Russo and Duci 2015) can be used to describe the problematic situation by including the Failure Modes. Through this tool, the user decomposes the time evolution of each Failure Effect though a sequence of temporal frames. Each frame represents a picture of what is happening in a specific instant of time, while the entire sequence builds a cause-and-effect chain. The aim of Film Maker is to help the problem solver in identifying the precise instant of time the Failure Effect occurred.

All critical photograms can be better analysed by "zooming inside" in order to increase the level of detail in order to better circumscribe the Failure Effect to a precise zone of occurrence as well as time of occurrence.

Figure 1 shows an example that explains how the presence of the perturbed element, "Presence of air bubbles in the paint," can make the adhesion of the paint on the surface insufficient (modification of the element of the functional analysis) by leading to the Failure Effect, "Peeling off the paint from the surface," after a certain time interval. The Failure Effect, when it occurs, can then be analysed by identifying its precise zone of occurrence, "The cavity between the beads."

Through this approach, the problem solver can identify the Failure Effects without using the entire bill of material for determining all the possible Failure Modes. In addition, the proposed method is able to

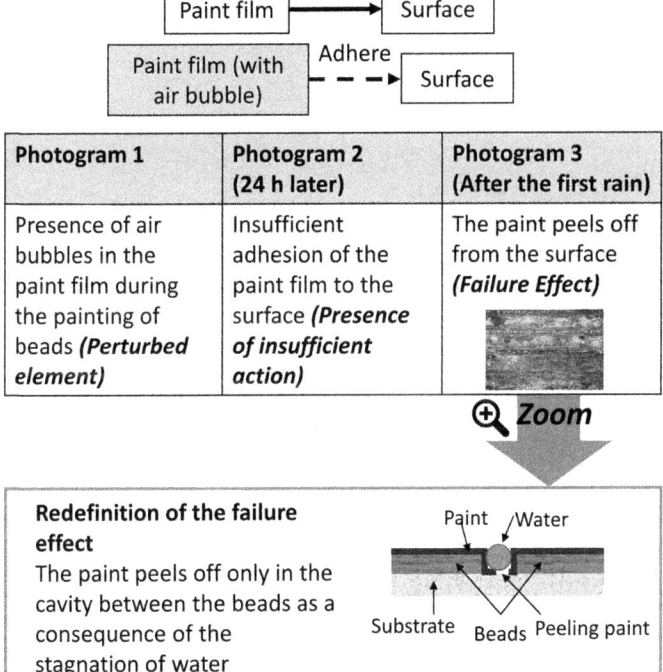

Fig. 1 Example of determination of modified actions between elements and Failure Effects through Film Maker

describe the Failure Effects in a more precise way, by identifying their time and space of occurrence.

3.6   Step 6. Failure Causes Determination and Problem Solving

Once all the Failure Effects have been determined, we evaluate them in order to identify the most critical ones, determining their Failure Causes and solving the related problem. For both cases, TRIZ tools are recommended, especially for Subversion Analysis suggested for identifying Failure Causes and problem solving.

Through Subversion Analysis, the problem solver tries to voluntarily provoke the Failure Effects, moving from a situation of defence to a new condition in which he has to use the resources at his disposal to attack. It is proved that this new condition is more powerful. Using Subversion Analysis in combination with Film Maker allows for exploitation of resources in a more precise time and space situation.

Furthermore, Subversion Analysis pushes users to use all the resources and the physical effects available in the system, increasing the probability of identifying all Failure Causes.

Once the Failure Causes are determined, we can solve the problem by using TRIZ tools.

4   Case Study

The proposed methodology has been used in previous collaborations with two multinationals firms. In this chapter, for the sake of brevity, we illustrate a case study regarding the improvement of an innovative vacuum cleaner including a dust compactor inside, developed during an industrial project with an Italian sector leader in the field of home appliances. The device compacts the aspirated dust in samples instead of collecting it inside the classical sachet, and it is activated manually by the user. The results of the project are patented in ITMI20131928. The results achieved through the proposed approach have been compared to those proposed by traditional FMEA for the same case study.

Step 1. Main Useful Function Determination

The main useful function of the system is to compact the dust sample.

Step 2. Main Elements Identification

Figure 2 represents the main elements of the system.

Step 3. System Map Through Functional Analysis

Figure 3 (left) represents the functional analysis of the main components of the system. In synthesis, the user pulls the dust from the collector compartment into the belt by using the piston. The dust is compacted against the matrix through the belt necking operated by the sprocket on which the belt is wound. The sprocket is actuated by the user through the lever (the suction pipe).

Step 4. Determination of the "Perturbed" Functional Analyses (PFA)

Among all the possible perturbation of the elements, we propose in Fig. 3 (right), as example, the modification of the joint and the related perturbed functional analysis. In this case, the "Deterioration of material" is

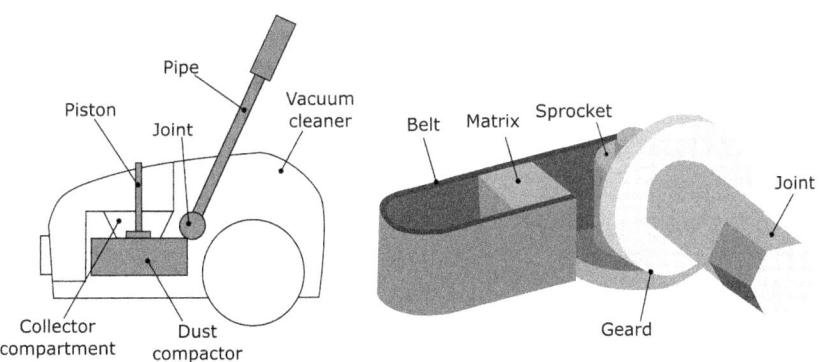

Fig. 2 Vacuum cleaner and dust compactor

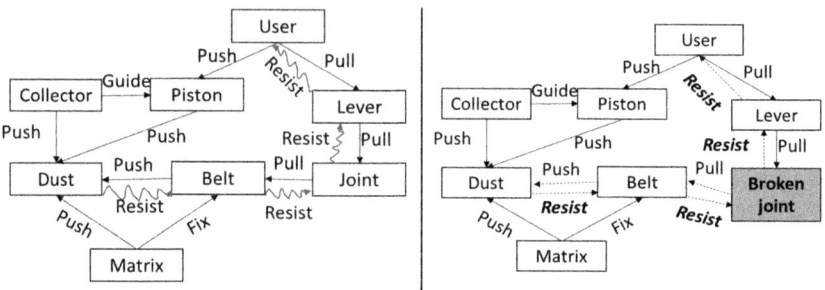

Fig. 3 Functional analysis (left) and perturbed functional analysis (right) for dust compactor obtained through the modification of the joint according to the noise factor "Deterioration of material"

the considered Noise Factor among the possible ones. As we can see, some of the previously described relations changed (bold).

Step 5. Failure Effects Determination

Based on the determined modified interactions highlighted by the perturbed functional analysis, the Failure Effects were determined. Compared to those determined by using traditional FMEA, the new effects are referred to the same Failure Mode or to new ones, and they focus on unedited aspects, such as user injury: whiplash injury due to the unexpected lessened resistance offered by the lever to the actuation, in turn caused by the low dust amount contained in the belt out of position. In particular, this aspect can be due to the ability of TRIZ functional analysis to better highlight the cause-and-effect chains, especially those that afflict the user, which is an integral part of the analysis.

Step 6. Failure Causes Determination

By using the proposed approach, an increased number of Failure Causes were determined, both for the already determined Failure Effects (by using traditional FMEA) and for the new ones. In this way, new aspects not previously considered for solving or improving the system are considered as possible dangerous physical effects, like the vibration of the engine.

The most critical Failure Effects and Failure Causes (evaluated through conventional RPN-index by the professionals of the company) were selected and analysed through problem-solving tools (e.g., Film Maker). In particular, a series of possible resolutive directions were determined for the different identified sub-problems.

The activity of robust design led to recreating the Failure Modes in a laboratory by using the prototype of the device in order to verify if this condition really led to the described fault. Various solutions were determined, such as using a different material for the belt and improving the prototype through the introduction of an opposing restraining force for containing the swelling of the belt by using an additional pin that keeps the belt stretched. These improvements were made to the device before the production phase.

4.1   Test

The proposed approach was tested by four MsD students in engineering and four PhD and academic researchers with previous knowledge in FMEA and TRIZ, and the achieved results were compared to those achieved by a control group who used traditional FMEA.

The specific objectives of this test are verified if the approach is able to achieve all the failures determined by traditional FMEA, identify new, unknown failures, and improve those already found through a better contextualization of the operative zone of occurrence.

Table 2 shows the total number of failures determined by the two groups.

Through the proposed approach, a greater number of failures were determined. In addition, those already found by the control group were also found by the proposed approach and some of the Failure Effects have been enriched with a better contextualization.

The entire analysis is proposed in Fig. 4, where the results achieved by using the proposed approach are highlighted.

Table 2 Comparison between the number of failures identified through traditional FMEA and the proposed approach

	Failure modes	Failure effects	Failure causes
Traditional FMEA	10	2	14
Proposed guidelines	12	7	20

Elements	Failure Modes (12)	Failure Effects (7)	Failure Causes (20)
Dust compactor	Perforated belt (1)	Sample with reduced density (due to the dust leakage) (1)	Stone or needle in the dust sample (1) / *Stone stuck between the belt and the side of the matrix (2)*
	Dilated belt (2)	Sample with reduced density (due to the overabundant dust mass collected) (1) / *Blocked belt between the sprocket and the containment wall due to the belt overlap on the sprocket (2)*	Belt material degradation (3)
	Belt out of position (closer to the matrix), (3)	Reduced dust quantity collected in the belt (3)	-Shifted matrix (4) / -Misaligned sprocket (5) / *Infiltration of the dust or stone between the belt and the sprocket during the welding (6)*
	Rotated matrix (4)	*User injury (whiplash injury) due to the reduced lever resistance offered by the reduced dust mass collected in the contract belt (4)* / Sample with reduced density (since the matrix is not able to constraints the belt as it should) (1) / *Blockage of the belt since it is pinched between the rotated matrix, the spilt dust and the wall (2)*	Broken fixing screw (7)
	Dirty matrix (adhesion with dust) (5)	*in the posterior surface of the sample; not cylindrical shape of the sample (3°)*	*Presence of lubricant or aspirated colloids (8)*
Transmission (gear, lever)	Misaligned gear (6)	Sample with reduced density (since the belt cannot be completely pulled) (1)	-Bumps on the lever during the use tilts the gear in perpendicular direction to that of actuation (9) / -Presence of compacted dust between the gear teeth (10) / *A stone in the dust sample overstressed the belt that tilts the sprocket (1)*
	Broken gear teeth (7)	*-The belt does not return to its original position consequently the next sample cannot accommodate the required dust amount in the anterior part (5)* / *-User injury (whiplash injury) due to the reduced lever resistance offered by the reduced dust mass collected in the contract belt (4)*	*Engine vibrations causes rubbing and breakage of the teeth (11)*
Collection compartment and piston	Misaligned compartment (8)	Sample with reduced density (1) / *on the top side since the piston does not reach the end position by blocking the dust leakage in the collection compartment (1*)*	-Frame deformation (12) / -Collection compartment deformation (13) / -Broken of the compartment fixation (14)
	Hole smaller diameter (9)	(1) / *(1*)*	-Particles bonding (15) / -Constructive errors (16) / -Collection compartment deformation (13)
	Hole larger diameter (10)	*Piston broken part of the edge of the hole by reducing the diameter if pulled too strong by the user. Part of the edge falls into the sample by compromising in part its density (6)* / (1) / *(1*)*	Particles pinched between the piston and the hole (17) / *The misaligned piston ruined the collection compartments (18)*
	Misaligned piston (11)	(1) / *(1*)*	Frame deformation (12)
	Piston larger diameter (12)	*The piston wears the walls of the compartment hole by enlarging its diameter (7)* / (1) / *(1*)*	*Particles bonding inside the sliding guide during the suction (19)* / Particles bonding on the piston border (20)

Fig. 4 FMEA analysis of the considered device enriched with the failures determined by the two groups, with the results found by the two (Cells with white background), the unknown failures determined through the proposed approach and the improvement of the already determined failures (Cells with grey background)

5   Discussion and Recommendations

This chapter proposes a new method to simplify and empower the traditional FMEA approach by introducing TRIZ tools and Subversion Analysis. The functional analysis has been used for mapping the main elements of the system, the perturbed functional analysis and Film Maker for determining the Failure Effects. Subversion Analysis is instead used for determining the Failure Causes. The advantages of this approach involved a better definition of the Failure Effects based on the time and space of occurrence. An exemplary case study shows how many failure modes, causes and effects can be added to results previously found with traditional FMEA.

References

Annamalai, N., & Muthukarapan, S. (2010). *Lean in burn-in preventive maintenance through TRIZ process analysis.* Proceedings of the 10th ETRIA world TRIZ future conference: TRIZ future conference 2010, Bergamo.

Augustine, M., Yadav, O. P., Jain, R., & Rathore, A. P. (2011, January). *An approach to capture system interaction failures of a complex system.* Proceedings of IEEE reliability and maintainability symposium (RAMS), Orlando.

Byrne, D. M., & Taguchi, S. (1987). Taguchi approach to parameter design. *Quality Progress, 20*, 19–26.

Denson, B., Tang, S. Y., Gerber, K., & Blaignan, V. (2014, January). *An effective and systematic design FMEA approach.* Proceedings of IEEE reliability and maintainability symposium (RAMS), Colorado Springs, CO, USA.

Dobrusskin, C. (2016). On the identification of contradictions using cause effect chain analysis. *Procedia CIRP, 39*, 221–224. Amsterdam: Springer.

Kaplan, S., Zlotin, B., Zusman, A., & Visnepolschi, S. (1999). *New tools for failure and risk analysis: An introduction to anticipatory failure determination (AFD) and the theory of scenario structuring.* Farmington Hills: Ideation International Inc.

Kara-Zaitri, C., Keller, A. Z., Barody, I., & Fleming, P. V. (1991, January). *An improved FMEA methodology.* Proceedings of IEEE reliability and maintainability symposium (pp. 248–252). Orlando, FL, USA.

Kmenta, S., & Ishii, K. (2000, September). *Scenario-based FMEA: A life cycle cost perspective.* Proceedings of ASME design engineering technical conference, Baltimore.

Koltze, K., & Souchkov, V. (2011). *Systematische Innovation: TRIZ-Anwendung in der Produkt-und Prozessentwicklung.* Munich: Carl Hanser Verlag GmbH Co KG.

Liu, et al. (2002). *Failure modes and effects analysis with Bayesian belief networks: Bridging the design-diagnosis modeling gap.* Stanford: Stanford University.

Liu, H.-C., Liu, L., & Liu, N. (2013). Risk evaluation approaches in failure mode and effects analysis: A literature review. *Expert Systems with Applications, 40*, 828–838.

Mader, R., Armengaud, E., Grießnig, G., Kreiner, C., Steger, C., & Weiß, R. (2013). OASIS: An automotive analysis and safety engineering instrument. *Reliability Engineering and System Safety, 120*, 150–162.

Mann, D. (2002). Evolving the inventive principles. *The TRIZ Journal.*

Price, C. J., & Taylor, N. S. (2002). Automated multiple failure FMEA. *Reliability Engineering and System Safety, 76*(1), 1–10.

Regazzoni, D., & Russo, D. (2011). TRIZ tools to enhance risk management. *Procedia Engineering, 9*, 40–51. Amsterdam: Springer.

Russo, D., & Birolini, V. (2011). Towards the right formulation of a technical problem. *Procedia Engineering, 9*, 77–91. Amsterdam : Springer.

Russo, D., & Duci, S. (2015). How to exploit standard solutions in problem definition. *Procedia Engineering, 31*, 951–962. Amsterdam: Springer.

Russo, D., Birolini, V., & Ceresoli, R. (2016). FIT: A TRIZ based failure identification tool for product-service systems. *Procedia CIRP, 47*, 210–215. Amsterdam: Springer.

Spreafico, C., & Russo, D. (2016). TRIZ industrial case studies: A critical survey. *Procedia CIRP, 39*, 51–56. Amsterdam: Springer.

Xiao, N., Huang, H.-Z., Li, Y., He, L., & Jin, T. (2011). Multiple failure modes analysis and weighted risk priority number evaluation in FMEA. *Engineering Failure Analysis, 18*(4), 1162–1170.

Xu, K., Tang, L., Xie, M., Ho, S., & Zhu, M. (2002). Fuzzy assessment of FMEA for engine systems. *Reliability Enginerering System Safety, 75*, 17–29.

Yang, C., Letourneau, S., Zaluski, M., & Scarlett, E. (2010, January). *APU FMEA validation and its application to fault identification.* Proceedings of ASME 2010 international design engineering technical conferences and computers and information in engineering conference (pp. 959–967). Montreal: American Society of Mechanical Engineers.

A TRIZ and Lean-Based Approach for Improving Development Processes

Jens Hammer and Martin Kiesel

1 Introduction

Evolving technologies and shorter product life cycles lead to increased speed of innovation. Regarding system development of software- and hardware-related products, the reduction of complexity and lead time, especially for the testing process, is mission critical for fast product delivery, the sustainable success of products and thus the survival of a company. A methodical approach must be applied to identify the key problems of the current process, achieve a clear understanding of the right target state, and elaborate the key levers for improvement.

Ikovenko and Bradley mentioned that there is a high relevance for combining the TRIZ algorithm with the Lean toolkit (Ikovenko and

J. Hammer (✉)
School of Business and Economics, Friedrich-Alexander-Universität Erlangen-Nürnberg, Erlangen-Nürnberg, Germany
e-mail: Jens-hammer@gmx.de

M. Kiesel
Siemens AG, DF MC TTI, Erlangen, Germany

L. Chechurin, M. Collan (eds.), *Advances in Systematic Creativity*,
https://doi.org/10.1007/978-3-319-78075-7_7

Bradley 2004). Merging these Lean tools together with TRIZ created several combined methodologies (Ikovenko and Bradley 2004). For instance, dramatic results were achieved in a multimillion-dollar Lean project, where at different phases of a Lean Approach several TRIZ tools were applied (Ikovenko and Bradley 2004). This leads to process simplification, considerable cost reduction, and an improved degree of reliability and safety (Ikovenko and Bradley 2004). As a result, both methods can benefit from their specific advantages. For that reason, we would like to present a case study within Siemens for improving the testing process. We apply TRIZ methods (i.e., Ideality, Cause Effect Chain Analysis) and combine these with the Lean Approach (i.e., Value Stream Analysis [VSA], A3 Report, PDCA) (Poppendieck 2013). The operational result is a set of action items that significantly reduces the testing time required by a factor of three. Furthermore, this specific case study is used to derive a framework for systematic improvement projects in system development.

2   Background

2.1   Lean and Agile in Development

The main goal of the Lean Approach in the development domain is to "sustainably deliver value fast" (Larmann and Vodde 2014). This implies a fast reaction to existing and new customer needs with a sustainable pace, without sacrificing quality.

According to Poppendieck, Lean software development established the following agile practices (Poppendieck 2016):

- Eliminate/reduce waste
- Support continuous learning
- Decide late
- Deliver fast to learn fast
- Enable the team and individuals
- Build quality
- See the whole picture

Moreover, in 2001, seventeen independent-minded software practitioners signed the Manifesto for Agile Software Development with four important statements (Highsmith 2002): (1) Individuals and interactions are more relevant than processes and tools; (2) A working software is more important than comprehensive documentation; (3) Contract negotiation is important, but customer collaboration is more important; and (4) It is good to follow a plan, but responding to change will lead to success in the end (Highsmith 2003).

In comparison to Lean manufacturing, software development has a major focus on optimizing the steps of design, implementation, integration, test and deployment (Poppendieck 2016). Major issues in this context are early tests to make defects visible as early as possible and to verify whether the implementation assumptions are right or wrong. To optimize the software development process, Lean development focuses on fast feedback cycles considering test-driven development (first define the tests, then implement the code to pass the tests), continuous integration (integrate small code changes frequently and run the tests to verify the result), iterations (develop software in iterations of two to four weeks) and cross-functional teams (cover all competences to specify the functionality and furthermore to implement and test the code) (Poppendieck 2016).

According to Poppendieck (2013), "lean software development focuses on the flow efficiency of the entire value stream." To optimize the flow efficiency in software development, the Lean Approach provides a set of tools that supports the operational improvement process:

- **Value Stream Analysis** (VSA) for modeling the current and target flow (Poppendieck 2013)
- **Hoshin Kanri** for setting targets in different time frames (i.e., north star, blue sky) (Kudernatsch 2013)
- **A3** for structured problem solving (Kudernatsch 2013)
- **PDCA** (plan–do–check–act) for incremental improvement (Kudernatsch 2013)
- **Set Based Design** for generating solution options and selecting the best solution (Singer et al. 2016)

2.2   TRIZ in Development

The Theory of Inventive Problem Solving (TRIZ) is called one of the most powerful inventive methodologies derived from the analysis of the world patent collection (Moehrle 2005; Muenzberg et al. 2016). It provides methods for problem-solving and innovating systems in a structured and methodical way (Muenzberg et al. 2016). TRIZ includes different methods to model the problem, create an abstract problem, derive solutions for that abstract problem, and afterwards create specific solutions using problem-solution methods including empirical knowledge about typical technical evolutions (Fig. 1) (Hammer and Kiesel 2017). No attempt will be made to explain the methods in detail in this chapter. Frequently used TRIZ tools and techniques are described in various scientific and application-specific contributions. The authors recommend collections from Moehrle, Ilevbare et al., and Muenzberg et al. (Moehrle 2005; Ilevbare et al. 2013; Muenzberg et al. 2016) to get an impression. Furthermore, the MATRIZ provides a structured educa-

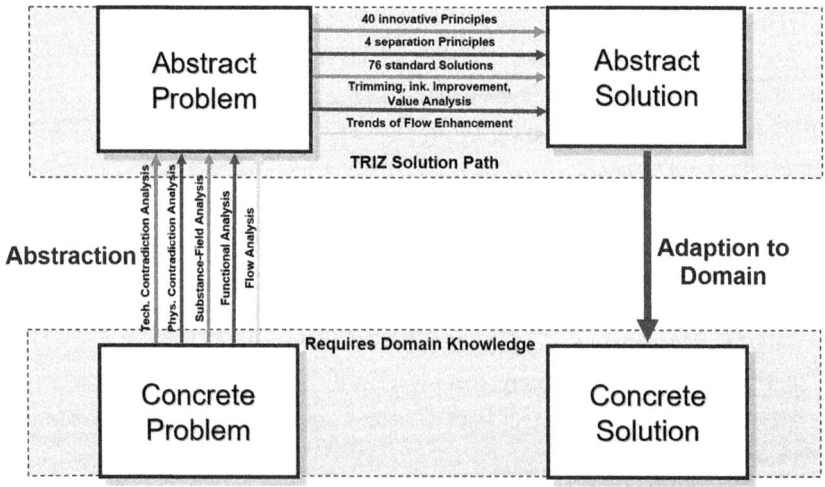

Fig. 1   General TRIZ way of thinking and acting (Hammer and Kiesel 2017)

tional approach to get used to a high amount of TRIZ tools (MATRIZ Level 1–3).

The methodology can be applied in certain fields along the whole value chain. It is suitable for different disciplines, such as research and development (R&D), quality management (QM), value engineering, business processing, crisis management (Muenzberg et al. 2013), and marketing and innovation management (Livotov 2008). Moreover, TRIZ has been combined with diverse methods to extend its scope and increase its usage. Among others, these combinations include TRIZ principles and system improvement; TRIZ and Quality Function Deployment (QFD); TRIZ and Failure Mode and Effect Analysis (FMEA); TRIZ and Lean; TRIZ, Statistical Process Control (SPC) and Lean; and finally, Six-Sigma and TRIZ (Bligh 2006).

2.3   TRIZ and Lean: Differences and Similarities

Both, TRIZ and Lean are ways to improve the operation of a system or process (Maia et al. 2015). TRIZ concentrates mainly on individual elements to optimize (with the exception of the Algorithm for Inventive Problem Solving [ARIZ]), while Lean evaluates the entire system/process to identify potential for increasing efficiency. Many TRIZ tools have a Lean equivalent (Maia et al. 2015). Moreover, both take a significant amount of time to define and analyze a problem (Maia et al. 2015). Within TRIZ, it is relevant to analyze and understand all the resources that exist to identify and solve the problem. It can be done by applying a 9-Screen Approach or by using other methods (i.e., innovation situation questionnaire, resource checklist, Function Analysis). Within Lean, the main objective is to analyze the whole system and the flow of materials and information (VSA for the current state) (Poppendieck 2013; Maia et al. 2015). In TRIZ, the problem solver will identify the key tasks or questions. In Lean, a common term is "go to gemba." For the Lean Approach it implies to go and see how a system really works, instead of just believing that the information that has already been collected is true (Bligh 2006; Maia et al. 2015).

Furthermore, both methods look towards the ideal future and have some sort of a methodology to create a vision of how a system or process should evolve. TRIZ uses Ideality of the system or process, 9-Screen Approach, the IFR and Trends of Technical System Evolution to understand the system's next steps and what an ideal solution should look like. Lean uses Value-Stream Mapping (VSM) (future state), Hoshin Kanri (North Star, blue sky), in addition to define a set of target states for dedicated timeframes that have to be achieved. With both, the target is to reach a more ideal state than the current one and both are totally akin to the eco-efficiency factors in sustainable development (Bligh 2006; WBCSD 2010).

As a final point, TRIZ and Lean try to increase and improve the use of available resources. With Lean, the goal is to optimize the complete value chain and eliminate waste (i.e., inefficiencies and non-productive actions). With TRIZ, the solution of the problem uses a resource that had previously been seen as waste, not/less useful (Bligh 2006) or used in another context.

These similarities support the deployment of Lean methods when implementing TRIZ and vice versa. Depending on the target of a project, it is very useful to apply the best suitable from both tool sets.

3   Case: Improvement of a Testing Process of a Motion Controller Product Comprising Hardware and Software Artefacts

In this chapter, we present a case study involving the development of automation products. The case study focuses on "the improvement of a testing process of a motion controller product comprising hardware and software artifacts." From development freeze to delivery of the product, the testing process for the motion controller and the associated engineering system has a lead time of six months. The system comprises a motion control device (with integrated motion control and drive control functionality) and an engineering suite comprising motion and drive engineering for configuring and programming the motion control device.

Phase 1	Phase 2	Phase 3			Phase 4	Phase 5
Setup and Planning	Kick-off	Workshop	Workshop	Workshop	Consolidation and preparation	Stakeholder Presentation

Fig. 2 Stepwise project approach

Based on the problem, we defined an improvement project with a stepwise approach (phase 1–5, see Fig. 2) to clarify the following questions:

- What are the main root causes for the current situation, especially time needed for the testing process?
- How can the testing time be significantly reduced?
- What is the expected result of the improvement measures?

3.1 Step 1: Project Setup and Planning

The project setup was defined by an architect (with TRIZ and Lean experience) in close cooperation with a TRIZ expert/moderator. We decided to use TRIZ and Lean methods for working on this project. We drew up a rough plan for the complete project and a detailed plan for the first workshop. We used a time-boxed approach with a clear expectation of the team effort required and the target results to get all the participants on board. Up front, we decided to run a kickoff meeting (one hour), three workshops (workshop sprints), each a half day with the team, and a summary meeting (two hours) to review and consolidate the results and a final presentation for the stakeholders.

The team setup was a cross-functional team (10 persons) covering the whole development and test chain (developer, tester, architects) to incorporate the relevant know-how and experience and to create a basis for the acceptance of the final measures.

3.2 Step 2: Project Kickoff

We conducted a project kickoff to present the problem, our approach, the input expected from each participant and project goals. VSM was presented to the participants to give them a basic understanding. Finally, we discussed the concerns and critical comments of the participants and came up with a working agreement for the project team.

3.3   Step 3: Workshops

We conducted three workshops using two sprints (1.25 hours) with two sub-teams running in parallel. Between the sprints, we presented the results to the whole team and defined the next steps. The following LEAN/TRIZ methods were used (see Table 1):

3.4   Step 4: Consolidation of Measures

Within the consolidation of the measures and preparation of the final presentation, there were two main questions to define the key message. What must be done? What improvement can be achieved? The prioritized list of measures included the aspects estimated impact and value, the effort for implementation and feasibility.

Table 1 Lean/TRIZ approach

1	**Value Stream Analysis (VSA)** for the current situation	Lean
2	**Analysis** of the test team "interruptions"	Lean
3	**Problem "Hotspot" Identification** based on the value stream and Prioritization	Lean
4	**Root Cause Analysis (RCA)** for the top three problems	TRIZ
5	**Ideality** for the top three problems	TRIZ
6	**Ideality** of the future value stream	TRIZ
7	**Idea Generation** for the future value stream	Lean/ TRIZ
8	Development of the **Future Value Stream**	Lean
9	**Aggregation of the working** result and creation of an **A3 report** (PDCA cycle) (i.e., current state, future state, RCA, measures for improvement)	Lean
10	**Identified measures** for improvement based on the different perspectives of the process, i.e., value stream perspective (current, future), problem perspective (root causes), necessary condition perspective, a collection of ideas that came up during the process.	Lean/ TRIZ

3.5   Step 5: Final Presentation to Stakeholders

Within the step "final presentation to stakeholders (to get their buy in)," a joint team presentation was given to the stakeholders. In general, within the project, we realized that the methods provided a holistic view of the problems and the measures that were identified cover totally different aspects, that is, degree of test automation, environmental issues (maturity of the test infrastructure), project prioritizing and planning issues, organizational issues (team stability, skill sets) and portfolio issues (complexity and variability). The estimated impact on the implementation of the proposed measures leads to a reduction of the testing time by a factor of three (see Figs. 3 and 4). Figure 3 shows an abstraction of the current state of the testing process. Development, integration and several levels of system tests are done in a sequential and iterative way.

The future state of the testing process is shown in Fig. 4. Feature and integration tests are done in parallel—the time periods needed for testing were significantly decreased.

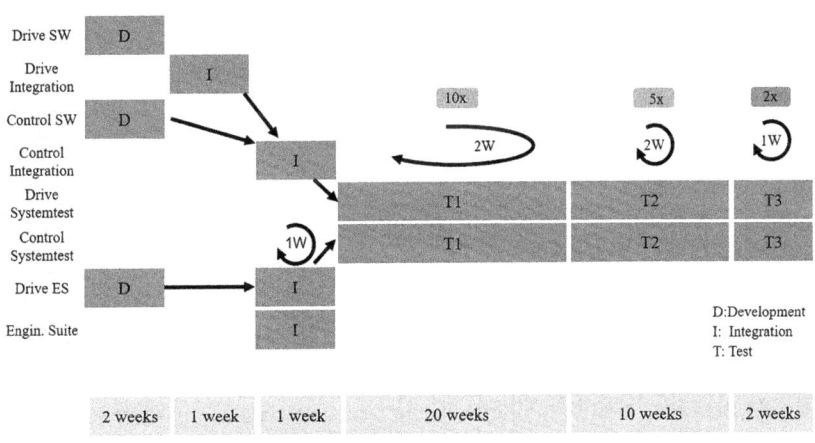

Fig. 3   Value stream analysis (current state)

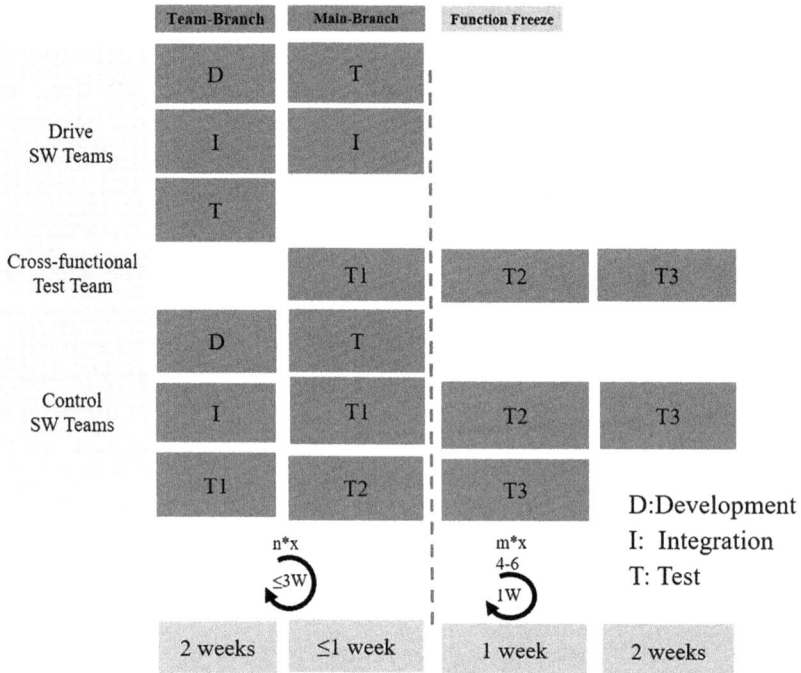

Fig. 4 Value stream analysis (future state)

4   Discussion and Recommendations

In terms of a general approach, we realized the following benefits of the Lean/TRIZ approach in methodology and working model.

4.1   Methodology

We realized that VSA is very powerful tool for these types of problems. The dedicated problem-specific "deep dive" with Root Cause Analysis (RCA) provides a holistic view of the problems and considering Ideality at different abstraction levels is helpful to generate ideas with different levels of detail. A pairwise usage of RCA and Ideality with a defined scope and at different abstraction levels has proved to be a good thinking tool and the A3 report is a good method to combine the different aspects of analytical current and future state.

4.2   Lean Working Model: Positive Experiences

Related to the Lean working model, we realized positive aspects regarding a clear scope and goal of the project. Cross-functional teams for generating a holistic view and as a base for the acceptance of the outcome is mission critical. A rough planning in the beginning and detailed before the next workshop fits to an agile thinking and working within specific time boxes with a predefined number of workshop sprints leads to success. This is a very useful Lean principal. Two sprints in each workshop proved to be a good compromise between (team) time invested and results achieved. Moreover, visual working in workshop sprints helps the participants to share information quickly and keep the overview. Digitalization, summarization, and consolidation of workshop results is a basis for the next sprint. Finally, working in small sub-teams (≤ 5 people) facilitates good collaboration and communication.

Based on the experience of the test improvement and different other projects, we derived a generic framework that allows effort to be scaled and adapted to the specific situation.

4.3   General Issues

The following describes a generic framework:

- Build a **cross-functional team** that covers the complete problem and solution space
- Work in **cross-functional sub-teams in the workshops** with ≤ 5 **people** to allow effective communication
- Let the **parallel team work** with the same or different methods/topics (selection depends on the specific situation)—both worked in our context

- **Train your technical leads in TRIZ/Lean Thinking** and let them guide how deep the discussion should go (when should we stop? In which cases is it valuable to allow an additional step?)
- **Work as an agile team in short sprints**, present the result and define your precise next steps incrementally
- **Invest five hours for each workshop with two sub-sprints** in the workshop (proven to be a very good approach in terms of time invested versus value generated)
- **Summarize and consolidate** the intermediate results after each workshop
- **Define a backlog of measures** and tag the measures with value, feasibility, effort and responsibility

4.4　Scaling Effort and the Level of Detail

The following set up is recommended for scaling the effort depending on the project (see Table 2).

Table 2　Recommendations for scaling project efforts

Small set up	Medium set up	Large set up
Use RCA and Ideality for a specific scope Derive improvement measures	Generate all information to cover a full A3 report Model current state (e.g., using VSM) Identify key problems Analyze the root causes using RCA Model the Ideality for the key problems Model the future state with different perspectives	Medium footprint with additional methods: 　Make clear what resources are available and which constraints must be regarded (innovation situation questionnaire) 　Generate information for an adequate ideality and future state using the 9-Screen Approach 　Use technical system evolution trends

5   Conclusion

The combination of TRIZ and Lean methods has proven to be very effective. Lean has a strong emphasis on human collaboration, flow efficiency, and provides a lot of methods, principles, and values that are very helpful. TRIZ brings in a structured approach for RCA and provides additional methods (e.g., 9-Screen Approach) and knowledge (Inventive Principles, Trends of Technical System Evolution) concerning typical technical evolution, which is extremely helpful to generate a promising picture of the future. This can be used for deriving ideas and for defining long-term or short-term goals.

Two years after the end of the improvement project, many of the proposed measures were implemented in the company's Lean transition context. The estimated target state (improvement by a factor of three) has almost been achieved. Finally, the company decided to set more ambitious goals to achieve a further reduction of the time "from order to cash."

References

Bligh, A. (2006). *The overlap between TRIZ and lean. manufacturing systems* (pp. 1–10). University of Rhode Island. Rhode Island, USA.

Hammer, J., & Kiesel, M. (2017). *Applying TRIZ to improve lean product lifecycle management processes.* TRIZFest 2017. Poland.

Highsmith, J. (2002). *Agile software development ecosystems.* Boston: Addison-Wesley.

Highsmith, J. (2003). *Agile project management: Principles and tools* (Vol. 4(2)). Arlington: Cutter Consortium.

Ikovenko, S., & Bradley, J. (2004). *TRIZ as a lean thinking tool.* 4th TRIZ future conference, Florence.

Ilevbare, I. M., Probert, D., & Phaal, R. (2013). A review of TRIZ, and its benefits and challenges in practice. *Technovation, 33*, 30–37.

Kudernatsch, D. (2013). *Hoshin Kanri Unternehmensweite Strategieumsetzung mit Lean-Management-Tools.* Stuttgart: Schäffer-Poeschel.

Larmann, M., & Vodde, B. (2014). *Scaling lean & agile development thinking and organizational tools for large-scale scrum* (p. 46). Westford.

Livotov, P. (2008). *TRIZ and innovation management.* www.tris-europe.com

Maia, L. C., Alves, A. C., & Leão, C. P. (2015). *How could the TRIZ tool help continuous improvement efforts of the companies ?* (Vol. 131, pp. 343–351). Procedia engineering.

Moehrle, M. G. (2005). *What is TRIZ? From conceptual basics to a framework for research* (Vol. 14, pp. 3–13). Creativity and innovation management.

Muenzberg, C., Hammer, J., Brem, A., & Lindemann, U. (2016). *Crisis situations in engineering product development: A TRIZ based approach* (Vol. 39, pp. 144–149). Procedia CIRP.

Poppendieck, M. (2013). *Lean software development an agile toolkit* (pp. 9–13). Crawfordsville: RR Donnelley.

Poppendieck, M. (2016). *Lean software development: The backstory*. Retrieved September 5, 2016, from http://www.leanessays.com/2015/06/lean-software-development-history.html

Singer, D. J., Doerry, N. C., & Buckley, M. E. (2016). *What is set-based design?* Retrieved September 5, 2016, from www.doerry.org/norbert/papers/SBDFinal.pdf

WBCSD. (2010). *Eco-efficiency and cleaner production: Charting the course to sustainability.* Switzerland.

A Method of System Model Improvement Using TRIZ Function Analysis and Trimming

Nikolai Efimov-Soini and Kalle Elfvengren

1 Introduction

The chapter concerns a function-based method for design improvement and new design development. According to Ullman, 75% of product cost is defined at the conceptual stage, and the cost of product improvement grows exponentially in the manufacturing stage, but the use of systematic methods makes it possible to minimize the funds lost (Ullman 2010). This means that these stages are extremely important in the product life-cycle, and the use of systematic methods at these stages is thus very useful for design development.

Several systematic approaches exist, for example, Axiomatic Design (Suh 1990), Unified Structured Inventive Thinking (USIT) (Sickafus 1997), and the Theory of Inventive Problem Solving (TRIZ) (Altshuller 1984). In this chapter, the TRIZ methodology is used for design

N. Efimov-Soini (✉) • K. Elfvengren
School of Business and Management, Lappeenranta University of Technology, Lappeenranta, Finland

© The Author(s) 2019
L. Chechurin, M. Collan (eds.), *Advances in Systematic Creativity*,
https://doi.org/10.1007/978-3-319-78075-7_8

development because it is easy to use and understand. TRIZ utilizes formal approaches and inventive tools, and it is widely used in science and industry (Luo et al. 2012; Di Gironimo et al. 2013; Chechurin 2016).

TRIZ, in Russian *Theoria Resheniya Izobretatelskih Zadech*, is an inventive method proposed by the Soviet inventor Genrich Altshuller in 1956 (Altshuller and Shapiro 1956). He studied about 40,000 patents and drew out the formal processes for some new ideas of the generation and the technical evolution trends. The method has 40 inventive principles, contradictions, ideality, and patterns of evolution.

This chapter concerns modern TRIZ tools, such as function analysis and trimming (Gadd 2011). The latter is a formal method for system development and improvement, based in the reduction of system complexity. Different types of this tool are used for patent-around design (Li et al. 2015), system improvement (Sheu and Hou 2013), and to form new design patterns (Efimov-Soini and Uzhegov 2017). In this chapter, the advanced method of trimming is used for system improvement and development in a formal manner. Function analysis is used as the input to the trimming process.

Previous trimming methods used formal rules for step-by-step improvement of the system (Ikovenko et al. 2005; Sheu and Hou 2013; Li et al. 2015). The functions were independent in these approaches, and the authors did not use a special formal ranking index to define the importance of the function in the system. In contrast to the previous methods, the new approach takes the relation between the functions and elements into account. The function analysis step is improved and a new operation, "creation and analysis of the function interaction matrix," is added before the trimming step. This improvement highlights the "function streamlines" in the system. This means that the functions are grouped in sets. This idea makes it possible to automate the trimming algorithm and to receive a new concept pattern.

The rest of the chapter is structured as follows: Sect. 2 is devoted to the state-of-the-art, Sect. 3 describes the method, Sect. 4 illustrates the method through an industrial case study, Sect. 5 consists of discussion, and Sect. 6 presents conclusions.

2   State-of-the-Art

The functional part of the presented method is based on TRIZ function analysis. There are several types of function presentation of the system model, such as the Functional Flow Block Diagram (FFBD) (Akiyama 1991), the Functional Analysis System Technique (FAST) (Cooke 2015), and the Integrated Computer-Aided Manufacturing Definition for Function Modeling (IDEF0) (Defense Acquisition University 2005). All these methods use the function approach for system model presentation. For example, FFBD is a function-oriented approach based on the sequential relationship of all system functions. FFBD develops the system from the top to the bottom and proposes a hierarchal view of the functions across a series of levels. The aim of each level is to identify a single task on a higher level by means of functional decomposition. In comparison to FFBD, the FAST diagram focuses on the product functions rather than a specific design. In contrast to FFBD and FAST, the TRIZ function modeling takes account of the physical interaction between the system elements and the type of this interaction. There are four interaction types: useful, harmful, insufficient, or excessive. FFBD and FAST use the static approach in the function analysis, meaning that the number of elements and functions, and the relation between the elements, are time-independent and do not change in time, whereas many TRIZ practitioners point out the need to identify problems clearly at each system level, and to solve them separately. This goal is achieved by integrating well-known models and instruments for system description and function representation. O. and N. Feygenson also suggest the Advanced Function Approach in the Modern TRIZ (Feygenson and Feygenson 2016), where they add some steps, such as: "Indicate the place the function is performed" and "Indicate the time the function is performed." Also, this approach has been used by Litvin et al. (2011) in their research in the application history and the function analysis evolution. The research indicates that the next logical step for enhancing the Function Approach is the introduction of two parameters: "time of performing a function" and "place of performing a function." The presented method combines previous works in the domain and proposes taking the physical relation and time-dependence of the system model into account.

The second part of the method consists of trimming. This is a formal process to improve the system model by means of system complexity reduction. There are different approaches in this case. For example, Miao Li's method (Li et al. 2015) is used in patent-around design; it takes account of the importance of each element. The Gen3 method (Ikovenko et al. 2005) is used for design improvement and development. This method uses the formal functional approach to rank the importance of each function in the system. The approach presented in this chapter combines and supplements these methods, and collects previous works in this domain.

3   Method Description

This chapter contains the method description by using a simple example. The method uses function modeling for system model improvement by means of system complexity reduction. The suggested method consists of two main parts: function analysis and trimming.

3.1   Function Presentation and Other Definitions

In TRIZ function modeling, the functions are presented in the following manner: the tool (the function carrier), the function, and the object. The function must create real action from the tool to the object, for example, "a helmet deflects a bullet" is a legitimate function, but "a helmet protects a head" is not a legitimate function. On the other hand, the function does not have to be declarative, for example, "a pill improves the health."

The function rank is a formal factor that defines the importance of the function in the function model. In this case, the rank is a positive integer number. As such, the argument of the rank function is inversely proportional to the importance of the function. For example, a function with rank three is more important than a function with rank five.

The ranking factor is a formal index that defines the function rank. It is a rational number and may be positive or negative. The latter is also inversely proportional to the importance of the function. Thus, a small value defines the most important function

The target element is called the main element in the system. This means that this element defines the purpose of the system in the initial function model.

The target function, on one hand, is a function that interacts with the target element. On the other hand, this function defines the main function of the initial system.

3.2   Function Analysis

The function analysis part consists of three main steps: component analysis, interaction analysis, and function modeling. Component analysis concerns decomposition of the system model to the main parts (e.g., to big assemblies). This step may be done by means of Computer-Aided Design (CAD) model analysis (Efimov-Soini and Chechurin 2017). In the interaction analysis step, the interaction between the elements is defined. In the final step, a function diagram (function model) is created and the rank (importance) of the functions in the system is defined.

Component Analysis

This step concerns system model decomposition. Complex systems are usually divided into big assemblies and simple systems to parts. This analysis includes not only system model parts but also external parts. For example, a car consists of an engine, a frame, a body, and so on, but in some cases a road may be included in the system model. The main goal of this step is to create a detailed system decomposition model in the area close to the target element.

Interaction Analysis

A special interaction matrix is used in the interaction analysis step. In this matrix, the interaction between the elements of the system model is denoted with a plus sign (+) and the lack of interaction with a minus sign (−). As an example, the interaction matrix of the system a cup with a cap—coffee—is presented in Table 1. This system model is used below to illustrate the method description. The case study is illustrated with a real industrial case.

Function Modeling

There are six substeps in this step: definition of the interaction as a function, initial ranking, initial function model creation, model selecting, final ranking, and function interaction analysis.

Firstly, each interaction is defined as a function in the substep called "interaction definition as a function." If interaction between the elements is not available, the function is not defined. The functions in the system model are presented in Table 2.

The target function and the target element are also defined in this step. In this case, the target function is "cup holds coffee" and the target element is "coffee."

Next, the initial function rank is defined. The rank (importance) of each function is defined on the interval $(1...+\infty)$. When the initial rank of the target function is 0, it is the highest in the system. The initial ranking of the suggested system is presented in Table 3.

Table 1 An interaction matrix

	Cup	Table	Coffee	Cap
Cup		+	+	+
Table	+		−	−
Coffee	+	−		+
Cap	+	−	+	

Table 2 Functions in the system model

Element 1	Function	Element 2	
Cup	Holds	Coffee	Target function
Table	Holds	Cup	
Cap	Holds	Coffee	
Cup	Holds	Cap	

Table 3 Initial ranking

Element 1	Function	Element 2	Rank
Cup	Holds	Coffee	0
Table	Holds	Cup	1
Cap	Holds	Coffee	1
Cup	Holds	Cap	1

After this, the function model is created by means of the previous step results. At this point, it is possible to create the function model simultaneously with the initial rank definition. In the function diagram, the elements are marked as rectangles. It is also recommended to place the target element on the right side. In addition, for elements that are impossible to modify or are not included in the model, create action (e.g., gravitation) is noted with a hexagon.

In the suggested method, two model types are presented: static and dynamic. In the static model, the function rank (importance), the interaction number in the system, and the number of the elements in the system are permanent. In the dynamic model, one or all parameters can be changed. The dynamic model is presented as a set of system snapshots. Each snapshot is a static state of the system, which means that the function rank, the number of interactions, and the number of elements are permanent in each state. For each static state, the time (duration) of each snapshot is defined by means of the following formula:

$$t_{ni} = \frac{t_i}{t_w} \tag{1}$$

where t_{ni} is the normalized time of the state i, t_i is the time of the state (in minutes, seconds, years, etc.), and t_w is the total observation time (in minutes, seconds, years, etc.).

The presented approach is used to calculate the final function rank. In this case, the special formal index (the ranking factor) is used. The latter is inversely proportional to the importance of the function. Thus, a small value of ranking factor defines the most important function. Using normalized time, the final ranking factor is defined as

$$FFR = \sum FR_i \times t_{ni} \qquad (2)$$

where FFR is the final ranking factor, FR_i is the ranking factor in state i, and t_{ni} is the normalized time of the state i. For a static system, $t_{ni} = 1$.

For the presented system, both the dynamic and the static approach may be used. For the dynamic approach, two states are considered, the cup is on the table ($tn_1 = 0.9$) and the cup is lying on its side ($tn_2 = 0.1$).

In the final ranking substep, the special index, called the ranking factor, is added to define the function rank. The following formula is used to calculate this:

$$FR_i = R - N_l - N_d \qquad (3)$$

where FR_i is the ranking factor in this state, R is the initial rank, N_l is the number of function carrier links, and N_d is the number of duplicated functions. The final ranking is presented in Table 4.

Additional subindexes are used in the suggested case, such as 2A and 2B. The letter is used to distinguish functions with an equal ranking factor. The function with subindex A is geometrically closer than the function with subindex B. This means that the element "cup" is closer to the element "coffee" than the element "cap" to the element "coffee."

The situation is different in the dynamic model, and the situation is different in the dynamic approach as well. There are two system states: the cup is on the table ($tn_1 = 0.9$) and the cup is lying on its side ($tn_2 = 0.1$). The latter state function model is equal to the model in the static approach.

Table 4 Final ranking

Element1	Function	Element2	R	Nl	Nd	FR	Final rank
Cup	Holds	Coffee	0	3	0	−3	1
Table	Holds	Cup	1	1	0	0	3
Cap	Holds	Coffee	1	2	0	−1	2B
Cup	Holds	Cap	2	3	0	−1	2A

The function "cap holds cup" is not available in this state because these elements do not interact here. The ranking table for this case is presented in Table 5.

Finally, sets of functions are defined after ranking. This approach makes it possible to improve the trimming process and to receive some new design patterns. In this case, the functions are divided into three types: independent (−), dependent (+), and equal (=). Independent functions do not interact, for example, in the fun system, functions installed in the wall "wall holds fun" and "fun moves air" are independent. In the dependent type, functions create the result "together," for example, "bolt holds nut" and "nut holds plate" are dependent functions. The equal functions create a similar result in the system, for example, functions "welding holds plate" and "bolt holds plate" are equal in many cases. In this case, independent functions trim in a separate manner, and dependent and similar functions in sets. The function interaction matrix for the presented model is shown in Table 6.

There are two sets of functions, the dependent set "cup holds cap" and "cap holds coffee" and the set of equal functions "cup holds coffee" and "cap holds coffee."

Table 5 Ranking in the dynamic approach

Element1	Function	Element2	R1	R2	NI1	NI2	Nd	FR1	FR2	Final rank
Cup	Holds	Coffee	0	0	3	3	0	−3 × 0.9	−3 × 0.1	1
Table	Holds	Cup	1	1	1	1	0	0 × 0.9	0 × 0.1	4
Cap	Holds	Coffee	1	NA	2	2	0	−1 × 0.9	−1 × 0.1	3
Cup	Holds	Cap	2	2	3	3	0	−1 × 0.9	−1 × 0.1	2

Table 6 The function interaction matrix

	Cup holds coffee	Table holds cup	Cap holds coffee	Cup holds cap
Cup holds coffee		−	=	−
Table holds cup	−		−	−
Cap holds coffee	=	−		+
Cup holds cap	−	−	+	

3.3   Trimming

Trimming is a formal process to improve the system model by means of system complexity reduction. There are three formal rules. The function may be trimmed if

- Rule A: An object of the Function does not exist;
- Rule B: An object of the Function performs the function itself;
- Rule C: Another Engineering System Component performs the useful function of the Function Carrier.

The trimming procedure starts with the function with a lower rank. If sets of function are defined in the system model, the trimming process starts with the last one. Three formal rules are used to trim the sets in the trimming process. This is a radical method, but it makes it possible to gain a new qualitative design pattern.

In the presented case the trimming process has the following steps:

- "Cup holds cap" and "cap holds coffee" (set "cup holds coffee") may be trimmed in case of the transfer function. For example, the function "holds" is transferred to the table and the new system is a table with a thermos. Another example: the function transfers to coffee itself to create a solid shell of coffee. In this case, the coffee holds itself. This trimming process is similar to the previous one.
- "Table holds cup" has a lower rank in the system (in the static and dynamic approaches) and may be trimmed by means of Rule B. It is possible to place the cap on the floor or hold it in the hand.
- "Cap holds coffee" may be trimmed if the function is transferred to the cap (similar to a baby cup). In this case, the element "cap" is trimmed, and the function "Cup holds cap" is trimmed as well by means of Rule A.

4   Case Study

Here, we present an industrial case. The case concerns a special tool for flow meter assembling. This procedure is used by the firm Termotronic (Saint Petersburg, Russia) in the manufacturing process. The presented mechanism is very complex, and therefore this chapter focuses only on the flow meter holding system. The step-by-step algorithm is presented below.

The holding system is presented in Fig. 1. The system is inspired with a linear actuator and it is used to hold the flow meter on a vertical axis. Considering that the various flow meter models have different tube diameters, the system must be adaptive. In this case, a system based in pinions is used. The user rotates a handle, this handle rotates the first pinion, then the pinion rotates the driven pinion, and the last one moves the thread. In this process, the thread moves the holder on the horizontal axis. The frame in this system holds the holder.

5   Function Analysis of the Holding System

The first step in the suggested approach is function analysis. This approach is based on previous developments (Efimov-Soini and Chechurin 2016, 2017; Efimov-Soini and Uzhegov 2017) and the Gen3 function analysis approach (Ikovenko et al. 2005).

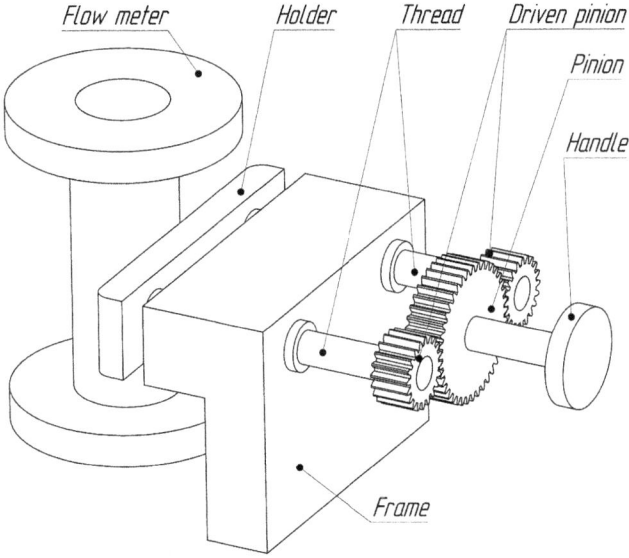

Fig. 1 Initial flow meter holding system

There are three parts to this step: component analysis, interaction analysis, and function analysis. Component and interaction analyses are accomplished by means of a special software (Efimov-Soini and Chechurin 2017) that uses a CAD model of the flow meter holding system.

The function model of the holding system is presented in Fig. 2.

5.1   System Function Interaction Matrix

The interaction between the functions and the interaction type are defined in this step. There are three types of the interaction in the suggested method: dependence (+), independence (−), and similarity (=). In Table 7, independent and dependent functions are distinguished. By means of this table, it is possible to divide the functions into two parts— the function "Frame holds Holder" and a set of the five remaining functions, named "Handle moves Holder."

5.2   Trimming of the System

The system simplification is completed by means of a trimming algorithm in this step. There the set of functions "Handle moves Holder" and

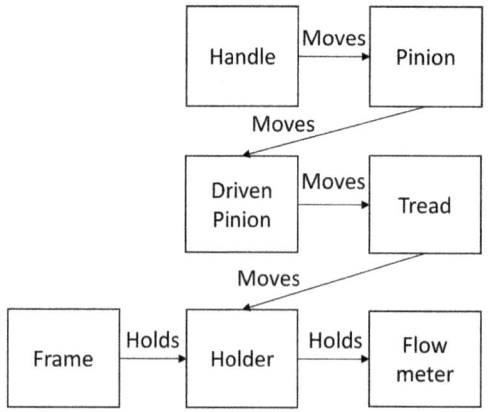

Fig. 2 Function model of the holding system

Table 7 Holding system function interaction table

	Handle moves pinion	Pinion moves driven pinion	Driven pinion moves tread	Tread moves holder	Holder holds flow meter	Frame holds holder
Handle moves pinion		+	+	+	+	−
Pinion moves driven pinion	+		+	+	+	−
Driven pinion moves tread	+	+		+	+	−
Tread moves holder	−	+	−		+	−
Holder holds flow meter	+	+	+	+		−
Frame holds holder	−	−	−	−	−	

the function carriers of this set are transformed to the system "Handle–Spring–Holder." This means that a spring is added into the system and the functions set "Handle moves Holder" is trimmed by means of Rule C. This system is self-adaptive, which means that the spring holds the flow meter with a different pipe diameter without any action. The function model of the improved system is presented in Fig. 3. The improved system is presented in Fig. 4.

6   Discussion

A special survey was performed to verify the presented algorithm. In this survey, 10 engineers from the firm Termotronic and Institut Telecomunicatsiy improved the existing systems by using the suggested

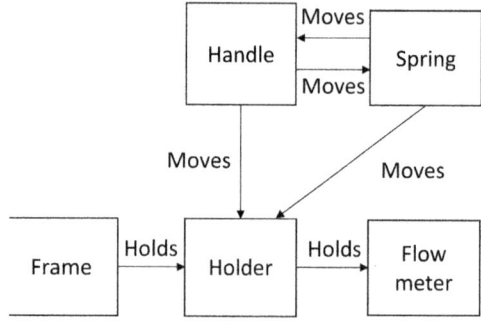

Fig. 3 Function model of the improved holding system

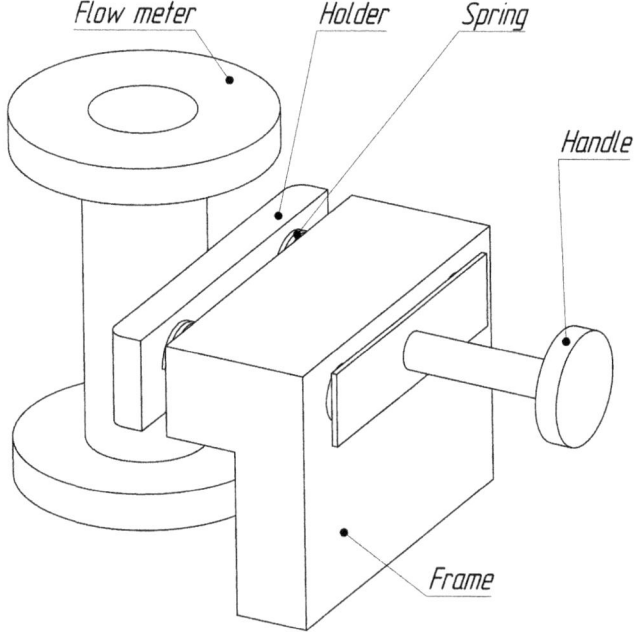

Fig. 4 Improved holding system

method. They worked manually without any special tool for system decomposition and function ranking to understand the weakness of this approach better. All specialists commended the new design pattern received by means of this method. Eight of the 10 engineers mentioned the difficulty of calculation in complex assemblies and seven of the 10 mentioned the ambiguity of the functional definition.

7   Conclusions

The presented algorithm makes it possible to improve a system model by means of system function presentation. It takes account of the evolution of the situation (and therefore its function model) and shows how the approach yields trimming ideas that are different from what can be derived from the standard static function ranking procedure. In addition, it is believed that the introduction of the time domain makes function analysis more accurate and realistic. The function analysis is formularized and requires a number of calculations, but makes it possible to create the improvement in a systematic manner.

Acknowledgments The authors would like to acknowledge the EU Erasmus plus program and its project Open Innovation Platform for University-Enterprise Collaboration: new product, business and human capital development (Acronym: OIPEC, Grant Agreement No.: 2015-3083/001-001) for the support.

References

Akiyama, K. (1991). *Function analysis: Systematic improvement of quality and performance*. Cambridge, MA: Productivity Press Inc.

Altshuller, G. (1984). *Creativity as an exact science*. New York: Gordon and Breach.

Altshuller, G., & Shapiro, R. (1956). Psychology of inventive creativity. *Issues Psychological, 6*, 37–49.

Chechurin, L. (2016). TRIZ in science. Reviewing indexed publications. *Procedia CIRP, 39*, 156–165. https://doi.org/10.1016/j.procir.2016.01.182.

Cooke, J. (2015). TRIZ-based modelling and value analysis of products as processes, *Trizfuture,* (2015): 1–11.

Defense Acquisition University. (2005). *System engineering fundamentals*. Virginia USA: Defense Acquisition University Press. https://ocw.mit.edu/

courses/aeronautics-and-astronautics/16-885j-aircraft-systems-engineering-fall-2005/readings/sefguide_01_01.pdf

Di Gironimo, G., Carfora, D., Esposito, G., Labate, C., Mozillo, R., Renno, F., Lanzotti, A., & Suiko, M. (2013). Improving concept design of divertor support system for FAST tokamak using TRIZ theory and AHP approach. *Fusion Engineering and Design, 88*(11), 3014–3020. https://doi.org/10.1016/j.fusengdes.2013.07.005.

Efimov-Soini, N., & Chechurin, L. (2016). Method of ranking in the function model. *Procedia CIRP, 39*, 22–26.

Efimov-Soini, N., & Chechurin, L. (2017). The method of CAD software and TRIZ collaboration. In A. Kravets, M. Shcherbakov, M. Kultsova, & P. Groumpos (Eds.), *Creativity in intelligent technologies and data science, Communications in computer and information science* (Vol. 754, pp. 517–527). Cham: Springer.

Efimov-Soini, N., & Uzhegov, N. (2017, May 22–25). The TRIZ-based tool for the electrical machine development. *Progress in electromagnetics research symposium*, St Petersburg, Russia.

Feygenson, O., & Feygenson, N. (2016). Advanced function approach in modern TRIZ. In *Research and practice on the theory of inventive problem solving (TRIZ)* (pp. 207–221). Cham: Springer International Publishing.

Gadd, K. (2011). *TRIZ for engineers: Enabling inventive problem solving.* Chichester: Wiley.

Ikovenko, S., Litvin, S., & Lyubomirskiy, A. (2005). *Basic training course.* Boston: GEN3 Partners.

Li, M., Ming, X., He, L., Zheng, M., & Zhitao, X. (2015). A TRIZ-based trimming method for patent design around. *CAD Computer Aided Design, 62*, 20–30. https://doi.org/10.1016/j.cad.2014.10.005.

Litvin, S., Feygenson, N., & Feygenson, O. (2011). Advanced function approach. *Procedia Engineering, 9*, 92–102. https://doi.org/10.1016/j.proeng.2011.03.103.

Luo, Y., Shao, Y., & Chen, T. (2012). Study of New Wall materials design based on TRIZ integrated innovation method. *Management Science and Engineering, 6*(4), 15–29. https://doi.org/10.3968/j.mse.1913035X20120604.635.

Sheu, D., & Hou, C. (2013). TRIZ-based trimming for process-machine improvements: Slit-valve innovative redesign. *Computers & Industrial Engineering, 66*(3), 555–556. https://doi.org/10.1016/j.cie.2013.02.006.

Sickafus, E. (1997). *Unified structured inventive thinking: How to invent.* Grosse Ile: Ntelleck.

Suh, N. (1990). *The principles of design.* New York: Oxford University Press.

Ullman, D. (2010). *The mechanical design process* (4th ed.). New York: McGraw-Hill.

Function Analysis Plus and Cause-Effect Chain Analysis Plus with Applications

Min-Gyu Lee

1   Introduction and Background

In this rapidly changing world, anyone—either a company or an individual—should be good at continuously innovating one's products, services and operations to survive. Among several stages of innovation, the earlier, conceptual stages usually have very high impact on the latter stages and outcome: (a) the stage of finding a good opportunity for innovation or a good problem to solve and (b) the stage of finding good conceptual solutions to the chosen problem. This chapter introduces an application of a central phase of the Define–Analyze–Solve–Execute (DASE) roadmap, which is a roadmap for stage (b)—finding good conceptual solutions for a given initial problem.

M.-G. Lee (✉)
Department of Industrial Engineering and Management,
Lappeenranta University of Technology, Lappeenranta, Finland

QM&E Innovation & Uni Innovation Lab,
Bundang-gu, Seongnam-si, Gyeonggi-do, South Korea

© The Author(s) 2019
L. Chechurin, M. Collan (eds.), *Advances in Systematic Creativity*,
https://doi.org/10.1007/978-3-319-78075-7_9

This methodology is based on the Theory of Inventive Problem Solving (TRIZ) (Litvin et al. 2012), and especially, the standard product value innovation methodology of the St. Petersburg TRIZ School, which is characterized by emphasis on practical implications of the theory, extensive use of functional approach and methodological recommendations, algorithmized in greater detail (Gerasimov 2010).

DASE methodology is a result of long efforts (Lee et al. 2016, 2017; Lee 2016, 2017) to improve existing TRIZ methods and roadmaps to be more useful and convenient while solving hundreds of real problems in various industries during TRIZ consulting in Korea by the author and two TRIZ masters, Vasily Lenyashin and Dr. Yury Danilovskiy, and Korean consultants, Sunghong Kim and Kyujin Jung.

Section 2 provides an introduction to help readers understand better how the two main methods, FA+ and CECA+, are used in the DASE roadmap, often complementing each other.

2   Introduction to the Methods: FA+ and CECA+

2.1   Introduction to FA+

Since FA+ is developed from function analysis methods used in TRIZ, the traditional definitions and methods are maintained as much as possible. However, changes are made when necessary to improve the method to be more widely applicable and convenient. Definitions and classifications for components and actions are rearranged as shown in (a) of Fig. 1. The shapes of components are similar to those in classical definitions, but we slightly changed their definitions to help direct visualization of the constraints—a rectangle is defined to be a controllable component instead of a component of the technical system, a hexagon as an uncontrollable component instead of a component of the super system. To simplify FA+ methods, a new category is generated to include all kinds of 'unsatisfactory action', which is useful because most model solutions coincide for those actions.

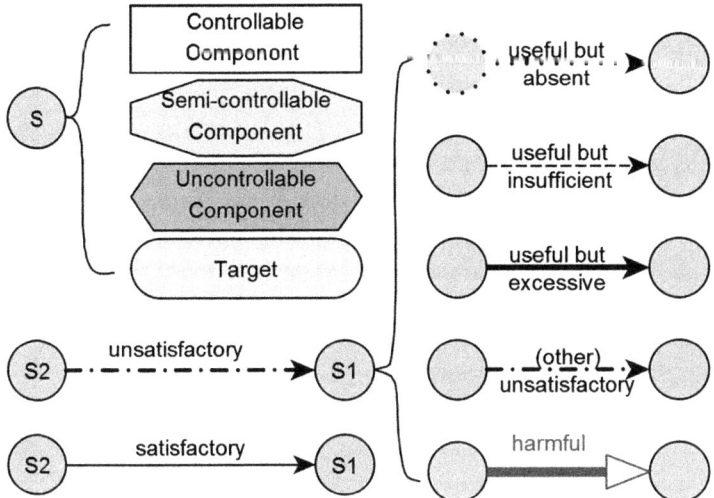

(a) Conventions for substances (components) and actions

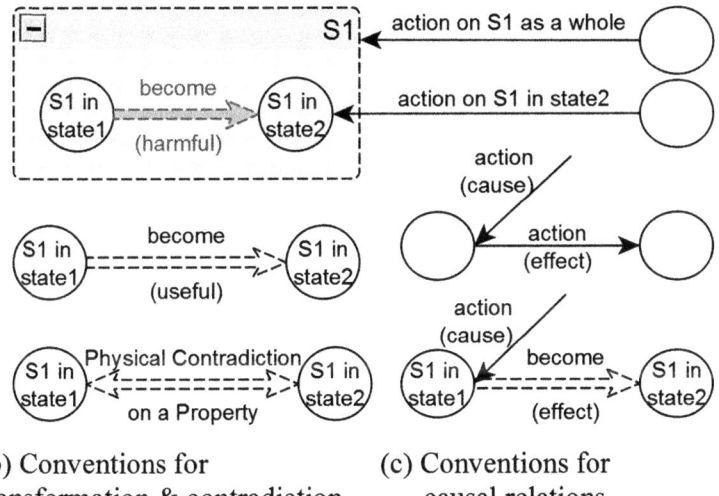

(b) Conventions for
transformation & contradiction

(c) Conventions for
causal relations

Fig. 1 Drawing conventions for function analysis plus (FA+)

Fig. 1b and c summarizes new drawing conventions added to supplement the missing notations in function analysis. The conventions (b) for transformation and contradiction can add ways to express situations from different times and different occasions in a diagram. The conventions in the middle of (c) were added to represent causality, which adds some features of CECA to function analysis.

Since some of the actions and transformations (in time) have causal relations with each other, connecting these arrows end to end can naturally make the flow of cause-effect actions more visible (Fig. 3).

The process of analyzing a problem and generating solving directions using the FA+ method can be summarized as follows:

1. Identify the technical system to improve and define the problem and goal, if not yet done in earlier steps of the roadmap.
2. Build an As-Is FA+ diagram for the problem situation. An FA+ diagram with new conventions in Fig. 1 is recommended, but a conventional FA diagram can also be used.
3. From the diagram, select one of the unsatisfactory actions (arrows) that are important for solving the problem and complete the following steps with this arrow.

   (A). Identify which of the two model problems (in Table 1) the selected action corresponds with.
   (B). Test whether each of the model solutions (MS1–1–2 or MS2–1–5 in Table 1) for the selected model problem is feasible. Often, the problem situation and constraints prohibit some of the model solutions. Then draw the feasible model solutions in the FA+ diagram as clouds with the name of model solutions and hollow arrows directing to the unsatisfactory actions that they are addressing to solve.

4. Repeat step 3 for all the other unsatisfactory actions that are important for solving the problem. This will complete the FA+ diagram with feasible model solutions (or 'solving directions').
5. Try to match suitable resources (substances and fields) for each feasible model solution. They can be searched initially from the FA+ diagram itself with the help of a few trends of evolution (e.g., MATChEm and Macro to Micro, etc.). A model solution (solving direction) with a set

of well-defined properties of substance and field can be called an idea. An Idea can also be drawn in the FA+ diagram as a second tier of clouds.

Tables 1 and 2 show the meanings of two model problems and five model solutions in FA+. They are designed to be as simple as possible so that an average TRIZ user can memorize and use in an FA+ diagram but are still helpful enough for the users to check main high-level solving directions existing in Inventive Standards or other TRIZ methods.

The maximum number of verbs, each of which represents a model solution, to be written near an arrow from a cloud is six (modify, replace, mediate, add field, remove and repair), but practically, four verbs are enough since the last two verbs are rarely used.

2.2   Introduction to CECA+

A CECA+ is done in four steps—building and verifying the as-is cause-effect chains and then generating and evaluating solving directions (Fig. 2). Although both FA+ and CECA+ models can be drawn with any diagramming tool, we strongly recommended using yEd Graph Editor (yWorks n.d.) to maximize drawing efficiency.

Since CECA+ was built on the merits of existing Root Cause Analysis (RCA) methods, especially those based on TRIZ, the basic cause-effect chain structure is similar to those. However, unlike those methods, the process and result of generating solving directions is explicitly integrated into the diagram, helping the user generate and interpret the solving directions easily.

It is important that once an As-Is CECA+ diagram is built, the third step of generating solving directions can be done automatically.

The process and result of verification of causes and solving directions are also incorporated into the diagram. The readability and ease of drawing was also much improved in CECA+ by using convenient diagramming software with intuitive design conventions. For more details of the conventions of drawing and processes, refer to the original paper (Lee et al. 2016).

Table 1 Model problems and model solutions for FA+ (in words)

Model PROBLEM (MP)	Model SOLUTION (MS)	Comment
MP1. Needed but absent action There is a needed but absent action on a substance S1. (Sometimes caused by the result of applying MS2-1-Removing) Noth-ing. needed but absent action → S1	**MS1-1.Modify or replace substance** Modify the properties of the substance S1 or replace it with other substance S_{add} so that the action is not needed any more	Modification (S1") or replacement (S_{add}) of S1 should no longer need the action
	MS1-2. Add a tool (a field adder) Add a new tool (field-adder) S_{add} that can provide the needed action on S1	The new tool S_{add} provides the needed action on S1
MP2. Unsatisfactory action A substance S2 has an unsatisfactory action (either harmful or useful) on another substance S1. Sometimes S2 could have other useful action that must be preserved unsatisfactory action S2 ⇢ S1 (harmful or useful)	**MS2-1. Remove substance** Just remove S2 (or S1, or both). If they have only harmful functions and are removable, just removing them can solve the problem	If a secondary problem MP1 (Needed but Absent Action) is generated by application of MS2-1 or 2-2, try application of MS1's
	MS2-2. Modify or replace substance Modify the properties of the substance S2 (or S1, or both), or replace it with another substance S_{add} so that all actions by or on it become satisfactory	The method of trimming corresponds to "Re-place" in MS2-2
	MS2-3. Add a mediator Add a substance S_{add} that can mediate the field (or action) from S2 to S1 (to amplify, attenuate, filter, stop, remove, control, transform or deflect it)	Both MS2-3 and 2-4 add a substance S_{add} in the interacting pair In 2-3, S_{add} is a field mediator, whereas, in 2-4,
	MS2-4. Add a tool (a field-adder) Make the action satisfactory by creating a helping field (or action) (on S1, S2 or both) by adding a tool S_{add} to the interacting pair	S_{add} is a carrier of a new field (which is not a modification of the original field)
	MS2-5. Repair S1 later Leave the unsatisfactory action change S1 to S1' (S1 in an unsatisfactory state) and repair S1' later by adding a tool (field-adder) S_{add} acting on S1' to change it to S1" (S1 in a repaired, satisfactory state)	Often, repairing (or post treatment) is the least efficient way of solving the problem but sometimes it's useful

Table 2 Model problems and model solutions for FA+ (in diagrams)

Model PROBLEM (MP)	Model SOLUTION (MS)	Meaning of MS
MP1. Needed but absent action There is a needed but absent action on a substance S1. (Sometimes caused by the result of applying MS2-1-Removing)	MS1 -1. 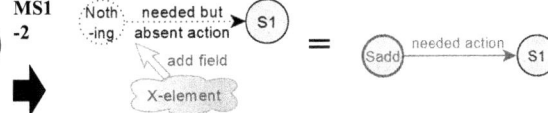	
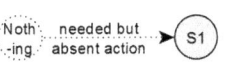	MS1 -2	
MP2. Unsatisfactory action A substance S2 has an unsatisfactory action (either harmful or useful) on another substance S1. Sometimes S2 could have other useful action that must be preserved.	MS2 -1 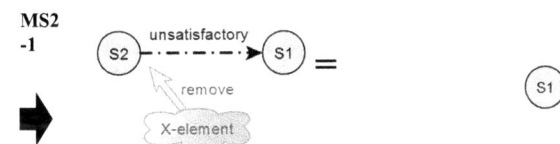	
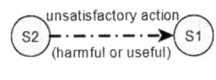	MS2 -2 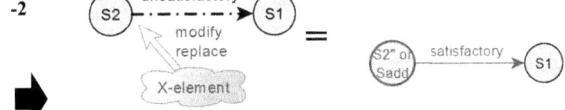	
	MS2 -3 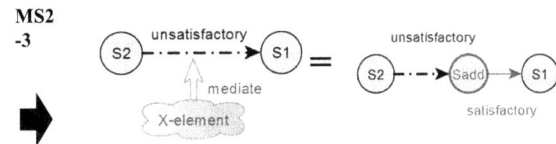	
	MS2 -4 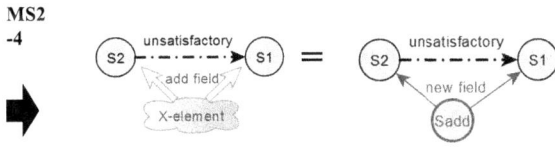	
	MS2 -5 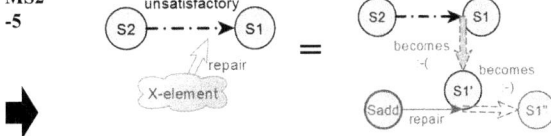	

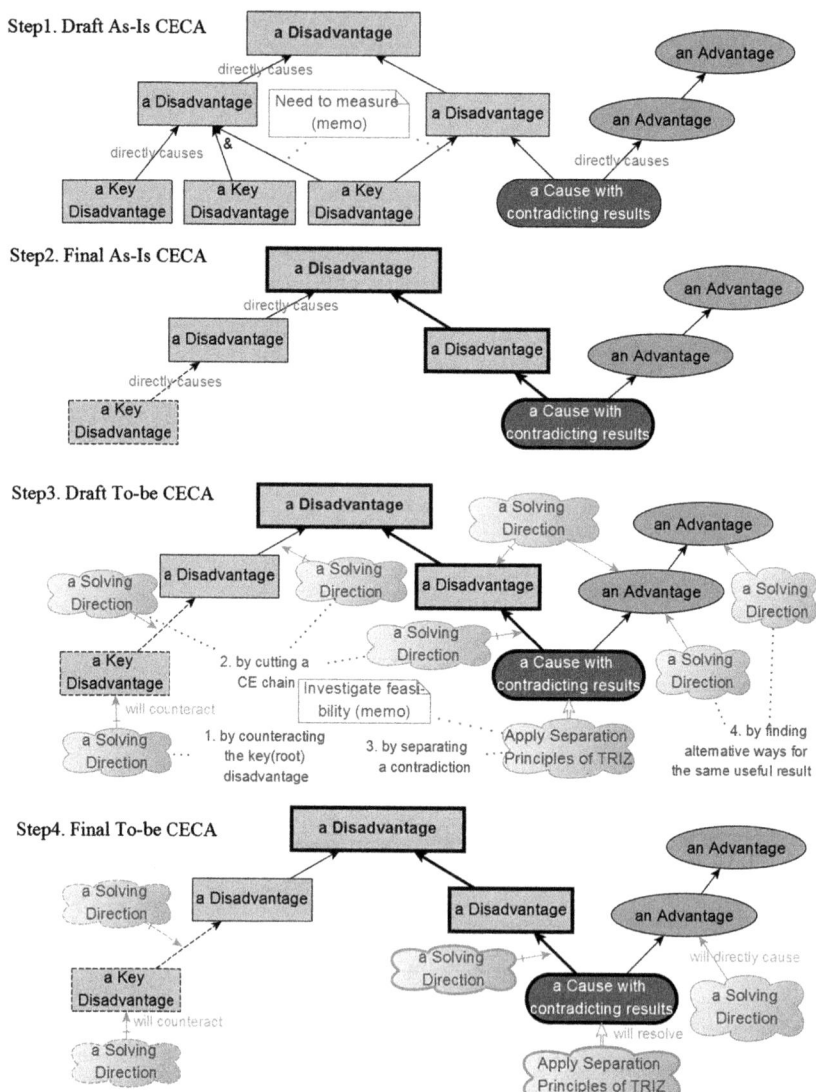

Fig. 2 Brief illustration of the four steps of CECA+

3 A Case Study of Application of FA+ and CECA+ Within the DASE Roadmap

Section 3 introduces how FA+ and CECA+ methods are used in the important latter part of the 'Analysis' stage of the DASE roadmap (Lee et al. 2017) for an artificial coaching project of reinventing a self-watering flowerpot. We chose this project mainly because it is simple and easy for readers while still being helpful for understanding how to use the methods in real projects. Since the basic conventions were already introduced in Sect. 2, here we will focus more on the context, thinking process and related tips. For ease of reading, we simplified some topics.

3.1 Initial Situation and Problem Definition

For many people, it is difficult to keep plants healthy in flowerpots. They often find their plants dried up or even dead. The common cause is incorrect watering. Both too much or frequent watering and too little or rare watering can cause the plant to die. In this chapter, ideas for a self-watering flowerpot are developed. Focus is on showing examples of using the methodology and not on the details of gardening knowledge or result of this reinvention. If the beginning part of the Analysis stage of the DASE roadmap were covered here, a benchmarking analysis and an analysis of the needs (and complaints) of the customers for self-watering flowerpots could have shown which product has the best potential to be the market winner and which features of it need to be improved. However, due to limited space, we skip these steps and choose the conventional flowerpot without a self-watering function as the technical system to analyze and improve further.

3.2 Application of the FA+ Method

After some preliminary information searches and investigation about this traditional flowerpot, a FA+ diagram is built as in Fig. 3. Here, several new conventions are used (Fig. 4).

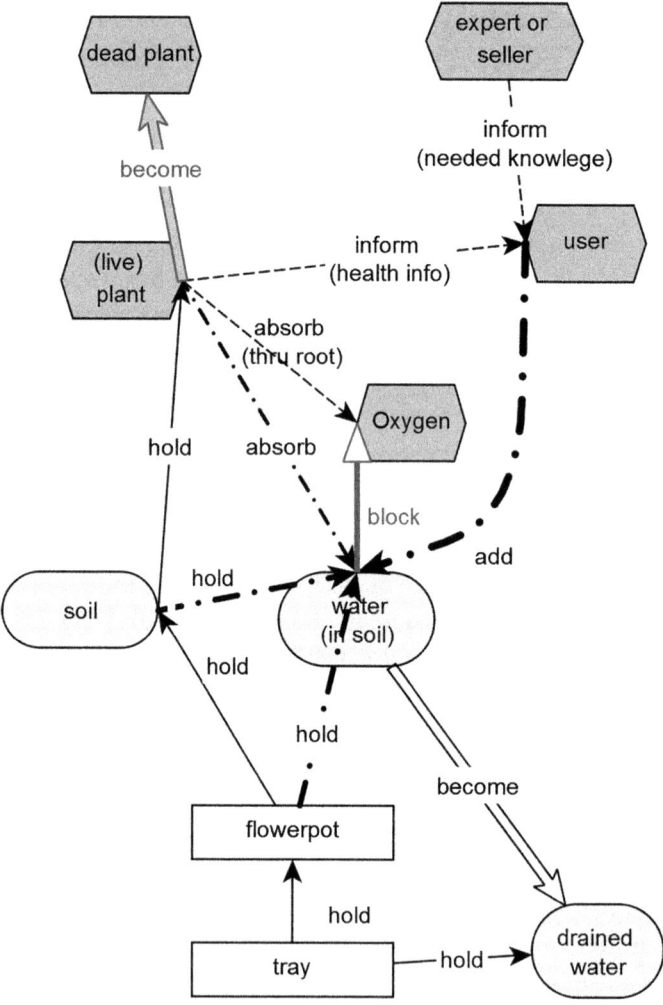

Fig. 3 As-Is FA+ diagram of the flowerpot problem

- The same component in two different times, 'live plant' vs. 'dead plant' and 'water in soil' vs. 'drained water', were expressed separately. All the other components appear only once because their change in time is small or not important for solving this problem.

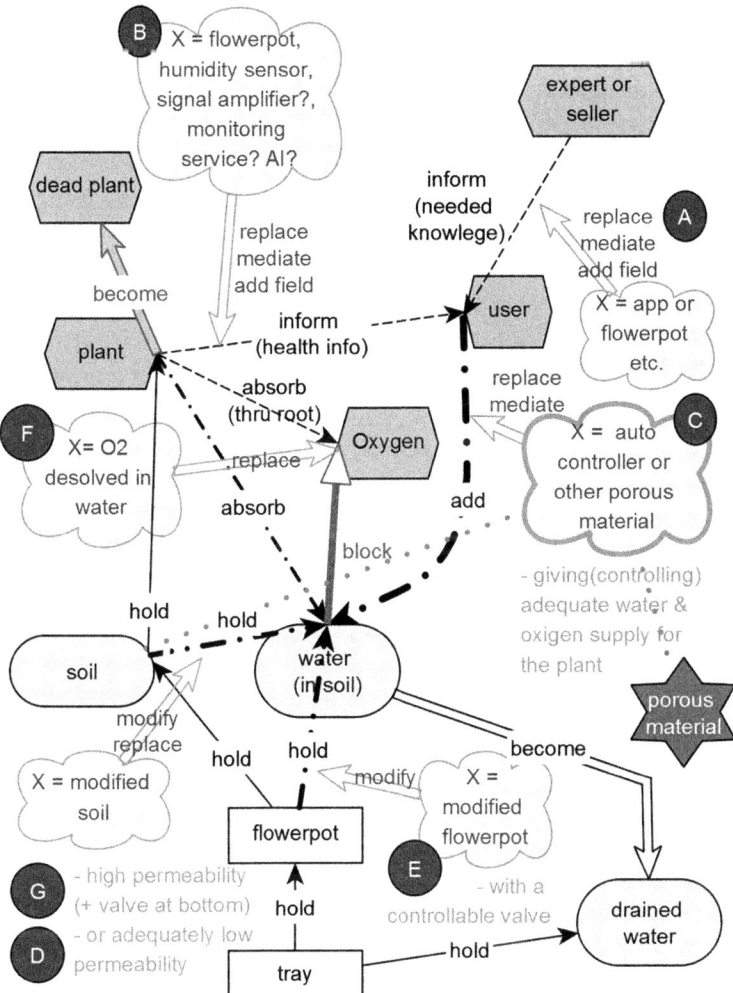

Fig. 4 FA+ diagram after adding feasible model solutions (verbs on arrows), solving substances (in clouds) and their needed properties (below the clouds)

- If two arrows (actions or transformations) meet exactly at the same point (on a boundary of a node), it means that there are **causal relations** between the connected arrows. Because they are connected, the 'chain' or 'network' of causally related actions is explicitly visible. Even though following the logical flow of causes could be easier in a CECA+ diagram, an FA+ diagram is still useful here to develop a proper understanding of the components and interactions before building a correct CECA+ to find out the most promising unsatisfactory actions to solve, generate model solutions and determine which components can be modified, replaced or introduced into the focusing action.

Model solutions are searched for all the unsatisfactory actions in the As-Is FA+ diagram because it is a quick process of selecting feasible model solutions (or solving directions) from four to six for each arrow, taking into account the impact and controllability of the related components or nearby components with needed properties (Fig. 3). In the cloud, list initial ideas for the solving substances as a memo. If possible, list as broadly as possible at this stage. Alternatively, one can focus on the first few model solutions that often have higher ideality. Or, one can reduce the scope of 'controllable' or 'addable' components to minimize cost or difficulties.

3.3   Application of the CECA+ Method

Since the FA+ diagram generated many ideas, the solver can move directly to the 'Solve' stage of the DASE roadmap. However, for more systematic analysis and resolution of the causes of the problem and contradictions, completing a CECA+ session is recommended. In the DASE roadmap, for an inventive problem with normal level of difficulty, either a FA+ or CECA+ is often enough to solve the problem. However, for problems with a bit higher level of difficulty or complexity, using both methods in correct sequence is recommended—FA+ then CECA+. On the bases of deeper and less biased understanding of the system and problem from creating an FA+, a CECA+ diagram (Fig. 5a) is drilled down, usually starting from top-level target disadvantages, which are often on the same level as the main parameters of value (MPVs).

The selection of the most ideal and practical directions and ideas can require several stages of logical thinking. For example, instead of going deep into the not-so-ideal direction of informing users to do better watering (directions A and B), one should select directions C and D, especially

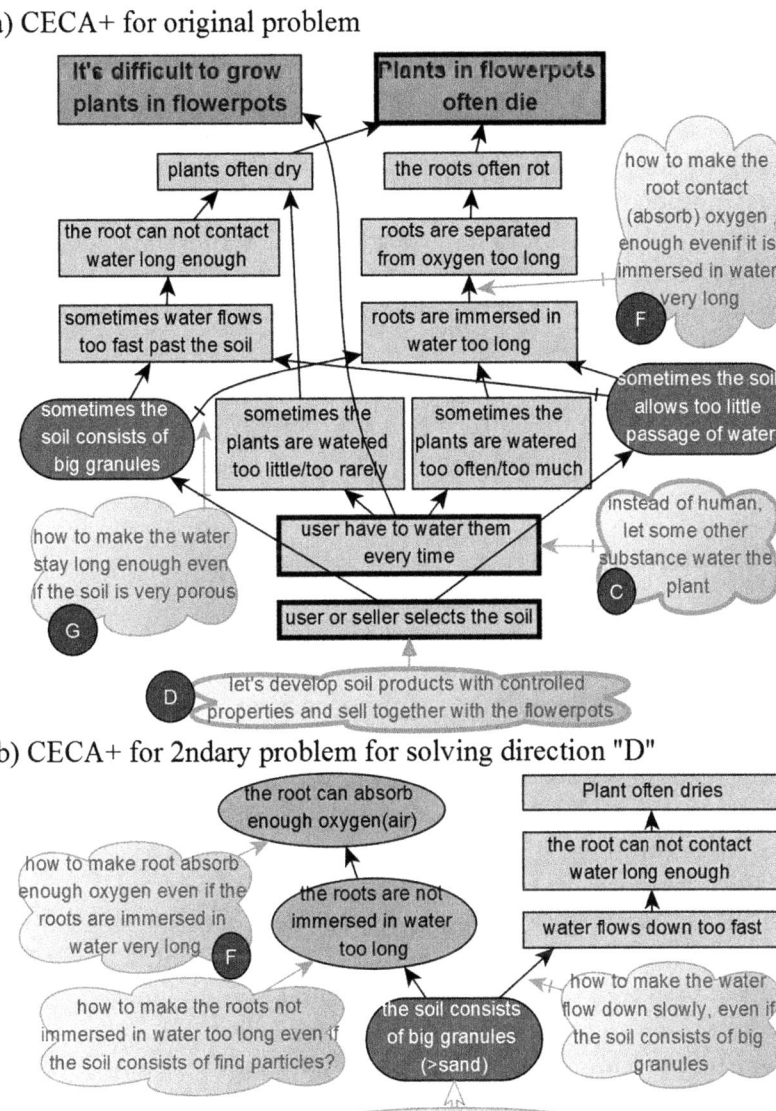

Fig. 5   CECA+ diagrams for the flowerpot problem

those that do not require electronic devices. For example, changing the water supplier from the user to a porous material (soil or artificial porous materials) that sucks water from a water tank and releases it toward the roots in an adequately low flow rate. Selecting this direction can lead to more ideas for informing the user of when to refill the tank or selling the soil as a product with an optimized level of porosity (and nutrition). At this stage, the solver confronts a contradiction on the level of porosity of the soil, which now becomes a controllable parameter. Figure 5b shows how solving directions can be semi-automatically generated from a contradiction with CECA+.

The stages of the DASE roadmap on how to generate concrete ideas from these solving directions and how to combine them to develop best designs are out of the scope of this chapter.

4   Conclusion

In this chapter, we provided a brief introduction to the FA+ and CECA+ methods and the DASE roadmap along with an example of the roadmap focusing on the synergetic use of the aforementioned methods. These methods and roadmap, including earlier versions, have been used in consulting, coaching and training for many years by the author, other consultants and students and have helped much in generating innovative ideas for problems of various levels and from various fields.

Other roadmaps are needed, being developed and used for more difficult inventive problem solving, development of design for new products, processes or services with new MPVs, or finding innovation opportunities. However, even for such different kinds of innovation tasks, DASE with FA+ and CECA+ plays a very important role as the engine of inventive problem solving, which the solver should use very often, very quickly and fluently.

Acknowledgements I would like to thank Professor Leonid Chechurin at Lappeenranta University of Technology, Finland, and TRIZ masters Vasily Lenyashin and Yury Danilovskiy at QM&E Innovation for their guidance and help in this exciting field of innovation.

References

Gerasimov, O. M. (2010). *Technology of selecting tools of innovation design based on TRIZ and VEA analysis.* TRIZ master dissertion. TRIZ Summit 2010. St. Petersburg, Russia.

Lee, M.-G. (2016). *How to generate simple model solutions systematically from function analysis diagram.* TRIZ future conference 2016. Wroclaw, Poland.

Lee, M.-G. (2017). *How to generate ideas systematically from function analysis of an inventive problem.* TRIZ future conference 2017. Lappeenranta, Finland.

Lee, M.-G., Chechurin, L., & Leniachine, V. (2016). *Improvement of cause effect chain analysis, CECA+, for systematic cause analysis and semi-automated idea generation for inventive problems.* Flexible automation and intelligent manufacturing 2016. Seoul, South Korea.

Lee, M.-G., Danilovskiy, Y., Lenyashin, V., Kim, S., & Jung, K. (2017). *Introduction to a mainstream Korean TRIZ methodology, DASE, and its application in Korea.* International conference on systematic innovation 2017. Beijing, People's Republic of China.

Litvin, S., Petrov, V., Rubin, M., & Fey, V. (2012). *TRIZ body of knowledge.* https://matriz.org/auto-draft/triz-body-of-knowledge-final/

Yworks. (n.d.). 2017. https://www.yworks.com/products/yed

Part II

Advances in Tools and Technologies for Creating New Innovations

The five chapters in Part II present novel tools and techniques for creating innovations that support and enhance the design or ideation process as well as discuss the education needed to generate new ideas.

Identification of Secondary Problems of New Technologies in Process Engineering by Patent Analysis

Pavel Livotov, Mas'udah, Arailym Sarsenova, and Arun Prasad Chandra Sekaran

1   Introduction

Efficient, cost-effective, and safe processing is of prime importance in production plants. Recently, there has been a growing trend of Process Intensification (PI) to overcome the problems faced by industries in process engineering (Wang et al. 2017). According to the research of Prof. Ramshaw and his colleagues, PI focuses on "the physical miniaturization of process equipment while retaining throughput and performance" (Cross and Ramshaw 1986) and on the associated cost savings (Reay et al. 2013). PI targets dramatic improvements in processing by developing and applying more precise and efficient existing operation schemes and promises to systematically solve the technological problems in process

P. Livotov (✉) • Mas'udah • A. Sarsenova • A. P. Chandra Sekaran
Faculty of Mechanical and Process Engineering, Laboratory for
Product and Process Innovation, Offenburg University of Applied Sciences,
Offenburg, Germany
e-mail: pavel.livotov@hs-offenburg.de

© The Author(s) 2019
L. Chechurin, M. Collan (eds.), *Advances in Systematic Creativity*,
https://doi.org/10.1007/978-3-319-78075-7_10

engineering primarily through new approaches, such as new methods, new processes, or new technologies.

As described in the research (Casner and Livotov 2017), the implementation of PI solutions leads to technologically new equipment in production plants and often causes secondary problems and/or significant investments, which results in engineering contradictions, that is, the intensification of one property may cause the worsening of another parameter (Kardashev 1990; Benali and Kudra 2015). There is undoubtedly a certain number of disadvantages in any novel solution concept or invention. As outlined in Zlotin and Zusman (2010), no invention can be implemented without solving a certain amount of secondary problems that increase dramatically with the level of invention. Therefore, the early identification of secondary problems and the prediction of engineering contradictions in novel technologies can enable a smooth loss-free shift to a new PI technology without "teething" problems. In the next step, the identified negative side effects and contradictions can be systematically limited with the inventive tools of the Theory of Inventive Problem Solving (TRIZ) methodology (Altshuller 1984; Hua et al. 2010), which helps to transform new PI technologies, to mobilize available resources, and to reach the maximum efficiency with a minimum of expenditures (Casner and Livotov 2017).

TRIZ is a comprehensive methodology for problem solving, analysis, and technological forecasting derived from the study of patterns of invention and technical evolution in the global patent literature (Altshuller 1984). The advantages of TRIZ methodology over conventional creativity techniques in practice are outlined in numerous interdisciplinary studies (Hua 2010; Ilevbare et al. 2013; Chechurin and Borgianni 2016), as well as in the field of process engineering (Casner and Livotov 2017; Pokhrel et al. 2015) and intensification of chemical processes (Srinivasan and Kraslawski 2006; Kim et al. 2009).

A typical scenario of introducing novel technologies in process engineering companies can be characterized by the lack of information about potential secondary problems that would appear during the implementation and operational phases of new processes and equipment. In practice, plant engineers often do not have enough data to evaluate the effectiveness of new technologies and do not possess specific experience or qualification to predict side effects or even to find any reliable information concerning potential concealed disadvantages of new technologies.

Numerous studies have shown that, depending on the years and the technical domain, 70–90% of technical information can be found in patent documents only. The rising number of patents or patent applications outlines their growing importance as the source of actual information (Asche 2017). As summarized in Livotov (2015), patent information enhanced through computer-aided classification and retrieval contributes significantly to the acceleration of innovation processes. It can be used for the automatic retrieval of problems (Souili et al. 2015) or technical contradictions (Cascini and Russo 2007), for the analysis of possible improvements of products (Verhaegen et al. 2009), for technology road mapping (Lee et al. 2008) and evaluation of potential sources for the external generation of technological knowledge, for competitor monitoring and R&D portfolio management (Ernst 2003), for the forecasting of technical and technological evolution (Daim et al. 2006) and developing acquisition strategies (Moehrle and Geritz 2004), for identification of new product features with high market potential (Livotov 2015), and other applications. The patent information is usually readily available as most patent applications are published 18 months after the first filing. Despite the efforts of numerous research groups, patent information remains an under-utilized resource at a practical level, especially in the conceptualization and detailed design phases (Kanda 2008). Since design problems and solutions in patents are not easily understood, even with the use of computer algorithms, Vasantha et al. (2017) discuss an affordable crowdsourcing approach for interpreting patents in the early phases of engineering design. Recent research analyzes numerous aspects of using patent citations: influence of patent citation on its market value (Patel and Ward 2011), patent citation differences across technologies (Noailly and Shestalova 2017), patent citation indicators and types (Karvonen and Kässi 2013), patent citations to non-patent and scientific literature (van Raan 2017), patent development paths and evaluation of the citation weight between patents (Choi and Park 2009).

In addition, the majority of patent documents in the field of process engineering provide information about problems of the technologies or equipment, which can be useful for PI and engineering design (Casner and Livotov 2017). In this context, this chapter introduces three methods to find secondary problems in patent literature: (a) direct knowledge-based identification of secondary problems in new technologies or equipment; (b) identification of secondary problems of prototypes mentioned in patent citation trees; and (c) prediction of negative side effects using the correlation matrix for invention goals and secondary problems in a specific engineering domain.

2   Methods for the Identification of Secondary Problems

The three proposed methods are illustrated with the example of the granulation process. For this purpose, we analyzed 150 patent documents from 15 countries with an application date between 2000 and 2015 and related to granulation in pharmaceutical and ceramic industries. We retrieved patent documents by using online search engines and databases of the German Patent and Trade Mark Office (DPMA), the European Patent Office (EPO), and the United States Patent and Trademark Office (USPTO). Granulation is an important process step in agrochemicals, pharmaceuticals, foods, chemicals, and minerals (Wang et al. 2017). The techniques can be divided into two types according to the process used to facilitate granulation: wet granulation that utilizes liquid in the process and dry granulation that requires no liquid (Wang et al. 2017; Shanmugam 2015). Granulation may be further classified as a batch or continuous process.

The granulation process and the proposed methods for problem identification are illustrated here using the analysis of the US Patent 6499984B1 "Continuous production of pharmaceutical granulation" (Ghebre-Sellassie et al. 2002), which replaces the traditional discontinuous batch process for granulation, negatively characterized by high cycle times, high production costs, and lower process efficiency. The benefits of the existing wet granulation technology include: (a) high shear mixer granulators, (b) single pot processing with a high shear mixer granulator and microwave drying, and (c) a high shear granulator integrated with a fluid bed dryer, such as a semi-continuous, multi-cell apparatus (Ghebre-Sellassie et al. 2002)

In this context, the US Patent 6499984B1 describes a novel "true" continuous granulation process starting with individual ingredients or a powder blend (Ghebre-Sellassie et al. 2002). As an alternative to the batch process used in pharmaceutical industries, the US Patent 6499984B1 provides a Twin Screw Wet Granulator-Chopper for continuous granulation (TSWGC). TSWGC is an automated, single-pass system, including process and apparatus, for continuous granulation.

Innovation technologies described in patent documents are always provided with information about their advantages (positive effects) but are almost never provided with information about the drawbacks (negative effects). Providing information about disadvantages of new inventions certainly is not in the interest of inventors. In process engineering in particular, it is often difficult for engineers to identify all problems in the new technologies or solutions described in the patent documents. In the following, we demonstrate the three proposed methods for the identification of secondary problems in new technologies with an example of the TSWGC patent.

2.1 Direct Knowledge-Based Identification of Secondary Problems in New Technologies or Equipment

The first proposed method for the identification of secondary problems can be performed by analyzing and comparing new technology with the advantages of its prior technology or cited prototypes. The idea of this approach is that the advantages of cited prior technologies that are not a part of new technology can be identified as one of the potential secondary problems in the future. The method is explained using an example of the TSWGC patent 6499984B1 (Ghebre-Sellassie et al. 2002), which describes the High Shear Mixer Granulator (HSMG) as a prior technology. In Table 1, all identified advantages of prior and new technologies are assigned to HSMG (prior) and TSWGC (new) with abbreviations Y,

Table 1 Comparison of the impact of the novel and prior technologies

N	Advantages of prior and new technologies (positive impact)	Prior technology (HSMG)	Novel technology (TSWGC)
1.	Increases overall process efficiency	N	P
2.	Reduces cycle time of granulation	N	Y
3.	Achieves product cost saving	N	P
4.	Enhances homogeneity of granulation	Y	Y
5.	Reduces energy consumption	Y	P
6.	Reduces residence time in drying apparatus	Y	Y
7.	Eliminates separate wet milling step	N	Y
9.	Measures parameters online	N	Y
10.	Provides feedback for process tuning	N	Y
11.	Enables densification and uniformity	N	Y
12.	Reduces cleaning/maintenance efforts	Y	P
13.	Enhances yield of product	Y	P
14.	Suitable for various materials	Y	P
15.	Provides unit operation sequentially	N	Y

Y provided by technology, *N* not provided by technology, *P* possible secondary problem

N, and P: Y–provided by technology; N–not provided by technology; P–possible secondary problem.

As shown in Table 1, some drawbacks of prior technology HSMG, marked with N, are eliminated in the novel technology TSWGC. However, not all advantages of HSMG are assigned to TSWGC. For example, process efficiency, product cost saving, energy consumption, maintenance, the yield of product, and applicability for various materials are unknown features in TSWGC. These features (marked with "P") could be considered as possible secondary problems of TSWGC in the implementation phase. In general, a checklist of typical technical problems in the specific domain (here, *granulation process*) could be helpful in this step. However, the application of such a relatively long checklist for the identification of secondary problems of a novel technology is not only time-consuming, but also requires specific expert knowledge and practical experience with the new technology, which is often not available. This challenge can be overcome by using the patent citation trees.

2.2   Identification of the Secondary Problems of Prototypes Mentioned in Patent Citation Trees

Each patented technology or equipment cites prior patents in its patent document (backward citation) and can be itself cited in the later patents (forward citation). Information about the disadvantages of the cited prior solution can be relatively quickly and precisely extracted from citing patent documents by using various online patent databases and advanced search and text-mining tools. The information about the disadvantages can be directly obtained in the sections *State of the Art or Background* of inventions. The information of the advantages is available both in the citing and cited documents as shown in the citation map in Fig. 1.

As the disadvantages of the analyzed patent can be extracted using forward citations in later patents only, the application of this method for new or other technologies without forward citations is limited. For a new technology without forward citations, one can use the backward citations to identify possible secondary problems systematically as described in Sect. 2.1. The method is also restricted to a limited number of patents. For example, not all inventions are patented in accordance with the intellectual property strategies of the companies. Furthermore, not all technologies fulfill the patentability criteria, and not all patent citations provide full information about the disadvantages of the prior technology.

Figure 2 shows a forward citation tree for the US Patent 6499984B1, which was cited by at least 26 patents. Figure 3 illustrates secondary

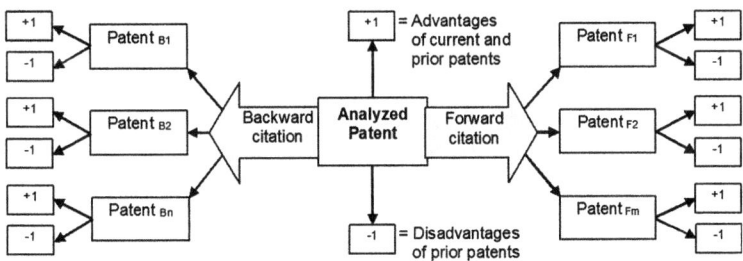

Fig. 1 Citation map of an analyzed patent with the backward (prior patents) and forward citations (later patents)

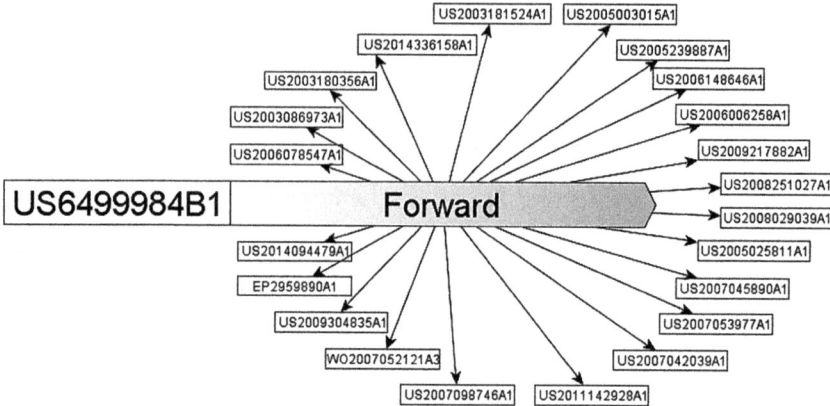

Fig. 2 Forward citation tree of the US Patent 6499984B1

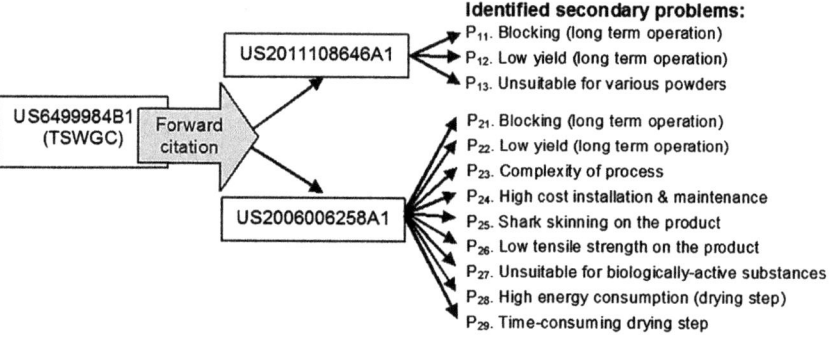

Fig. 3 Examples of the secondary problems of TSWGC (US6499984B1) found in the forward citations

problems extracted from two exemplary patent documents of the presented citation tree.

2.3 Prediction of Negative Side Effects Using the Correlation Matrix of Invention Goals and Secondary Problems

The analysis of a high number of patent documents allows not only identification of secondary problems of technologies but also delivers a basis for prediction of possible secondary problems in the future.

In the presented case study, the analysis of 150 patents resulted in the identification of 208 invention goals, formulated as solution-neutral requirements for PI (Table 2). This checklist of requirements can be in general used for an easier identification of the possible negative effects or secondary problems of the inventions as presented in Sect. 2.1. The numbers in the table indicate the impact of patented solutions on the PI requirements, that is, the invention solves the problem (positive effect: +1), the invention causes a secondary problem (negative effect: −1), or there is no effect identified (=0).

The obtained information about the relationship between the initial problems solved by inventions and the corresponding secondary problems can be used for the formulation of the correlation matrix to predict engineering contradictions in the field of analysis. In the correlation matrix, the previously identified numerous requirements for PI can be combined into a finite number of categories or problem clusters.

In the granulation case study, almost all extracted 208 requirements for granulation process were combined into 27 categories as presented in Table 3.

The smaller number of categories allows a fast application of the matrix in practice. A fragment of the correlation matrix for granulation process is presented in Table 4, where "−1" indicates a possible contradiction

Table 2 PI requirements for granulation process, extracted from 150 patents (fragment)

PI solution-neutral requirements	Category	Patent 1	Patent 2	Patent 3	Patent 4	...	Patent 150
1. Reduce water consumption	Water consumption	+1	+1	−1	0	...	0
2. Reduce energy consumption	Energy consumption	0	+1	+1	0	...	+1
3. Reduce efforts for cleaning, maintenance	Cleaning & maintenance	−1	0	+1	0	...	0
4. Reduce air pollution	Environmental concerns	0	0	+1	0	...	0
...	...	...	...	...	...	...	...
208. Reduce additives consumption	Additives consumption	+1	0	0	−1	...	0

Table 3 Categories of PI requirements for granulation process extracted from 150 patents

N	Category	N	Category
1	Process duration, time expenditures	15	Binder or additive consumption
2	Quality of product (cracks...)	16	Size or volume of equipment
3	Energy consumption	17	Solid handling efforts (transporting ...)
4	Water consumption	18	Controllability of the process
5	Uniformity of granulation product	19	Adaptability of equipment
6	Complexity of process or equipment	20	Production capacity
7	Costs (investments, installation...)	21	Product composition
8	Maintenance and cleaning	22	Process efficiency (mass balance...)
9	Productivity or yield	23	Mechanical properties (hardness...)
10	Reliability of equipment and process	24	Chemical properties (bioavailability ...)
11	Disintegration, solubility, dispersion...	25	Physical properties (density ...)
12	Agglomeration (solidification...)	26	Homogeneity of product
13	Moisture content	27	Replaceability of equipment
14	Environmental performance	28	Others

between two requirements categories, "+1" indicates a possible complementary amplification of two categories, and "0" indicates a neutral or unknown counteraction. The engineering contradictions marked with "−1" are not deterministic and should be interpreted as an indication of a possible secondary problem only.

The method is explained by using an example of the TSWGC patent 6499984B1 (Ghebre-Sellassie et al. 2002), which claims to reduce the granulation process duration. In accordance to the correlation matrix, the TSWGC implementation can face the following secondary problems: reduction of product quality (cracks, brittleness, etc.), higher energy and water consumption, reduction of product uniformity (size distribution, pore size, content, blend, etc.), higher costs (investment, installation, human resources, etc.), and additional maintenance efforts. A comparison with the results achieved with other methods of problem identification (Sects. 2.1 and 2.2) shows a very good agreement with the recommendations of the correlation matrix.

Table 4 Correlation matrix for the prediction of secondary problems of new granulation technologies

Parameter to be improved	Secondary impact of new technology									
	Process duration, time expenditures	Quality of product (cracks, ...)	Energy consumption	Water consumption	Uniformity of granulation product	Complexity of process or equipment	Costs (investment, installation, ...)	Maintenance and cleaning		Replaceability of equipment
	1	2	3	4	5	6	7	8	...	27
1 Process duration, time expenditures		−1	−1	−1	−1	+1	−1	−1	...	0
2 Quality of product (cracks, etc.)	−1		−1	−1	+1	−1	−1	−1	...	0
3 Energy consumption	−1	−1		0	−1	−1	+1	+1	...	0
4 Water consumption	−1	−1	+1		−1	+1	+1	+1	...	0
5 Uniformity of granulation product	−1	+1	−1	−1		−1	−1	−1	...	0
6 Complexity of process or equipment	+1	−1	+1	+1	−1		+1	+1	...	+1
7 Costs (investment, installation, etc.)	+1	−1	−1	−1	0	−1		−1	...	0
8 Maintenance and cleaning	+1	+1	+1	+1	0	+1	−1		...	0
...	...	...	...	...	...	...	...	...		...
27 Replaceability of equipment	+1	−1	−1	0	0	+1	+1	−1	...	

−1 possible contradiction, +1 possible enhancement, 0 neutral

The additional benefit of this prediction method is that it not only helps to identify the categories of secondary problems but also forecasts possible positive side effects of technologies. For example, higher productivity, reliability, better environmental performance, and other effects belong to the expected additional positive effects of TSWGC anticipated by the correlation matrix.

3   Concluding Remarks and Future Perspectives

Thorough analysis of patent documents, performed in the presented research, reveals multiple engineering contradictions and secondary problems in novel process engineering technologies and equipment. The anticipatory identification of the negative side effects will essentially contribute to a faster and more cost-efficient innovation. The ability to forecast negative effects of new technologies objectively, comprehensively, and systematically is one of the crucial preconditions for successful PI. The proposed approach contains three complementary methods, which can be easily conducted by the engineers.

The graph in Fig. 4 illustrates a general algorithm for secondary problem identification. First, it should be checked which negative side effects or problems of a new technology are already known. If no information about the secondary problems is available, the recent patents or patent applications protecting a new technology should be identified in the patent databases. In the next step, the patent documents describing prior technologies should be identified using backward citations. Then, the methods A, B, and C (see Sects. 2.1, 2.2, and 2.3) can be applied separately, in combination, or in succession according to the obtained information. If the checklist of typical technical problems (Table 3) or the correlation matrix (Table 4) for a specific engineering domain are already available, the prediction of the secondary problems can also be performed without patent analysis. The correlation matrix offers the most efficient and rapid forecast of secondary problems, whereas the analysis of citation trees delivers more precise and detailed results. It is noteworthy that a very good agreement was observed between the results achieved

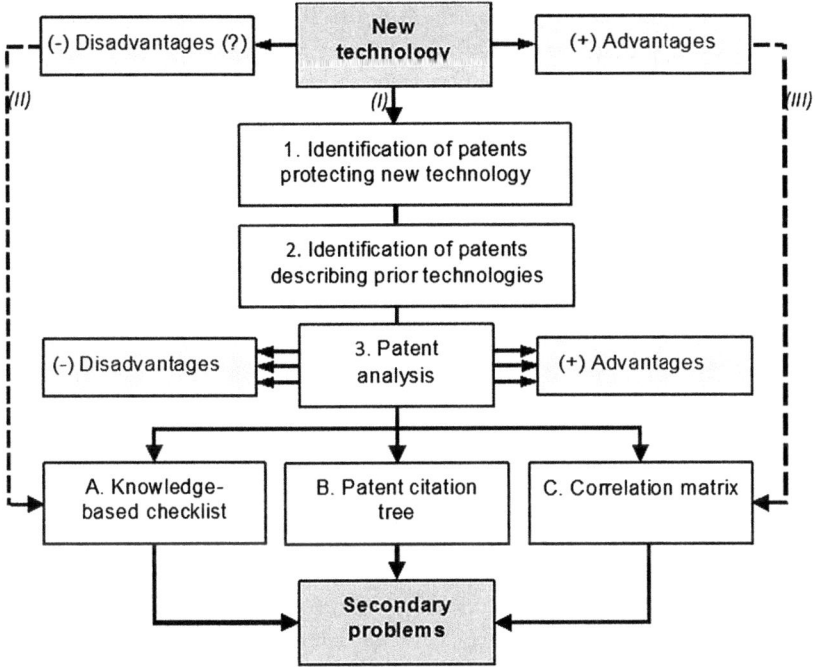

Fig. 4 General algorithm for secondary problem identification by patent analysis: (I) no initial information about secondary problems available; (II) checklist of typical problems available; (III) correlation matrix available

with different methods applied for identification of potential secondary problems in novel granulation technologies.

Furthermore, the suggested methods effectively support the advantageous linking of TRIZ methodology to the PI approach. The identified problems and contradictions can be inventively solved with TRIZ tools, such as 40 inventive principles, standard solutions, databases of physical, chemical, and geometrical effects, and others. Lastly, the proposed methods can also be used to identify additional benefits of new technologies.

In the future, the methods can be partially or completely automated by means of computer-aided data mining and processing. This will replace the currently time-consuming generation of citation trees and correlating parameters of engineering contradictions in the correlation matrix. Thus, for each technological or industrial domain, the analysis of patents can be performed continuously and will result in more universal and statistically more precise correlation matrices, which could be applicable in a wider range of fields in process engineering, for example, for all PI tasks involving solids.

Acknowledgments The authors wish to thank the European Commission for supporting their work as part of the research project "Intensified by Design® platform for the intensification of processes involving solids handling" within international consortium under the H2020 SPIRE programme.

References

Altshuller, G. S. (1984). *Creativity as an exact science. The theory of the solution of inventive problems.* Gordon & Breach Science Publishers, issn 0275-5807. Amsterdam.

Asche, G. (2017). "80% of technical information found only in patents" – Is there proof of this? *World Patent Information, 48*, 16–28. https://doi.org/10.1016/j.wpi.2016.11.004.

Benali, M., & Kudra, T. (2015). *Drying process intensification: Application to food processing.* Retrieved September 11, 2016, from https://www.researchgate.net/publication/266211018

Cascini, G., & Russo, D. (2007). Computer-aided analysis of patents and search for TRIZ contradictions. *International Journal of Product Development, 4*(1/2), 52–67.

Casner, D., & Livotov, P. (2017, August 21–25). *Advanced innovation design approach for process engineering* (pp. 653–662). Proceedings of the 21st International Conference on Engineering Design (ICED 17). Vol 4: Design methods and tools, Vancouver. isbn 978-1-904670-92-6.

Chechurin, L., & Borgianni, Y. (2016). Understanding TRIZ through the review of top cited publications. *Computers in Industry, 82*, 119–134. https://doi.org/10.1016/j.compind.2016.06.002.

Choi, C., & Park, Y. (2009). Monitoring the organic structure of technology based on the patent development paths. *Technological Forecasting and Social Change, 76*, 754–768.

Cross, W., & Ramshaw, C. (1986). Process intensification: Laminar flow heat transfer. *Chemical Engineering Research and Design, 64*(4), 293–301.

Daim, T. U., Rueda, G., Martin, H., & Gerdsri, P. (2006). Forecasting emerging technologies: Use of bibliometrics and patent analysis. *Technological Forecasting and Social Change, 73*(8), 981–1012.

Ernst, H. (2003). Patent information for strategic technology management. *World Patent Information, 25*(3), 233–242.

Ghebre-Sellassie, I., Jr Mollan, M. J., Pathak, N., Lodaya, M., & Fessehaie, M. (2002). Continuous production of pharmaceutical granulation. US Patent 6499984B1.

Hua, Z., Yang, J., Coulibaly, S., & Zhang, B. (2010). Integration TRIZ with problem-solving tools: A literature review from 1995 to 2006. *International Journal of Business Innovation and Research, 1*(1/2), 111–128.

Ilevbare, I. M., Probert, D., & Phaal, R. (2013). A review of TRIZ, and its benefits and challenges in practice. *Technovation, 33*(2), 30–37.

Kanda, A. S. (2008). *An investigative study of patents from an engineering design perspective.* PhD. Clemson University, Clemson.

Kardashev, G. A. (1990). *Physical methods of process intensification in chemical technology.* Moscow: Khimia. 208p. (in Russian)

Karvonen, M., & Kässi, T. (2013). Patent citation analysis as a tool for analysing industry convergence. *Technological Forecasting and Social Change, 80*, 1094–1107.

Kim, J., Kim, D., Lee, Y., Lim, W., & Moon, I. (2009). Application of TRIZ creativity intensification approach to chemical process safety. *Journal of Loss Prevention in the Process Industries, 22*(6), 1039–1043.

Lee, S., Lee, S., Seol, H., et al. (2008). Using patent information for designing new product and technology: Keyword based technology road mapping. *R&D Management, 38*(2), 169–188.

Livotov, P. (2015). Using patent information for identification of new product features with high market potential. World conference: TRIZ future, TF 2011–2014, Elsevier, Germany. *Procedia Engineering 131*, 1157–1164.

Moehrle, M., & Geritz, A. (2004). *Developing acquisition strategies based on patent maps* (pp. 1–9). Proceedings of the 13th international conference on management of technology, Washington, DC: R&D Management.

Noailly, J., & Shestalova, V. (2017). Knowledge spillovers from renewable energy technologies: Lessons from patent citations. *Environmental Innovation and Societal Transitions, 22*, 1–14.

Patel, D., & Ward, M. R. (2011). Using patent citation patterns to infer innovation market competition. *Research Policy, 40*(6), 886–894.

Pokhrel, C., Cruz, C., Ramirez, Y., & Kraslawski, A. (2015). Adaptation of TRIZ contradiction matrix for solving problems in process engineering.

Chemical Engineering Research and Design, 103, 3–10. https://doi.org/10.1016/j.cherd.2015.10.012.

Reay, D. A., Ramshaw, C., & Harvey, A. (2013). *Process intensification: Engineering for efficiency, sustainability, and flexibility* (2nd ed.). Oxford: Butterworth-Heinemann.

Shanmugam, S. (2015). Granulation techniques and technologies: Recent progresses. *BioImpacts: BI, 5*, 55–63.

Souili, A., Cavallucci, D., Rousselot, F., & Zanni, C. (2015). Starting from patent to find inputs to the problem graph model of IDM-TRIZ. *Procedia Engineering, 131*, 150–161. https://doi.org/10.1016/j.proeng.2015.12.365.

Srinivasan, R., & Kraslawski, A. (2006). Application of the TRIZ creativity enhancement approach to design of inherently safer chemical processes. *Chemical Engineering and Processing: Process Intensification, 45*, 507–514. https://doi.org/10.1016/j.cep.2005.11.009.

van Raan, A. F. J. (2017). Patent citations analysis and its value in research evaluation: A review and a new approach to map technology-relevant research. *Journal of Data and Information Science, 2*(1), 13–50. https://doi.org/10.1515/jdis-2017-0002.

Vasantha, G. V. A., Corney, J. R., Maclachla, R., & Wodehouse, A. J. (2017). The analysis and presentation of patents to support engineering design. *Design Computing and Cognition, 16*, 209–226. Retrieved August 10, 2017, from https://www.researchgate.net/publication/312026500

Verhaegen, P. A., D'Hondt, J., Vertommen, J., et al. (2009). Relating properties and functions from patents to TRIZ trends. *CIRP Journal of Manufacturing Science and Technology, 1*(3), 126–130.

Wang, H., Mustaffar, A., Phan, A. N., Zivkovic, V., Reay, D. A., Law, R., & Boodhoo, K. (2017). A review of process intensification applied to solids handling. *Chemical Engineering and Processing: Process Intensification, 118*, 78–107. https://doi.org/10.1016/j.cep.2017.04.007.

Zlotin, B., & Zusman, A. (2010). *Addressing secondary problems – The last obstacle on the way of successful problem solving with TRIZ. TRIZCON.* St. Petersburg: St. Petersburg Polytechnic University.

Control Design Tools for Intensified Solids Handling Process Concepts

Markku Ohenoja, Marko Paavola, and Kauko Leiviskä

1 Introduction

The Theory of Inventive Problem Solving (TRIZ) can be applied to generate new concepts for process intensification (PI). In order to meet the target performance of the intensified process and to avoid design bottlenecks due to process operation, the suggested concepts need to comprise a feasible control system. Therefore, a design step, where a systematic procedure for variable selection is performed, available measurement devices are mapped, and the control design is initialized, is needed. This chapter presents a systematic approach to tackle these issues in a structured manner in order to enable a smooth transfer from new innovative ideas into feasible process design from operation point of view.

M. Ohenoja (✉) • M. Paavola • K. Leiviskä
Control Engineering, Faculty of Technology, University of Oulu,
Oulu, Finland
e-mail: markku.ohenoja@oulu.fi

© The Author(s) 2019
L. Chechurin, M. Collan (eds.), *Advances in Systematic Creativity*,
https://doi.org/10.1007/978-3-319-78075-7_11

167

2   Background

Process intensification (PI) is process development leading to substantially smaller, cleaner, safer, and more energy-efficient technology (Reay et al. 2013). The best known examples come from, on the one hand, chemical engineering and miniaturization, such as miniaturized reactors, fuel processing systems, power sources and, on the other hand, integrating unit operations, such as reactive distillation and dividing-wall columns, see (Baldea 2014; Cremaschi 2015). Indeed, a European research programme has identified that while the highly intensified equipment is largely restricted to gas\liquid and liquid\liquid systems also the processes involving solids could benefit especially from continuous mode of operation. However, the intensification of solids handling processes is challenging due to an apparent risk of particulate fouling (scaling, caking, clogging) leading to operational and even safety problems. Taking into account also the complexity of the processes, the intensification project should consider the whole process chain, not only an isolated problem. The Intensified by Design® (IbD®) platform aims to facilitate a set of design tools aimed for PI and optimization of solids handling processes.

TRIZ can be applied to generate new concepts for PI. As a part of the IbD® project, Livotov et al. (2017) have identified cases where TRIZ and PI methods can be linked efficiently using the platform-supported PI database and TRIZ components. According to the Design of Six Sigma (DFSS) approach, the output of the concept development phase needs to lead to a sound, invulnerable, and robust system (do the right things, do them all the time) (Yang and El-Haik 2003). As process control is an intrinsic part of process engineering, without a process control assessment, the target of DFSS, and new concepts, is unreachable. The theoretical increment in process efficiency gained through PI might be compromised if the plant is difficult to control and therefore cannot be operated at its nominal operating point (Mauricio-Iglesias et al. 2013).

The integrated process and control design can be considered as an essential part in transferring a new idea into a product or a process. The integrated process and control design has been reviewed in several studies (Ricardez-Sandoval et al. 2009; Hamid 2011; Yuan et al. 2012;

Sharifzadeh 2013; Vega et al. 2014; Huusom 2015). In the intensified processes, the design of process monitoring and control encounters new challenges (Nikačević et al. 2012). Therefore, the integrated design approach is preferable, and it can also lead to novel sensing and actuating solutions in the intensified processes.

In the early stage of control design, it is crucial to identify the aims of process monitoring, which of the monitored variables are to be connected to closed-loop control, should the control be automated, and could the process be optimized during the operation with the proposed control structure. Integrating such an analysis into the process design stage, the intensified process will also have a fit-for-purpose, intensified process monitoring and control solution. Next, a systematic approach is presented to tackle these issues in a structured manner in order to enable a smooth transfer from innovative ideas into a feasible process design from process operation point of view. Finally, the tools are demonstrated with a case study example.

3 Systematic Approach

Naturally, there exist a number of tools for the integrated design at the early stage. For example, Hopkins et al. (1998) have presented a design procedure aimed at eliminating all uncontrollable flowsheet designs based on a qualitative structural controllability analysis. The systematic approach developed here aims to be an accessible tool for initial process control evaluation starting from information collection. It requires performing tasks that ensure that all the reasonable interconnecting elements are taken into consideration before proceeding to more detailed design. The approach assists the user in screening and solving the monitoring and control issues of the process design in hand. The systematic approach is divided into three main steps: (1) initial process analysis, (2) monitoring advisor, and (3) control advisor. Here, we focus on steps 1 and 3: initial process analysis and control advisor. The flowsheet of the approach is presented in Fig. 1. The associated tools are presented in the following subsections.

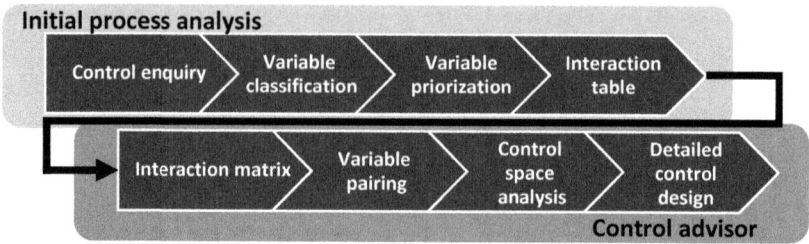

Fig. 1 Systematic approach for the initial control design for new process concepts

3.1   Initial Process Analysis

The initial process analysis step has various objectives. First, the process boundaries and the list of potential monitored variables need to be identified. Then the variables are classified into controlled, manipulated, and disturbance variables. Moreover, the critical process parameters and critical quality attributes are specified, and the relationship between the selected variables are evaluated. The identification of process boundaries is needed as the control design typically has slightly different boundaries from the process design due to auxiliary equipment, energy and material exchange, and disturbances. This step can be conducted with purely qualitative information and is therefore suitable for being performed for any new process concept. The systematic procedure for the initial process analysis mainly follows the principles of the design steps for the basic control scheme presented in Roffel and Betlem (2006).

Control Enquiry and Variable Classification

The aim of the control enquiry step is to study the operation of the process in a structured manner with the available data, such as process flow-sheets, PI&C charts, process operational points, and process input (raw materials) and output (product) specifications. The enquiry requires the user to define the goals of operation (target properties of the intermediate products and the final product), and to investigate the process boundaries as well as the external disturbances. If possible, the process should be divided into independent sub-processes, and the following analysis should be applied separately for every sub-process.

For each sub-process, the controlled (process conditions, material contents, qualities) and manipulated (correcting) variables are defined in a variable classification step. Moreover, the process variable(s) that determine the throughput (load) and the recycle flows of the process are specified. In addition to the manipulated and controlled variables, variables may also belong to disturbances, observed, or not applicable\ignored classes. Only one type can be selected for one variable.

Variable Prioritization

Here, expert knowledge is used for identifying the critical process variables from the comprehensive variable list provided by the previous step. Eventually, a decision is made on which variables are to be considered when the possible monitoring and control solutions are screened. This step can also be executed quantitatively if model-based tools or experimental set-ups are available. In this case, a sensitivity analysis (Saltelli et al. 2006) can be performed and the decision can be based on numerical data. One example of a model-based procedure is presented in Singh et al. (2009).

Interaction Table

In the final step of initial process analysis, the identified (most important/sensitive) controlled variables, manipulated variables, and disturbances are arranged in an interaction table for the selected sub-process. Then, a qualitative information of the power and speed of the control between the variables in the interaction table is estimated. The scale is dependent on the fastest/largest response. The interaction table acts as a basis for variable pairing and control design in the control advisor step.

3.2   Control Advisor

The control advisor allows for evaluating the desired and available control space with the selected variables, their desired range, and expected disturbances. The control advisor also presents a procedure for process identification, control strategy development, implementation, and verification. In order to evaluate the control space, the qualitative information used in the previous step needs to be converted into quantitative information. Naturally, this information is difficult to obtain in the conceptual design phase. The tools can, however, be used also for inaccurate data, allowing the user to inspect the sensitivity of changing process interactions and disturbance space to the design space and control space.

Interaction Matrix

The interaction table for the selected sub-process generated in the initial process analysis step need to be updated with quantitative information. If quantitative information is not available, or cannot be estimated, the user can use the qualitative interaction table and the guidelines presented in Roffel and Betlem (2006) for the variable pairing and selection of basic control schemes. If quantitative data are available, the interaction table corresponds with an interaction matrix, or a steady-state gain matrix between the selected inputs (manipulated variables, disturbances) and outputs (controlled variables). In addition, inaccurate estimates can be used. In this case, the matrix represents one possible process expression and the analysis should focus on the sensitivity of different interaction gain estimates to control space and variable pairing. If results are found to be highly sensitive, the proposed control strategy does not have generalizable features, and is not feasible for the process concept studied.

Variable Pairing

The selection of inputs and outputs (IO) in a small system may sometimes be intuitive. The larger the system, the more difficult the pairing of variables, and performance, complexity, and costs of the system may be far from the optimized solution. Systematic methods for variable selection, or IO selection, have, for example, been reviewed in van de Wal and de Jager (2001). In the initial phase, a simple criteria based on linear models and IO controllability measures can be recommended. One of the most popular methods is the relative gain array (RGA) (Bristol 1966), requiring only a steady-state gain matrix (that is, the interaction matrix) as an input, and the RGA matrix as an output. The RGA matrix can then be used with some basic rules to find recommended IO pairs for univariate control loops. Alternatively, the RGA matrix can indicate the need for multivariate control strategies instead of distributed univariate control strategies.

Control Space Analysis and Detailed Design

The interactions can be inspected more closely with more sophisticated controllability measures. Due to its visual interface, the output controllability index (OCI) developed in Vinson and Georgakis (2000) is used here. OCI indicates whether the allowable input space (range of manipulated variables) can satisfy the desired design space (range of controlled variables) with or without the presence of the expected range of disturbances. In this simplest form, the control space evaluation is made with steady-state linear models and is therefore well suited for the initial process control design. In the detailed design, the same analysis can be conducted with non-linear and dynamic models (Lima and Georgakis 2010).

The concept phase evaluation is followed by a detailed control design step. It supports, for instance, process identification, control algorithm development, and controller tuning tasks for particular application. In this step, it is necessary to use detailed dynamic simulation models and/or real process environment, and is therefore beyond the concept phase evaluation.

4   Case Example

The systematic approach described in Sect. 3 has been applied for a flash flotation cell, a unit process used in an intensified mineral beneficiation process (Newcombe et al. 2013). Table 1 presents a fictitious interaction

Table 1 Qualitative interaction table for the case example

	CV1	CV2
MV1	Moderate	Small
	Fast	Slow
MV2	Nil	Small
		Fast
DV1	Large	Small
	Fast	Slow
DV2	Moderate	Small
	Fast	Fair

Table 2 Quantitative values for the case example

	CV1	CV2	Nominal	Minimum	Maximum
Target	1700	7	n.a.	n.a.	n.a.
Minimum	1675	6	n.a.	n.a.	n.a.
Maximum	1725	8	n.a.	n.a.	n.a.
MV1	0.51	−0.08	90	0	500
MV2	0	0.15	300	260	340
DV1	0.85	−0.04	1700	1600	1900
DV2	0.71	−0.10	500	500	700

CV1 is the feed density (kg/m³), CV2 is the froth depth (cm), MV1 is the water addition rate (L/h), MV2 is the air addition rate (L/h), DV1 is the raw material density (kg/m³), and DV2 is the feed flowrate (L/h)

table with qualitative information for a small-scale flash flotation cell. Based on the initial analysis, major disturbances arise from upstream raw material density variations (DV1) and feed flow variations (DV2). One possible control strategy could involve the feed density (CV1) and cell froth depth (CV2) as controlled variables, and water addition rate (MV1) and air addition rate (MV2) as manipulated variables. Ideally, product throughput and quality (mineral concentration and particle size distribution) are controlled, but they cannot be effectively measured online. Table 2 depicts the quantitative strength of interactions, the range of studied variables, and their target values in this case example. The analysis is based on uncertain interactions around the nominal and target values of the process.

For the variable pairing, the interaction matrix (G_{ss}) between the manipulated and controlled variables can be written as in Eq. 1:

$$G_{ss} = \begin{bmatrix} 0.51 & 0 \\ -0.08 & 0.15 \end{bmatrix} \tag{1}$$

The corresponding RGA matrix is presented in Eq. 2:

$$RGA(G_{ss}) = \begin{bmatrix} 1 & 0 \\ 0 & 1 \end{bmatrix} \tag{2}$$

In the RGA matrix, there is one dominating entry to each column. Hence, the RGA matrix indicates that there are no severe interactions and two univariate control loops could be established; the first one between MV1 and CV1, and the second one between MV2 and CV2. From the process engineering point of view, utilizing the water addition rate (MV1) for controlling the feed slurry density (CV1) and adjusting the froth depth (CV2) with the air addition rate (MV2) is reasonable.

For the control space analysis, the variable ranges and disturbances also need to be evaluated. In Fig. 2, the control space without accounting for

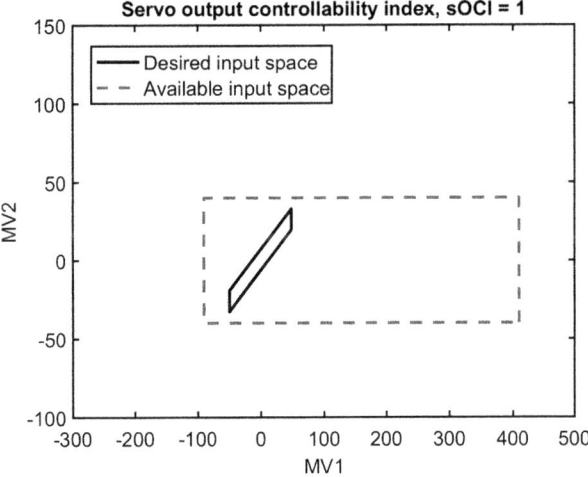

Fig. 2 Control space analysis without taking into account the disturbances. MV1 is the water addition rate (around the nominal value of 90 L/h), MV2 is the air addition rate (around the nominal value of 300 L/h)

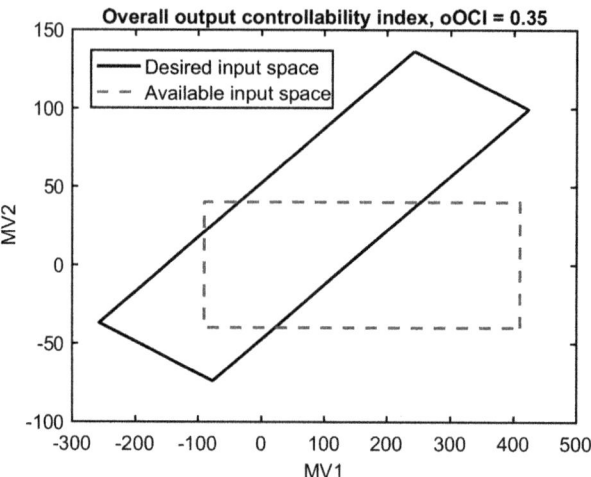

Fig. 3 Control space analysis when disturbance variables are taken into account. MV1 is the water addition rate (around the nominal value of 90 L/h), MV2 is the air addition rate (around the nominal value of 300 L/h)

the disturbances is presented. In this case, the control space analysis indicates that the range of MVs is feasible to cover the design space (sOCI = 1). It can also be seen that MV1 has unnecessarily large range. In Fig. 3, the analysis involves the effect of expected disturbances as well. Clearly, the available control space cannot cover the design space and expected disturbances (oOCI < 1). Therefore, the process and control design is not feasible unless the process design is altered. Without the identification of disturbance variables and taking into account their prevailing effect on process behavior, a false justification of the controllability and therefore the feasibility of the proposed process concept could be made.

How can the process design be altered in order to meet the controllability requirements? There are several alternatives to change the process requirements. The first option is to allow more variation in the design space (range of CVs). However, this option is often the last choice or compromise, as the intensification targets cannot be fully achieved. The second option is to expand the ranges of the MVs. It should be noted that there might be physical limitations to the manipulated variables, which eventually determine the allowable range. For example, in Fig. 3, it can be seen that the range of MV1 should be expanded by setting the lower limit to −280 with respect to the nominal value. However, Table 2 depicts that the nominal value is 90 and the current minimum value is zero. Evidently, a negative water addition rate (L/h) cannot be established and the expansion of MV1 range is not possible in this case. The third option is to control the upstream disturbances by some other means, hence reducing their effect in the inspected sub-process. This approach requires availability and expertise of plant-wide interactions, and may be beyond the process concept evaluation. The fourth option is to discard the proposed set of variables and try a different set of MVs (and possibly CVs). If none of these lead to a controllable process, the process concept studied cannot be seen as feasible from the process operation point of view.

5 Discussion and Recommendations

As presented, the proposed procedure offers a systematic way for designing control systems for intensified process concepts generated using the TRIZ approach. Together with the tools for choosing the suitable measurements (monitoring advisor), it will make it easier to achieve the feasible monitoring and control solutions in different PI cases. Therefore, for its part, the approach paves way for the industrial adoption of intensified processes. As a limitation, the proposed approach focuses on designing control systems for unit process level. The intensification, in turn, is likely to have an effect on one or more control subsystem(s) consisting of several unit processes, and maybe even plant-wide planning, scheduling, and optimization. However, it is an industrial practice to evaluate these aspects during a detailed control design and implementation process. In the detailed control design, dynamic behaviour (time-dependency and transitions between different steady states) of the system need to be considered as well. Dynamic behaviour is an essential feature in control design, but dynamic analyses require detailed models or experimental data, which are likely out of reach during process concept design. Another limitation of the proposed tools is that the control space analysis can efficiently be visualized only to small systems. Larger systems require interpreting several combinations, or evaluating only the numerical controllability measure.

It will be interesting to observe if TRIZ can produce inventive solutions for integrated process and automation solutions as well. The discussion has been initialized in Chechurin et al. (2016). Intuitively, the systematic approach presented here could benefit from internal TRIZ exercise targeted to monitoring solutions, for example. With the new monitoring solutions, a different set of variables and therefore new control strategies could be considered. Moreover, the results from this systematic procedure may initialize a need for TRIZ exercise, for instance, to overcome the physical limitations of manipulated variables or to control upstream disturbances.

Acknowledgments The present work has been developed under the financial support received from The EU Framework Programme for Research and Innovation – H2020-SPIRE-2015 (IbD® – Intensified by Design. GA – 680565).

References

Baldea, M. (2014). Multum in parvo: A process intensification retrospective and outlook. In J. D. Siirola, G. P. Towler, M. R. Eden (Eds.), *Computer aided chemical engineering* (Vol. 34, pp. 15–24). Proceedings of the 8th international conference on foundations of computer-aided process design. Elsevier. https://doi.org/10.1016/B978-0-444-63433-7.50003-1.

Bristol, E. (1966). On a new measure of interaction for multivariable process control. *IEEE Transactions on Automatic Control, 11*(1), 133–134. https://doi.org/10.1109/TAC.1966.1098266.

Chechurin, L., Berdonosov, V., & Yakovis, L. (2016). *Heuristic problems in automation and control design: What can be learnt from TRIZ?* TRIZ future conference. Wroclaw.

Cremaschi, S. (2015). A perspective on process synthesis: Challenges and prospects. *Computers & Chemical Engineering*, Special issue: Selected papers from the 8th international symposium on the Foundations of Computer-Aided Process Design (FOCAPD 2014), July 13–17, 2014, Cle Elum, Washington, DC, 81 (October): 130–37. https://doi.org/10.1016/j.compchemeng.2015.05.007.

Hamid, M. K. A. (2011). *Model-based integrated process design and controller design of chemical processes*. Ph.D. thesis, Kgs. Lyngby: Department of Chemical and Biochemical Engineering, Technical University of Denmark, Copenhagen, Denmark.

Hopkins, L., Lant, P., & Newell, B. (1998). Output strutural controllability: A tool for integrated process design and control. *Journal of Process Control, 8*(1), 57–68. https://doi.org/10.1016/S0959-1524(97)00027-9.

Huusom, J. K. (2015). Challenges and opportunities in integration of design and control. *Computers & Chemical Engineering*, Special issue: Selected papers from the 8th international symposium on the Foundations of Computer-Aided Process Design (FOCAPD 2014), July 13–17, 2014, Cle Elum, Washington, DC, 81 (October): 138–46. https://doi.org/10.1016/j.compchemeng.2015.03.019.

Lima, F. V., & Georgakis, C. (2010). Input–Output operability of control systems: The steady-state case. *Journal of Process Control, 20*(6), 769–776. https://doi.org/10.1016/j.jprocont.2010.04.008.

Livotov, P., Chandra Sekeran, A., Law, R., Mas'Udah, & Reay, D. (2017). *Systematic innovation in process engineering: Linking TRIZ and process intensification*. TRIZ future conference. Lappeenranta.

Mauricio-Iglesias, M., Huusom, J. K., & Sin, G. (2013, May). Control assessment for heat integrated systems. An industrial case study for ethanol recovery. *Chemical Engineering and Processing: Process Intensification*, Special issue: Hybrid and reactive separations, 67: 60–70. https://doi.org/10.1016/j.cep.2012.11.003.

Newcombe, B., Bradshaw, D., & Wightman, E. (2013). The hydrodynamics of an operating flash flotation cell. *Minerals Engineering, 41*, 86–96. https://doi.org/10.1016/j.mineng.2012.11.007.

Nikačević, N. M., Huesman, A. E. M., Van den Hof, P. M. J., & Stankiewicz, A. I. (2012). Opportunities and challenges for process control in process intensification. *Chemical Engineering and Processing: Process Intensification, 52*, 1–15. https://doi.org/10.1016/j.cep.2011.11.006.

Reay, D. A., Ramshaw, C., & Harvey, A. (2013). *Process intensification: Engineering for efficiency, sustainability and flexibility, Isotopes in organic chemistry* (2nd ed.). Amsterdam/Boston: Elsevier/BH, Butterworth-Heinemann is an imprint of Elsevier.

Ricardez-Sandoval, L. A., Budman, H. M., & Douglas, P. L. (2009). Integration of design and control for chemical processes: A review of the literature and some recent results. *Annual Reviews in Control, 33*(2), 158–171. https://doi.org/10.1016/j.arcontrol.2009.06.001.

Roffel, B., & Betlem, B. H. (2006). *Process dynamics and control: Modeling for control and prediction.* Chichester/Hoboken: Wiley.

Saltelli, A., Ratto, M., Tarantola, S., & Campolongo, F. (2006). Sensitivity analysis practices: Strategies for model-based inference. *Reliability Engineering & System Safety*, the 4th international conference on Sensitivity Analysis of Model Output (SAMO 2004), 91(10): 1109–25. https://doi.org/10.1016/j.ress.2005.11.014.

Sharifzadeh, M. (2013). Integration of process design and control: A review. *Chemical Engineering Research and Design, 91*(12), 2515–2549. https://doi.org/10.1016/j.cherd.2013.05.007.

Singh, R., Gernaey, K. V., & Gani, R. (2009). Model-based computer-aided framework for design of process monitoring and analysis systems. *Computers & Chemical Engineering, 33*(1), 22–42. https://doi.org/10.1016/j.compchemeng.2008.06.002.

van de Wal, M., & de Jager, B. (2001). A review of methods for input/output selection. *Automatica, 37*(4), 487–510. https://doi.org/10.1016/S0005-1098(00)00181-3.

Vega, P., de Rocco Lamanna, R., Revollar, S., & Francisco, M. (2014). Integrated design and control of chemical processes – Part I: Revision and classification. *Computers and Chemical Engineering, 71*, 602–617. https://doi.org/10.1016/j.compchemeng.2014.05.010.

Vinson, D. R., & Georgakis, C. (2000). A new measure of process output controllability. *Journal of Process Control, 10*(2–3), 185–194. https://doi.org/10.1016/S0959-1524(99)00045-1.

Yang, K., & El-Haik, B. (2003). *Design for six sigma: A roadmap for product development.* New York: McGraw-Hill.

Yuan, Z., Chen, B., Sin, G., & Gani, R. (2012). State-of-the-art and progress in the optimization-based simultaneous design and control for chemical processes. *AICHE Journal, 58*(6), 1640–1659. https://doi.org/10.1002/aic.13786.

Anticipatory Failure Determination (AFD) for Product Reliability Analysis: A Comparison Between AFD and Failure Mode and Effects Analysis (FMEA) for Identifying Potential Failure Modes

Renan Favarão da Silva and Marco Aurélio de Carvalho

1 Introduction

To be successful, especially in the field of product development, many skills are required from engineers due to the increasing complexity of technology. Failures during product design development can lead to disastrous results (Atman et al. 2008). Therefore, the use of tests and detection reviews as the main methods to ensure the reliability of products is no longer sufficient nowadays (Carlson 2014). Traditionally applied in the advanced stages of product development, these methods imply a greater restriction on design change possibilities.

Designing a product is a creative process, experimental and with no predefined result. This combination of factors makes the process complex and often poorly structured (Song et al. 2016). Consequently, most

R. F. da Silva (✉) • M. A. de Carvalho
Federal Technological University of Paraná (UTFPR), Curitiba, Brazil
e-mail: marcoaurelio@utfpr.edu.br

© The Author(s) 2019
L. Chechurin, M. Collan (eds.), *Advances in Systematic Creativity*,
https://doi.org/10.1007/978-3-319-78075-7_12

product failures have premature origin in the development cycle, being only detected in late stages (Bruch 2004). Using proactive approaches from the beginning of the process can reduce the amount of these project redefinitions in addition to lowering costs and anticipating factors detrimental to the product.

One of the most used amongst reliability tools is the Failure Mode and Effects Analysis (FMEA). This is the main method used in many industries. However, FMEA also has its failure modes and is not very reliable itself (Jenkins 2013). Even in the numerous satisfactory applications, FMEA results tend to be limited to known expectations of failure occurrence (Hipple 2006). Another problem with FMEA is that the analysis focuses on parts or subsystems, and problems related to the connections between those parts or subsystems tend not to be included in the analysis. A critical view of FMEA deems it as a rudimentary process routinely used in the industries that is not able to find all the potential faults of the components (Hiltmann 2015).

Allegedly, Anticipatory Failure Determination (AFD) is a method whereby the user can minutely analyze fault mechanisms, obtain an exhaustive set of potential failure scenarios, and develop inventive solutions to prevent, neutralize, or minimize the impact of failure scenarios (Kaplan et al. 2005). Anticipatory Failure Determination is the most common name, but the same method can also be found in TRIZ literature as Anticipatory Failure Identification (Livotov 2008), Subversion Analysis (Regazzoni and Russo 2010), Anticipatory Failure Prediction (Thurnes et al. 2012), and Diversionary Method or Diversionary Analysis (Frenklach 1998). Although not a widely recognized method in industry, the systematic approach of AFD can predict future problems and develop ways to prevent them (Bruch 2004).

A search in TRIZ Future Conference (TFC) papers published between 2001 and 2015 revealed that only two articles contained the keywords Anticipatory Failure Identification, Anticipatory Failure Determination, Anticipatory Failure Prediction, Diversionary Method, Diversionary Analysis, or Subversion Analysis. Besides the lack of publications, it was found that AFD-related literature tends to emphasize the advantages of the method. There is virtually no published critical analysis of the method. The few publications discussing AFD also tend to be theoretical and dissociated from case studies.

Therefore, this chapter critically compares AFD and FMEA as tools for predicting potential failures in product design. Its main contribution is an evaluation of a non-traditional method for identifying potential failure in product development that was not found in literature. As design fault solutions are mostly related to inexistent, insufficient, or inadequate methodological support (Bruch 2004), testing a different method helps to identify improvements and complementarity to traditional methods as FMEA.

2 Identification of Potential Failure Modes During Product Development

Functional product failures have always been a problem, however, it seems that with the complexity of current systems, the number and severity of incidents has increased (Ungvari 1999). This reinforces the popularity of the risk scenario analysis, motivated mainly by the search for product and process reliability, which is currently undergoing growth without signs of stagnation. Nevertheless, it usually focuses on learning and explaining the failures that have already occurred rather than preventing them from repeating themselves or identifying those that have not yet occurred (Kaplan et al. 2005).

To effectively minimize the occurrence of failures, designers should have excellent knowledge of fault mechanisms (Regazzoni and Russo 2010). It is assumed that the correction of a product failure during the planning, production, and use stages of the product costs, respectively, 10, 100, and 1000 times more than the correction of the same failure during the conceptual design step (Bruch 2004). Therefore, methods of anticipating failures are growing in importance (Livotov 2008).

Developed to identify potential product failure modes, FMEA and AFD provide advantages for their use. Although FMEA is a linear, well-accepted, and standardized process, AFD is an example of an inventive process of failures that is still little explored (Thurnes et al. 2012). Thus, this chapter focuses on exploring AFD's potential in relation to FMEA.

2.1   Failure Mode and Effects Analysis (FMEA)

FMEA is a process originally developed by engineers to study failures in military systems. Introduced in the 1940s by the US military, the technique was subsequently enhanced for the aerospace and automotive industries (Jenkins 2013). Currently, there are many relevant standards and guides that specify the use of the technique. Some of the most common are: SAE J1739, Potential Failure Mode and Effects Analysis in Design (FMEA Design), Potential Failure Mode and Effects Analysis in Manufacturing and Assembly Processes (Process FMEA), and the Automotive Industry Actions Group's (AIAG) Potential Failure Mode and Effects Analysis (FMEA) Reference Manual, Fourth Edition (Carlson 2014).

Despite the variations, the primary objective of FMEA remains in design improvement (Carlson 2014). In other words, FMEA basically consists of a qualitative method that identifies and lists the potential failure modes during the product development process. The tool is a systematic method for identifying and preventing problems in products and processes before they occur (McDermott et al. 2009).

Although the application of FMEA in existing products can also produce beneficial results, the analysis is conducted at the product design stage (McDermott et al. 2009). The tool is developed for the initial stages of product design development because it aims to identify and correct problems before they reach the consumer (Carlson 2014).

The most common types of FMEA are system FMEA (SFMEA), design FMEA (DFMEA), and process FMEA (PFMEA). DFMEA is characterized as a type of FMEA whose focus is on product design, typically at the subsystem or component level (Carlson 2014). Therefore, this type of FMEA is the one that is applicable to the investigation of this chapter.

DFMEA is mainly a structured approach to the prevention of problems related to product design, its causes, and its defects (Santana and Massarani 2005). The technique involves all items of the product structure (subsystems and components), identifying the potential modes and causes of failures, current controls, and effects. In other words, it is intended to discover problems in the products that will result in safety risk, dysfunctions, or shortening life cycle.

Each PFMEA/DFMEA uses as a communication tool a worksheet to document and store all important information of the steps. Its process can be summarized in ten steps: 1. Review the process or product, 2. Brainstorm the potential failure modes, 3. List potential effects for each failure mode, 4. Assign severity ratings to each effect, 5. Assign occurrence ratings for each failure mode, 6. Assign detection ratings for each failure mode or effect, 7. Calculate the risk priority number (RPN) for each effect, 8. Prioritize failure modes per action, 9. Take actions to eliminate or reduce high-risk failure modes, and 10. Calculate the resulting RPN with reduced or deleted failure mode (McDermott et al. 2009).

Therefore, FMEA is an analytical methodology used to ensure that potential problems are considered throughout the product development process. However, high industry adherence does not guarantee that these methods are employed properly or that they provide their true benefits (Santana and Massarani 2005). In this way, it is possible to investigate other analogous tools to contribute to the development of products, such as AFD.

2.2   Anticipatory Failure Determination (AFD)

The acronym TRIZ (in russian *Teória Rechénia Izobretátelskih Zadátchi*) was introduced by Genrich Altshuller in the former USSR (Sarno et al. 2005). From the outset, the approach differed from methods developed in the West, such as brainstorming, morphological method, and value analysis. Altshuller focused not on the study of the problem-solving process or the creative personalities, but on the products of the creative process: patents (De Carvalho and Hatekeyama 2003).

In this way, the development of TRIZ by its founder, along with its associates, studied a vast base of technological solutions, patents, and inventions. TRIZ originated in the mid-twentieth century to develop a method that would support the process of inventive generation of ideas and solutions in a systematic way. Before that, there was no systematic method to support this process beyond brainstorming, which is basically grounded in trial and error (Souchkov 2014).

With the end of classic TRIZ dating to the death of Altshuller in 1998, the globalization of TRIZ's knowledge begins. In this context, Ideation

International develops I-TRIZ, a methodology based on classic TRIZ and advances from practical experience and research in the United States. Through it, AFD was structured as a TRIZ-based application to analyze, predict, and eliminate product and process failures (IDEATION 2012).

AFD is a methodology that helps to reveal potential failure modes in systems, manufacturing processes, and products, among others, before problems appear with their harmful and undesirable effects (Kaplan et al. 2005). The methodology is a practical application of I-TRIZ and a systematic and structured process that proposes to reveal the causes of failures and to develop simple and effective ways to eliminate the associated problems (Clarke et al. 1999).

It may sound simplistic that AFD is structured in the reversal of a basic question of "What can fail?" to "How can I make it fail?" This change puts people's brains in different quarters: they are now allowed to be villains and do things that are generally not allowed. In experiences with projects such as bank fraud, food contamination, and chemical release, AFD has produced non-obvious responses, significantly increased possibilities, or improved previous responses (Hipple 2015).

The tool is used for the solution search after identifying the potential faults. In the AFD method, the faults are "invented" in a subversive way, by which it is known. Once the list of the scenarios of invented failures is complete, the problem needs to be reinvented and failures must be prevented from ever happening (Livotov 2004).

Therefore, AFD has in its concept the use of TRIZ in an inverse way in order to find all the ways to produce failures in the system and later to eliminate them (Ungvari 1999). The uniqueness of the technique is the intentional process of invention of failures (Jenkins 2013). This is claimed to assist in the search for causes by means of a desired potential disability, which differs from traditional methods (Kohnhauser 1999).

Although this TRIZ-based fault prediction process is known by other nomenclatures such as Subversion Analysis, Subversive Analysis, Anticipatory Failure Identification (AFI), and Anticipatory Failure Prediction (AFP), since AFD is the name registered by Ideation International, its essence is preserved. Even so, when comparing the different nomenclatures, small variations are noticed in the form of the presentation of the method (Silva 2017).

The script of AFD of Ideation International provides ten steps for failure prediction: 1. Original problem statement, 2. Identifying the success scenario,

3. Formulating the inverted problem, 4. Identifying obvious flaws, 5. Identifying available resources, 6. Using base knowledge, 7. Inventing new solutions, 8. Intensifying and camouflaging, 9. Analyzing harmful effects revealed, and 10. Preventing and eliminating failures and harmful effects (Kaplan et al. 2005).

Among the main objectives of AFD is not only to reveal or predict failure modes, but to eliminate them effectively with an appropriate time demand. The ideal way to prevent failures is to eliminate the causes. However, there are countless causes why this may not be possible since it can be very expensive, late, out of your responsibilities, and so on. However, the solutions should be sought from the most ideal solution: eliminate the cause of the failure, then eliminate the failure, and, ultimately, eliminate the effects (Kaplan et al. 2005).

Finding and preventing possible and future failures is not automatism but a process that requires, in addition to a systematic approach, a lot of creativity and inventive talent. AFD encourages the questioning of "how" failures can occur while other methods, such as FMEA, focus on "what" may fail (Thurnes et al. 2012). This discussion among methodologies is developed and explored in more detail in the following section.

3   AFD and FMEA Comparison

AFD and FMEA can be used during product design development to identify and prevent potential failures. Both methodologies have established parallels between them; the main difference lies in the orientation of the AFD to a more proactive approach (Regazzoni and Russo 2010). In this way, this section presents a theoretical comparison about both methods followed by an evaluation of practical results of AFD and FMEA applications.

3.1   Theoretical Comparison

The logic of a Subversion Analysis lies in determining all modes that destroy the system being designed. As a consequence, systems designed with this approach are less vulnerable to unforeseen failures. After this process, it is easier to design the system in which those failure modes are eliminated or at least taken into account when implementing corrective actions (Regazzoni and Russo 2010).

The TRIZ-based method for predicting and eliminating potential failures structures an effective and creative process that complements other methods of fault prevention and risk analysis. The methodology broadens the field of search and application when compared with other tools such as checklist and FMEA. AFD excels in analyzing previous failures that occurred for no apparent reason, as well predicting hidden sources of scenarios of potential failures or damages (Livotov 2008).

The main difference between the AFD and the traditional methods diffused in the West, such as FMEA, is the perspective by which the system has its certain failure modes. In traditional techniques, the fault prediction process develops linearly from the articulation of system functions to what may occur if there is a failure to deliver these functions. In other words, the logical line of analysis follows the design intent. It is given the potential failure, the effect of failure, the probability of occurrence, and its ability to be determined (Ungvari 1999).

The traditional approach seems logical, however, it carries structural weaknesses that may compromise the analysis. The process of identifying failures is essentially an exercise in brainstorming initiated by probing how the system might fail (Ungvari 1999). In comparison with the AFD, the process of identifying failures of traditional methods carries the Psychological Inertia syndrome. The analyses of identification of potential failures are performed in the same mental context in which the system was created (Ungvari 1999). The AFD methodology is able to overcome this psychological inertia (Proseanic et al. 2000) and prevent "mental blocks," motivating the user to find inventive solutions (Livotov 2008), thus being able to provide support needed to identify product barriers.

Besides, the characterization of failures is established in the absence of an intended or projected function. However, the characterization of failures should not only be analyzed in the absence of function perspective, but also include the insufficient and excessive conditions of the function. Also, "prohibited" or malicious functions are not considered as sources of potential failures in traditional methods, leading the analysis according to the original purpose of the project (Ungvari 1999).

Finally, traditional tools do not solve problems in an inventive way or in their root causes (Ungvari 1999). If the evaluated system is very dangerous, corrections are suggested with reengineering and insertion of security systems, barriers, or detection improvement (Ungvari 1999).

The deficiencies pointed out were overcome during the elaboration of the AFD methodology. The identification of failures occurs in the inverse of those used in conventional processes by carrying the power of the tool. This perspective allows us to fully exploit the weaknesses of the system by turning the user into an offensive mode in fault searches (Ungvari 1999).

Although the discussion about the potential of methodologies for product development brings interesting information, it was presented within the framework of the theoretical comparison. The reviewed literature compared the methodologies based on the scripts from guides and norms and on the expectations of results of the methods. No references were found in the literature comparing practical cases or perceptions of users based on similar applications.

This opportunity guided the development of this chapter in order to contribute to a practical evaluation of AFD in relation to FMEA. Applications of the tools with engineering students made it possible to collect, compare, and discuss the processes of both methods. Also, this content was finally confronted with the theoretical analyses available in the literature for the first time.

3.2   Practical Comparison

The methodological procedure of this comparison was divided in three stages: course planning for engineering students, data collection, and analysis of results. Therefore, the roles assumed by the researchers in each of these stages were specialist, facilitator, and analyst, respectively.

In the planning of the FMEA and AFD course, the activity to be performed with engineering students was structured to provide the data collection of the research. This process included the following points: scope and format of the course, theoretical material of the course, practical cases, evaluation forms, representation and sampling of participants, and schedule.

The data were then collected during the course sessions through the forms completed by the participating students throughout the activities. For scientific rigor improvement, it was defined a mixed method to include collections and analyses of quantitative and qualitative data in order to deepen the conclusions through triangulation. These forms were composed of the different data formats: sample, qualitative, quantitative, categorical, and practical results.

The sample data were characterized as the students' personal information. For the qualitative data, open questions recorded the opinions and comments of the participants. For quantitative data, evaluations given by participants on a numerical scale from 0 to 5, and for categorical data, the multiple-choice question format.

Finally, the data collected in the research was analyzed in order to support the discussions and the considerations about the research questions. Three processes were used to compose the set of analyses: exploratory analysis, statistical analysis, and comparative analysis.

In the scope of the course, an introductory approach was considered about product development, followed by an explanation of the FMEA methodology, and, finally, AFD, interspersed with the exercises of practical application in groups. This format, therefore, consisted of a theoretical-practical model with eight hours of duration, limited to 25 participants per session. As a requirement for participation, students should have, at least, already completed at least one half of the undergraduate engineering course.

The use of novices for design development experiments is considered sufficient for validations of research on creative methods, since the methods that are beneficial to them will also be to the practitioners, though not necessarily to the same degree. In addition, professional engineers tend to centralize the inspirations for problem solving in their previous experiences, which, for the scope of this research, would hinder the impartiality of evaluations (Atman et al. 2008; Yuan et al. 2014).

For the evaluations, four forms were developed for application with the students (see Table 1). They were intended for the closure of each section in order to record participants' conclusions. The set of answers formed the database (qualitative and quantitative) for analysis and discussions.

Table 1 Course evaluation forms

Form	Description
I. Presentation and expectation	Used at the end of the introduction section to record students' sample data
II. FMEA evaluation	Applied after completion of the FMEA section (theoretical and practical). The effectiveness of training, facilitation, and didactic material was evaluated (scale 0–5)
III. AFD evaluation	Applied after completion of the AFD section (theoretical and practical). The effectiveness of training, facilitation, and didactic material was evaluated (scale 0–5)
IV. Comparison FMEA and AFD	Evaluated the robustness of each tool, the ease and effectiveness in the identification of failure modes, as well as considerations of use recommendations for product design and complex cases

For the practical sections of FMEA and AFD, five product designs were selected for group activities: a frying pan, a notebook, school scissors, a pair of glasses, and a vegetable broom. These products were chosen for the low number of components, low complexity, and great familiarity to students.

The minimum sample of students participating in the research was calculated as 96, so that a standard level of significance (5%) and a moderate margin of error (10%) could be attained (Laboratório de Epidemiologia e Estatística 1996). According to the format proposed for the course, mechanical and industrial engineering students from five universities in the Curitiba-PR (Brazil) area participated in the course sessions, which were conducted between August and October 2016.

In total, 105 students participated in the courses. All of them were senior students, from the third, fourth, and fifth year. In Brazil, an engineering bachelor's degree is a five-year course. As for students' previous familiarity with FMEA and AFD, only 23% of the participating students claimed to have applied or participated in a FMEA or AFD application.

Following the conclusion of each practical session of FMEA and AFD, the practical results and qualitative evaluations were collected. In both methodologies, above-average evaluations were found for the effectiveness of the course to solve the practical case and the qualification of the teacher for training and facilitation during course and didactic material used. With FMEA, the means were 4.3, 4.6, and 4.8, while with AFD they were 4.0, 4.7, and 4.4, respectively.

Finally, students were asked to individually fill out the FMEA and AFD comparison form. In general, it has been observed that there was no tendency of total exclusion of either tool. Both methods stood out in different criteria.

In evaluating the robustness of the methodologies, AFD was the preferred method. The main reasons provided were: it allows more opportunities to identify failures (59%) and helps identify failures that are more unusual (17%). For the process of identification of failures, specifically, the evaluations of the methodologies were very similar: 42% emphasized FMEA, and 39% AFD. The main justifications were that FMEA is a more direct (22%), more organized (27%), and more linear (14%) method, whereas those who chose AFD mentioned that this tool provides greater comprehensiveness in identifying failure modes (30%), is visually clearer (18%), and stimulates the identification of more unusual failure modes (18%).

Later, the ease of use of the methods was evaluated. For 60% of undergraduates, FMEA excelled in this aspect. The main reasons for this choice were that the FMEA system was considered clearer and more definite (25%), more linear (19%), and easier or more practical (15%).

For 61% of the students, FMEA is best for application to cases of greater complexity, mainly because it is more organized (27%), easier to understand (14%), and less complex than AFD (14%). On the other hand, for 59% of the students, AFD is the most recommended tool for failure prevention in product development. The main reasons for this conclusion were the ability to expand the range of identified failure modes (44%), the higher probability of identifying factors that are less usual (18%), and a process more visual or less restricted than FMEA (11%).

The various sessions of the course carried out aimed not only at meeting the planned sample number, but also in contributing to the validity of the research through data triangulation. This technique limits personal and methodological biases and increases the generalization of a study as it uses more than one data source. The use of different methods of data collection also characterizes another type of triangulation: methodological triangulation (Azevedo et al. 2013).

Table 2 Course participant's perceptions about FMEA and AFD on robustness, fault identification and ease of use

Robustness			Fault identification			Ease		
Answers	%	Qtd.	Answers	%	Qtd.	Answers	%	Qtd.
Both	28.5	30	Both	12.4	13	Both I	6.7	7
FMEA	12.4	13	**FMEA**	**41.9**	**44**	**FMEA**	**60.0**	**63**
AFD	**49.5**	**52**	AFD	39.1	41	AFD	16.2	17
Other	9.5	10	Other	6.7	7	Both II	17.2	18

Table 3 Course participants' perceptions about FMEA and AFD on complex cases and product development

Complex cases			Product development		
Answers	%	Qtd.	Answers	%	Qtd.
FMEA	**60.9**	**64**	FMEA	25.7	27
AFD	17.2	28	**AFD**	**59.0**	**62**
No preference	26.7	13	No preference	15.2	16

In order to guide the answers to the questions, the data was tabulated from the total of participants, according to Tables 2 and 3.

The data tabulations reinforced the robustness of AFD: 49.5% of participants considered it to be more robust than the FMEA, since, in addition to identifying more potential failures, it allows users to find less usual failure modes. In contrast, for 60% of the participants, the tool is more difficult to use than FMEA, because it is less linear, less creative, and less restricted.

This conclusion was confronted with the potentialities promised in the literature of the methodologies in order to evaluate them where it is observed that both have their advantages. FMEA is characterized by a more linear, standardized, and well-accepted process (Thurnes et al. 2012), which meets the main annotations of the participants in its forms. Regarding the potential of AFD, the analytic steps lead to a more in-depth and inventive analysis (Ungvari 1999), which was confirmed by student responses.

The disadvantages reported by the course participants for AFD were similar to those found in the literature. The basic tools of TRIZ are not easy to learn with routine habits of thought and require more effort

(Thurnes et al. 2012). This was confirmed, with 60% of participants choosing FMEA as the easiest tool and almost 20% of students reporting that AFD was more confusing and required more conversations with the instructor during the application. The pattern for FMEA is well structured, differing from the lesser known and more comprehensive AFD (Thurnes et al. 2012).

Finally, the results of all practical applications of the two methodologies were analyzed. In all, there were 24 practical applications of FMEA and AFD, in which more failures and more potential causes were identified in 71% of them. In contrast, the AFD failure tree diagrams have made it more limited for immediate understanding due to freer and more connected organization. Compared to the FMEA deliveries, structured in spreadsheet form, the diagrams became more visually polluted.

In order to visualize the deliverables of two application cases by students, Figs. 1 and 2 exemplify the results of FMEA and AFD for the same products.

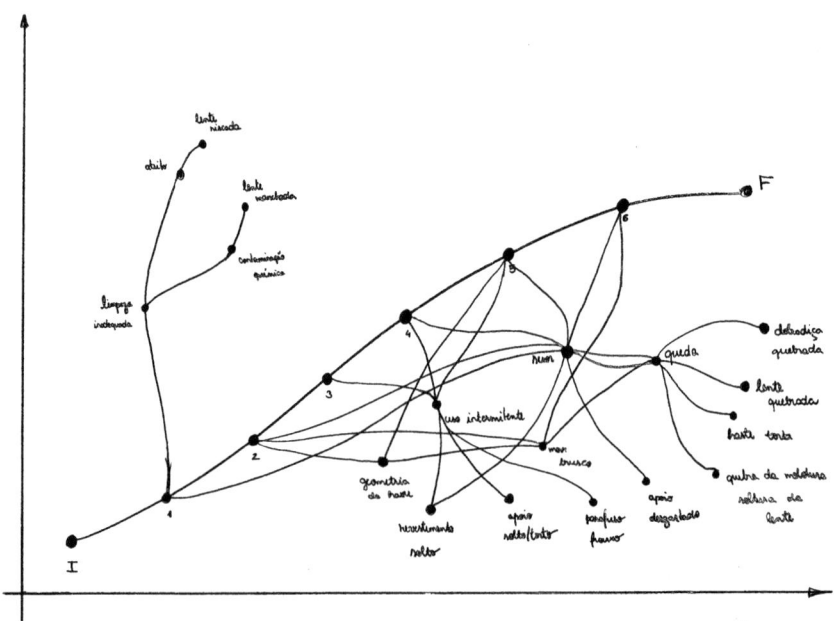

Fig. 1 Example of AFD application result for the practical case "glasses"

Análise de Modos e Efeitos de Falha
FMEA - Failure Mode and Effects Analysis

Facilitador FMEA: Eng. Renan Favardo
Equipe Titular: Carla, Rafael, João Gabriel, Taísse, Ariel
Outros Envolvidos:
Descrição do Item: Óculos de grau
Outros Detalhes:

FMEA Tipo: DFMEA
Data e horário inicial: 04/08/16 21:25
Data e horário final: 04/08/16 21:
Número do Grupo: 4
Universidade: UFPR

Item	Função	Modo de Falha Potencial	Efeito de Falha Potencial	s	Causa de Falha Potencial	o	Controle Atual Preventivo	Controle Atual Detectivo	d	RPN	Ação Recomendada
Lentes	Ajustar a imagem de acordo com o grau do cliente	Não ajustar a imagem	Não enxergar	8	Grau errado	1	Prova				
					Baixa resistência mecânica	4	Teste e Simulação				
Apoio nariz	Sustentar o óculos na posição correta confortavelmente	Não sustentar o óculos	Impossibilidade de uso	8	Dimensão errada	1					
		Não causar o conforto	Incomodo e insatisfação do cliente	4	Material inadequado	3					
Articulação	Conectar as pernas à armação do óculos.	Não conectar as pernas na armação	Impossibilidade de uso	8	Quebra	6					
Pernas	Sustentar o óculos sobre a orelha com o comprimento	Não sustentar o óculos sobre a orelha	Impossibilidade de uso	8	Dimensão errada	1					
	adequado para que o mesmo não se mova	Não causar conforto	Incomodo	4	Material e design inadequado	3					
					Dimensão errada	1					

Baseado SAE J1729

Fig. 2 Example of FMEA application result for the practical case "glasses"

4   Conclusions

Different from other studies, this chapter highlighted the comparison of AFD with FMEA through a theoretical and practical evaluation. As its main contribution, practical results of both tools were analyzed in parallel with theoretical evaluations in the literature. No previous work was found in the literature with this practical approach.

For practical comparison, the development of a course on the two tools of fault identification was the main research instrument. Student perceptions for AFD were positive and, for use in product design development, more recommended than FMEA. With a significance level of 5% and accuracy of 9.5%, 59% of participants would recommend AFD for use of product development over FMEA, recommended by 25.7%. For 49.5% of the participants, AFD has a more robust process of identifying potential failures. However, the FMEA was the main recommendation for 61% of users for complex cases and highlighted in the facility between the methods for 60%.

In the qualitative analyses, the participants classified the FMEA methodology as linear, simpler, and more organized, also emphasizing the practicality of the risk priority number (RPN) for directing the future failure prevention actions. For AFD, the strengths came through as a creative method, freer and with greater exploratory potential.

As for the weaknesses, for FMEA, the difficulty came from the concepts needed to develop the practical case, being classified as a more bureaucratic method. For AFD, students have pointed out difficulties in conducting a freer method and in drawing the most connected faulty scenario diagram, which sometimes becomes confusing to interpret. These considerations were in agreement with some studies present in the literature (Thurnes et al. 2012; Masin and Jirman 2012). Based on the results, AFD could be an alternative method for product development reliability analysis while a complement to FMEA for complex cases.

Therefore, it was concluded that the AFD has a large, and little explored, potential in identifying anticipated failures during the product development process. The tool performed satisfactorily during practical applications and had good assessments by participating students.

Acknowledgments The authors thank UTFPR, FUNTEF-PR, Fundação Araucária, CAPES, and CNPq for providing the infrastructure and financial support for this research.

References

Atman, C. J., Yasuhara, K., Adams, R. S., et al. (2008). Breadth in problem scoping: A comparison of freshman and senior engineering students. *International Journal of Engineering Education, 24*(2), 234–245.

Azevedo, C. E. F., Oliveira, L. G. L., Gonzalez, R. K., et al. (2013). A Estratégia da Triangulação: Objetivos, Possibilidades, Limitações e Proximidades com o Pragmatismo. In *Encontro de En-sino e Pesquisa em Administração e Contabilidade*. Rio de Janeiro: Associação Nacional de Pós-Graduação e Pesquisa em Administração.

Bruch, C. (2004). *Handling undesired functions during conceptual design – A state – and state – Transition – Based approach*. International design conference, 8. https://www.designsociety.org/download-publication/19843/handling_undesired_functions_during_conceptual_design-a_state-and_state-transition-based_approach. Accessed 4 Dec 2016.

Carlson, C. S. (2014). *Understanding and applying the fundamentals of FMEAs*. Annual reliability and maintainability symposium. http://www.reliasoft.com/pubs/2014_RAMS_fundamentals_of_fmeas.pdf. Accessed 4 Dec 2016.

Clarke, D. W., Vishnepolschi, S., & Zlotin, B. (1999). *Case study: Application of AFD – Failure analysis for the analysis of power transfer between a transmission sun gear and associated satellites*. Ideation International Inc. http://www.ideationtriz.com/new/materials/WalkingBearingCaseStudy.pdf. Accessed 5 Dec 2016.

De Carvalho, M. A., & Hatekeyama, K. (2003). Solução Inventiva de Problemas e Engenharia Automotiva: A Abordagem da TRIZ. In *Revista Engenharia Automotiva e Aeroespacial* (Vol. 3(13), pp. 34–37). Brasil: SAE. http://www.aditivaconsultoria.com/artigoengautomotivaeaerospacial-marcoekazuo.pdf. Accessed 4 Dec 2016.

Frenklach, G. (1998). Diversionary method. *TRIZ Journal*. https://triz-journal.com/diversionary-method/. Accessed 23 Aug 2017.

Hiltmann, K. (2015). *Predicting unknown failures* (Vol. 132, pp. 840–849). Triz future conference. http://www.sciencedirect.com/science/article/pii/S1877705815042769. Accessed 04 Dec 2016.

Hipple, J. (2006). Predictive failure analysis™: How to use TRIZ in "Reverse". *TRIZ Journal.* https://triz-journal.com/predictive-failure-analysis-use-triz-reverse. Accessed 5 Dec 2016.

Hipple, J. (2015). *Predictive failure analysis™: Planning for your worst business nightmare by figuring out what it is.* New & Improved White Paper Archive. http://newandimproved.biz/newsletter/2015.php. Accessed 4 Dec 2016.

IDEATION. (2012). *History of TRIZ & I-TRIZ.* http://www.ideationtriz.com/history.asp. Accessed 5 Dec 2016.

Jenkins, B. (2013). *Automating FMEA: Next-generation failure analysis.* Naneva Orare-search. http://oraresearch.com/wpcontent/uploads/2014/11/Ora_FMEA_130812. Accessed 5 Dec 2016.

Kaplan, S., Visnepolschi, S., Zlotin, B., et al. (2005). *New tools for failure and risk analysis: Anticipatory Failure Determination (AFD) and theory of scenario structuring.* Southfield: Ideation International Inc.

Kohnhauser, V. (1999). Use of TRIZ in the development process: Zero-defect-development for customer centered innovative products. *TRIZ Journal.* https://triz-journal.com/use-triz-development-process-zero-defect-development-customer-centered-innovative-products. Accessed 4 Dec 2016.

Laboratório de Epidemiologia e Estatística. (1996). *Estimação de Uma Proporção.* http://www.lee.dante.br/pesquisa/amostragem/calculo_amostra.html. Accessed 5 Dec 2016.

Livotov, P. (2004). *The undervalued innovation potential: Industrial application of TRIZ delivers more breakthroughs to less cost – If the right tools are applied at the right time and place.* Research Gate. https://www.researchgate.net/publication/237686175_The_undervalued_innovation_potential_industrial_application_of_TRIZ_delivers_more_breakthroughs_to_less_cost_if_the_right_tools_are_applied_at_the_right_time_and_place. Accessed 5 Dec 2016.

Livotov, P. (2008). *Innovative product development and theory of inventive problem solving. Triz and innovation management, Innovator.* http://www.triz.it/triz_papers/2008%20TRIZ%20and%20Innovation%20Management.pdf. Accessed 4 Dec 2016.

Masin, I., & Jirman, P. (2012). *Production problems solving based on negative effect and non-desirable component identification.* TRIZ future conference, Lisboa.

McDermott, R. E., Mikulak, R., & Beauregard, M. R. (2009). *The basics of FMEA* (3rd ed.). New York: CRC Press.

Proseanic, V., Tananko, D., & Visnepolschi, S. (2000). The experience of the Anticipatory Failure Determination (AFD) method applied to an engine concern. *TRIZ Journal*. https://triz-journal.com/experience-anticipatory-failure-determination-afd-method-applied-engine-concern. Accessed 5 Dec 2016.

Regazzoni, D., & Russo, D. (2010). *TRIZ tools to enhance risk managemente* (Vol. 9, pp. 40–51). TRIZ future conference. http://www.sciencedirect.com/science/article/pii/S1877705811001160. Accessed 04 Dec 2016.

Santana, A., & Massarani, M. (2005). *Engenharia do Valor Associada ao DFMEA no Desenvolvimento de Produto* (pp. 1–12). Society of Automotiv Engineers Inc. http://www.mecanica-poliusp.org.br/05pesq/cont/pdf/712.pdf. Accessed 4 Dec 2016.

Sarno, E., Kumar, V., & Li, W. (2005). A hybrid methodology for enhancing reliability of large systems in conceptual design and its application to the design of a multiphase flow station. *Engineering Design, 16*, 27–41. http://faculty.washington.edu/vkumar/microcel/linkfiles/labpapers/15.pdf. Accessed 4 Dec 2016.

Silva, R. F. (2017). *Determinação Antecipada de Falhas (AFD) para a Identificação de Falhas Potenciais no Projeto de Produtos: Uma Comparação com a Análise de Modo e Efeitos de Falhas (FMEA)*. Repositório Institucional da Universidade Tecnológica Federal do Paraná (UTFPR). http://repositorio.utfpr.edu.br/jspui/bitstream/1/2625/1/CT_PPGEM_M_Silva%2C%20Renan%20Favar%C3%A3o%20da_2017.pdf. Accessed 6 Jan 2018.

Song, T., Becker, K., Gero, J. et al. (2016). Problem decomposition and recomposition in engineering design: A comparison of design behavior between professional engineers, engineering seniors, and engineering freshmen. *Journal of Technology Education, 27*(2), 37–56. https://www.researchgate.net/publication/302418470_Problem_Decomposition_and_Recomposition_in_Engineering_Design_a_Comparison_of_Design_Behavior_Between_Professional_Engineers_Engineering_Seniors_and_Engineering_Freshmen. Accessed 8 Jan 2018.

Souchkov, V. (2014). *Breakthrough thinking with TRIZ for business and management: Na overview*. ICG training & consulting. http://www.xtriz.com/TRIZforBusinessAndManagement.pdf. Accessed 4–05 Dec 2016.

Thurnes, C. M., Zeihsel, F., & Visnepolshi, S., et al. (2012). *Using TRIZ to invent failures: Concept and application to go beyond traditional FMEA* (Vol. 131, pp. 426–450). TRIZ future conference. http://www.sciencedirect.com/science/article/pii/S1877705815043313. Accessed 04 Dec 2016.

Ungvari, S. F. (1999). The anticipatory failure determination fact sheet. *TRIZ Journal.* https://triz-journal.com/anticipatory-failure-determination-fact-sheet. Accessed 5 Dec 2016.

Yuan, J. T. J., Kong, K. Y., Parveen, H., et al. (2014). *An overview of design cognition between experts and novices.* International conference on advanced design research and education. https://www.researchgate.net/profile/Hashina_Parveen/publication/264545745_An_Overview_of_Design_Cognition_between_Experts_and_Novices/links/53e4702b0cf21cc29fc8f781.pdf. Accessed 04 Dec 2016.

Computer-Aided Conceptual Design of Building Systems: Linking Design Software and Ideas Generation Techniques

Ivan Renev

1 Introduction

The early design stage in architectural and construction projects is a crucial part of sophisticated long-term design process. This stage is also known as Conceptual Design (CD) and here many fundamental and critical solutions are taken into account. The smarter and less trivial the solutions developed during the CD are, the more technological, effective, and less costly the design is. Those solutions can be found by using different techniques of ideas generation, such as morphological charts, synectics, brainstorming, Theory of Inventive Problem Solving (TRIZ) tools, and so on. TRIZ is believed to be one of the most effective and well-structured problem-solving techniques (Altshuller 1999; Salamatov 2005) and it is well applicable to architecture and construction (Altun 2011; Chiu and Cheng 2012; Conall Ó Catháin 2009; Lee and Shin

I. Renev (✉)

School of Engineering Science, Lappeenranta University of Technology (LUT), Lappeenranta, Finland

e-mail: Ivan.Renev@student.lut.fi

© The Author(s) 2019

L. Chechurin, M. Collan (eds.), *Advances in Systematic Creativity*,

https://doi.org/10.1007/978-3-319-78075-7_13

2014; Mohamed and AbouRizk 2005; Mohamed and AbouRizk 2003; Renev and Chechurin 2016; Teplitskiy 2005a, b; Lin 2005). In our digital century, it is reasonable to link modern construction design software with ideas generation techniques in order to enhance and automate design creativity. Nowadays, Building Information Modeling (BIM) is a popular stream in construction design. Existing BIM software have a range of instruments enabling designers to bring all their knowledge and experience into projects, however, the software do not support users in searching for nontrivial conceptual ideas for design. That is why the ideas generation stage is still a separate, not automated, and human-dependent part of construction design.

2   Problem Definition

It is without a doubt that the conceptual design stage is one of the most important parts in a complex decision-making process in Architecture–Engineering–Construction (AEC) design at early stages. However, there is no professional software for automating the decision-making process at early design phases, including for searching for not only optimal and reliable solutions but also for inventive ones. Thus, BIM and graphical programming for design are state-of-the art in modern construction design, and Computer-Aided Invention (CAI) software is becoming more popular in different disciplines. Merging this with existing inventive techniques could add artificial intelligence to AEC design software and enhance and automate design creativity in the conceptual design stage.

3   State of the Art

Linking engineering design software with ideas generation techniques and development of CAI systems is still a new research topic. Ikovenko (2004) mentions that merging TRIZ with other methods gave birth to several integrated methodologies based on TRIZ and it opened new horizons for CAI development to cover all the parts of those methods, both analytical and concept generating. In addition, Bakker et al. (2011)

explain a link that is missing between CAI and Computer-Aided Design (CAD) software. Furthermore, the authors proposed integration of CAI and CAD software. León-Rovira (2001) suggested integrating TRIZ and CAD in order to increase design effectiveness and productivity. The review of existing literature in the field of architecture and construction showed that new technological advancements in AEC design have brought the "level of automation" as a pivotal factor in the success of projects. The article by Abrishami et al. (2014) shows that extant literature has identified a significant knowledge gap concerning the key impact links and support mechanisms needed to overtly exploit computational design methods, especially BIM, throughout the conceptual design stage. Moreover, most of the respondents studied in the paper highlighted several deficiencies in the existing tools, whilst they asserted that such a purposeful BIM interface can offer comprehensive support for automation of the entire AEC design and implementation phases, and particularly enhance the decision-making process at early design phases.

4 Building Information Modeling (BIM) and Leading Design Software

In current construction design practice, BIM is a popular process. The US National Building Information Model Standard Project Committee provides the following definition of BIM: "Building Information Modeling (BIM) is a digital representation of physical and functional characteristics of a facility. A BIM is a shared knowledge resource for information about a facility forming a reliable basis for decisions during its life-cycle; defined as existing from earliest conception to demolition" (NIMBS Committee 2007). Despite all BIM's advantages, compared to 2D CAD design, it has no solutions for helping designers automate the conceptual design stage. According to business software review site G2 Crowd, (2015), Autodesk Revit® was determined as a leader among the best BIM software products by customer satisfaction (based on user reviews) and scale (based on market share, vendor size, and social impact). Autodesk Revit® is a BIM software for architects, structural engineers, MEP engineers, designers, and contractors. It allows users to design buildings and structures and their components in 3D, annotate the model with 2D drafting elements, and access building information from the building model's database. Autodesk Revit® is 4D BIM capable with tools to plan and track various stages in the building's lifecycle, from concept to construction and later demolition. Based on the preceding, we selected Autodesk Revit® as the basic and most promising software for realization of a proposal for conceptual design stage automation in AEC projects. Moreover, this is the only software that has a built-in open-source graphical programming tool for design that extends BIM with the data and logic environment of a graphical algorithm editor and enables users to significantly expand functionality of the software without having special knowledge of programming. The tool is called Dynamo (Open source graphical programming for design n.d.).

5   TRIZ

Genrich Altshuller and his colleagues in the former USSR developed the TRIZ method between 1946 and 1985. TRIZ includes a practical methodology, tool sets, a knowledge base, and model-based technology for generating innovative solutions for problem solving. The approach includes a number of tools. Some of the most used are the Ideal Final Result and Ideality; Functional Modeling, Analysis, and Trimming; the 40 Inventive Principles of Problem Solving; Laws of Technical Systems Evolution and Technology Forecasting; and 76 Standard Solutions. The method is universal and finds its application in different fields. During the last decade, there were a number of attempts to apply TRIZ to non-technical areas such as business, service, art, and so on.

6   Proposal of a Design Algorithm

In this work, we suggest using computer-aided automation of searching for new ideas and nontrivial solutions during conceptual design in construction in order to simplify and automate complex decision-making processes in early design stages. We propose three steps of the automation process in terms of inventiveness: elementary (SoA1), medium (SoA2), and advanced (SoA3). Step 1 is not supposed to use any problem-solving techniques; hence, it is more about looking for optimal solutions but still has significant meaning for design process. Here, graphical programming for design (Dynamo) and extended built-in databases of optimal and well-tested design solutions are promising tools to be used. In Step 2, software users receive more inventive prompts in a semi-automated way. Here, I suggest implanting the TRIZ Contradiction Matrix into the software. In this case, users formulate technical contradictions and automatically gain inventive principles and examples on how those contradictions can be solved. Further and final design solutions are up to engineers. In Step 3, software is supposed to self-analyze BIM and suggest inventive solutions in order to improve the system. For that purpose, TRIZ Functional Modeling and Trimming can be used. SoA1 is optimization (optimal solution), and SoA2 and SoA3 are inventions (inventive solutions). The suggested design process scheme is shown in Fig. 1.

6.1   Steps of the Automation Process

Step 1 (SoA1)

For instance, a simple search for an optimal shape of a building by changing only its initial parameters (size, high, number of floors, etc.) instead of performing a completely new design with new parameters would allow designers to significantly reduce time when project schedules are tight. The architecture profession is not known for being quick to change, but as the design process goes digital and clients demand more value from their projects, graphical programming in this case could notably automate the conceptual design process. Once a graphical program for a building or a structure is created, a designer can "play" with its initial parameters in order to obtain optimal ones.

Graphical programming is a new approach in AEC design and it brings completely new possibilities for designers and customers and allows for significant automation in the early design stage. We propose to expand this approach with built-in databases of optimal and well-tested design

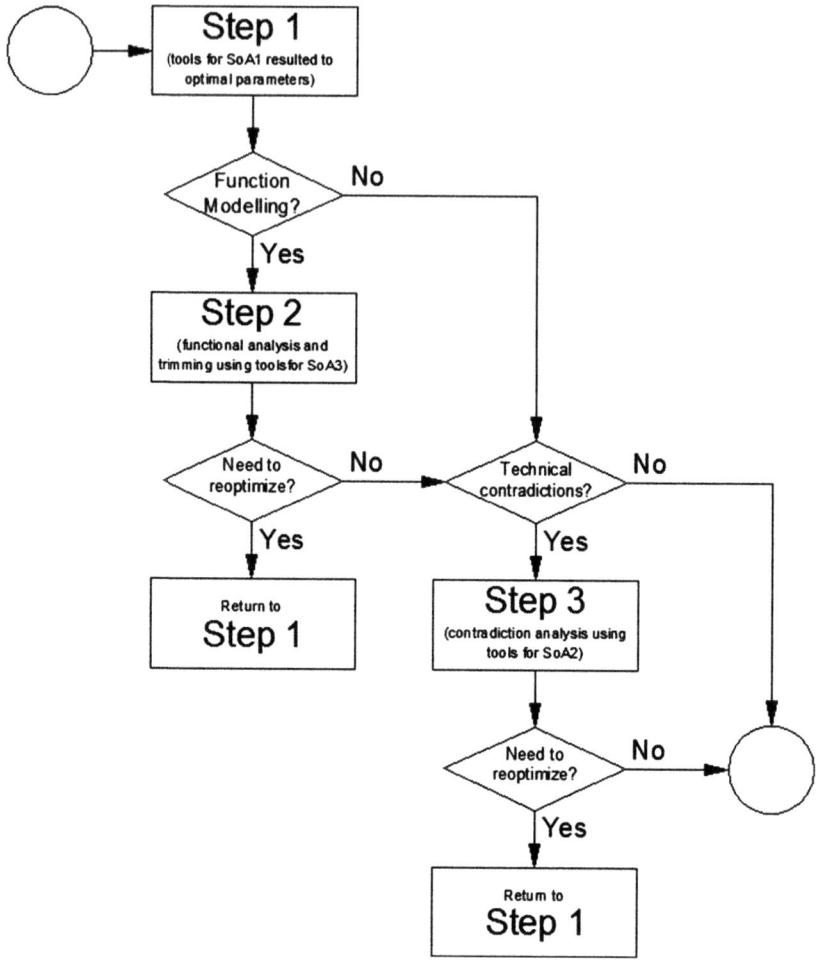

Fig. 1   Block diagram of the analysis process

solutions. In this case, software will also assists users by giving them technical prompts during design. For instance, it will automatically suggest an optimal cross-section of a structure depending on its structural parameters, material, boundary conditions, and so on. Thus, in case of, for example, floor slab design, a program suggests a range of different floor slab structures (concrete, steel, standing on columns or suspended, etc.). All that would support engineers during design, shorten the conceptual design stage, and accelerate the decision-making process.

Step 2 (SoA2)

When optimal parameters are obtained and there are technical contradictions to be solved in the system, users go to the next step, which is the contradiction analysis. Here, we suggest using the TRIZ Contradiction Matrix with 40 Inventive Principles. By clicking on an element, users go to the "contradiction matrix" menu and formulate technical contradictions for the chosen part of the design.

After that, designers use a number of principles in order to solve those contradictions. For instance, for horizontal bearing structures, a quite common contradiction is between their strength and weight. As a rule, strength of a structure is increased by increasing dimentions of its cross-section, which inevitably leads to use of extra materials and, hence, the heavier structure. In this case, a sufficient inventive principle is "Composite Materials"—change from uniform to composite (multiple) materials. Indeed, using composite materials in construction (reinforced concrete, bi-steel beams and trusses, etc.) is a quite common approach that helps to improve strength without affecting weight of structural elements.

However, it is up to engineers' decision and experience to either accept or decline suggested inventive principles in order to improve the design. If new, inventive conceptual solutions were gained during such analysis, designers can go back to the first step for re-optimization.

Step 3 (SoA3)

The final step in our approach is Function Modeling, Analysis, and Trimming of a BIM. An important part here is building a hierarchy of system elements and identifying the functions they perform. As all system elements and their functions are identified by the BIM software, ranking is performed and the best candidates for possible trimming are automatically suggested to a designer who either accepts or declines those proposals. A similar procedure can be done on a subsystem level when every single separate element of a system and subsystems can be analyzed. During such analysis, for instance, different materials, types of cross-sections, and so on of an element are automatically tested.

7   Common Description of the Functional Analysis

Functional Analysis is the most complex part in the conceptual design process according to our methodology, so we will describe it in more detail. In our approach, software is supposed to self-analyze BIM and suggest solutions in order to improve the system. For that purpose, we suggest using TRIZ Functional Modeling and rules of Trimming. In order to teach software to extract a function model from a BIM model, we have built a special graphical program or script with the help of Dynamo. The script first automatically detects elements/components in the BIM model and places them into an interaction matrix. The interaction matrix defines those elements that interact with each other and shows all interaction between those elements of the system. In the next step, the software defines functions of elements in the interaction matrix. In building structures, these functions are usually "holds." However, such functions as "bends," "expands," "compresses," "twists," and so on may take place. For identification of interactions, the script also applies special rules. Next, a function model diagram is automatically generated from the interaction matrix. This diagram shows a hierarchical structure of the components and the functions between them. Such function analysis helps to eliminate mental and thinking inertia since attention of designers is put on elements and functions. In addition, the software helps to achieve a more complete and convenient workflow for design engineers as all is done within the BIM environment. Finally, having this overview is a prerequisite for performing the other TRIZ tools, such as Function Ranking and Trimming. Trimming is a method from TRIZ used to reduce the amount of system components without losing system functionality. The method is based on transferring functions performed by a component that should be trimmed to another component. The software uses rules of Trimming for finding other components to perform this functionality; the component no longer performing any functions can be removed (trimmed) from the function model without losing any functionality.

As a result, the software highlights the best candidates for trimming in the BIM model and the design engineer can either accept or decline those proposals. In Function Ranking, functions of elements are ranked according to their level of usefulness. In order to perform ranking, we first have to identify a "target function" (for instance, "carry live load"). The higher rank belongs to the functions that are closer to the target function. Therefore, the software chooses the furthest from the target functions as the candidates for trimming. There are also so-called harmful functions like "bends" or "twists" since bent or twisted elements require more materials in order to stay stable rather than tensioned ones and it may be wise to eliminate such functions if there is a demand for a cheaper but equally stable design. Function Ranking offers the user a quick overview on the structure of a system and on the importance of distribution of functions. Such analysis during conceptual design enables engineers to automatically analyze the BIM model and easily obtain nontrivial design, avoiding complex processes of topology optimization and structural analysis, which are issues for further detailed design. As a further case study, we analyze a simple beam structural system.

8   Function Analysis of a BIM Model

Detailed logic of the step-by-step Function Analysis of BIM models with full description of its technical realization is provided in our previous work (Renev et al. 2017). However, we provide a brief description of its main pillars in the section that follows.

8.1   Interaction Matrix

Building the interaction matrix is the first step in the Functional Analysis. Here, the position of elements in space, their geometric characteristics, and functions are not taken into account. The matrix of interactions represents the model as a system of individual elements interacting with each other in order to perform the functions assigned to them. Identifying the presence of interaction between the elements of the system is the goal of this type of analysis. The aim of this part of the study is to automatically construct the interaction matrix based on the result of the BIM model analysis. In order to do this, the following tasks must be performed:

1. Analyze the BIM model for presence of physical interaction between its components;
2. Output the result as the final interaction matrix in a user-friendly way.

8.2   Defining Functions of Elements

The second step in the Functional Analysis is the construction of a matrix of functions of elements. When constructing the interaction matrix, we did not take into account the geometric characteristics of elements, their position in space. The goal of constructing the interaction matrix is to reveal the fact of physical interaction of the elements. The next step in the analysis is definition of functions of interacting elements relative to each other. Thus, we gradually go deeper into the analysis of the model, moving from general to particular.

8.3   Function Diagram

Construction of a function diagram is the third step in the Functional Analysis. The functional diagram displays the 3D model in two-dimensional form where each element is presented in the form of a block with the name of the element. Let us call it "block of the element." The interaction between them is displayed in the form of an arrow. Hereinafter "arrow of interaction." Presence of an arrow between the blocks will indicate the presence of interaction between the elements and direction of the arrow indicates direction of the action. The nature of interaction (functions), as a rule, is written above the arrows. Construction of the functional diagram is especially important for the analysis of complex systems with a large number of elements and functions.

Construction of the functional diagram is convenient for monitoring unwanted functions and state of the model after trimming when the function analysis is repeated.

8.4   Ranking

Ranking is an analysis that precedes the main objective of Functional Analysis—trimming. It implies a discrete examination of elements of the model under a number of criteria according to which the elements are assigned a rank. The higher the rank of the element, the higher its significance in the model and the higher the chance of its remaining in the model after trimming.

According to different ranking methods, the criteria for evaluating the model have different scales of evaluation: alphabetic, numerical, and so on. The numbers obtained as a result of the ranking for the element are summed up; the letters are added to the number and also determine the significance level of the element.

8.5   Formation of the First Ranking Rule

Closeness to a Target Function

Work of the elements in the framed building system is, as a rule, reduced to one final goal. For example, the work of the beams is reduced to keeping the slab plate in place. In turn, these beams are supported by columns. From the preceding, we conclude that the work of beams and columns of the first floor is reduced to providing a stable position in the space of the first-floor slab. The floor slab is needed in order to carry a live load, which includes the weight of people, equipment, and so on. Thus, it can be said that the slab is the key element necessary to achieve the ultimate goal of design and operation.

We have combined all the elements into three main groups according to the degree of closeness to a target function:

1. Elements of category "A" – high importance of elements;
2. Elements of category "B" – the average significance of elements;
3. Elements of category "C" – low significance.

In the framed systems, the following elements, the function of which is targeted, have been identified and cannot be trimmed:

1. Foundations
2. Floor slabs
3. Roof slab

The listed elements belong to the first group and are marked with the letter "A."

Elements of the second group are elements whose work is aimed at helping to achieve a target function. These elements are marked with the letter "B." These elements also have high significance in the model. Such elements can be identified through the following criteria:

1. The element interacts with the element that performs the target function;
2. The Z coordinate of the bottom point of the element is below the coordinate of the lowest point of the element that performs the target function.

Elements that do not interact with the "target elements" or have the Z coordinate higher than the Z coordinate of the target element refer to the third group and are marked with the letter "C."

8.6   Formation of the Second Ranking Rule

Harmful Functions

An important factor in the work of a structure is the function of its individual elements. Along with target functions, there are so-called harmful functions caused by certain factors. For example, the rigid connection of two columns provokes the appearance of compression and bending forces, and the hinged one—compression. The bending force will be harmful. The presence of harmful functions will also cause a decrease in the rank of the element. The following types of functions are assigned a certain number of points:

1. Compress – (−1);
2. Bend   (−2);
3. Torque – (−3)

8.7   Formation of the Third Ranking Rule

Useful Functions

The third rule is opposite to the second rule and takes into account only positive functions, which include:

1. Stretch – 1;
2. Hold – 2.

The rank of the elements is the sum of the scores based on the results of evaluating the elements on three grounds. The rank of the element is represented as follows:

$$R = (-X)N(+Y) \tag{1}$$

Where, X is the total score of harmful functions performed by the element, N is the letter designation of the group, and Y is the total score of positive functions performed by the element.

8.8   Trimming

Trimming is the main goal of Functional Analysis. At this stage, non-functional elements are "cut off" and the useful functions of the elements are transferred to other elements of the model.

The rules of trimming are as follows:

1. Removal of elements occurs only if they fall into category "C"
2. Elements with harmful functions are highlighted in the model

9   A Case Study

In order to test the algorithm, we developed a program using the visual programming tool Dynamo realized in the Autodesk Revit® software environment. After a user creates a BIM of a framed building, this program performs the following algorithm of actions: construction of the intersection matrix, identification of functions of the model's elements, construction of the functional model, and ranking of elements and trimming. In this research, we obtained a result that allows designers to automatically analyze and exclude non-functional elements from the model and further propose a solution to prevent unfavorable functions of its elements. The key advantage is that this analysis is being done in the early design stage before deep structural analysis, which is time- and cost-consuming. We provide a description the software prototype based on a frame building design case study.

9.1   Experimental Model Descriptions

In order to carry out the experiment, a two-story framed structure (Fig. 2) was constructed in Autodesk Revit® software. It includes eight columns, four foundations, one slab, and twelve beams. The choice of this composition of structures is justified by better visibility for representing the results. Despite the fact that the selected model does not include a large number of elements, it includes all the main categories of elements that are widely used in framed structural systems.

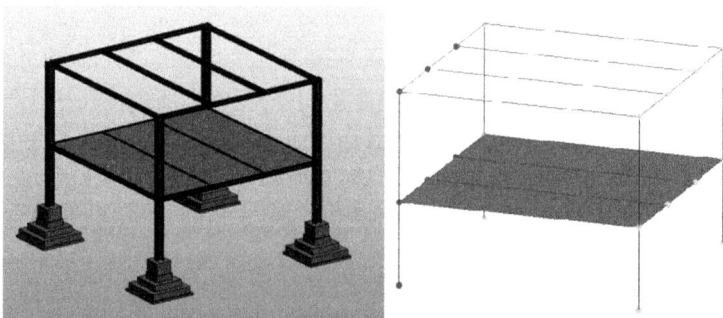

Fig. 2 Experimental model – physical (left). Experimental model – analytical (right)

Geometrical characteristics (cross-section of beams and columns, thickness and spatial dimensions of the slab and foundations) are not taken into account and do not influence the results of the experiment. We consider functions that are performed by the elements. In order to present this idea as clearly as possible, we decided to analyze the analytical model where each element is represented as a primitive. The nodes of connection of individual elements are represented in simplified form. Yellow nodes are hinged connections and violet nodes are rigid ones.

9.2 Interaction and Function Matrixes Construction

The script analyzes the elements of the model and creates a matrix. The interaction matrix for the analyzed model is shown in Fig. 3.

	1. Foundation	2. Foundation	3. Foundation	4. Foundation	5. Column	6. Column	7. Column	8. Column	9. Beam	10. Beam	11. Beam	12. Beam	13. Beam	14. Beam	15. Slab	16. Column	17. Column	18. Column	19. Column	20. Beam	21. Beam	22. Beam	23. Beam	24. Beam	25. Beam
1. Foundation					+																				
2. Foundation						+																			
3. Foundation							+																		
4. Foundation								+																	
5. Column	+								+	+					+	+									
6. Column		+							+	+					+		+								
7. Column			+								+	+			+			+							
8. Column				+								+		+	+				+						
9. Beam						+		+		+		+	+	+	+	+			+						
10. Beam						+	+		+		+				+	+	+								
11. Beam						+	+		+			+	+	+	+		+	+							
12. Beam							+	+	+		+				+	+									
13. Beam									+		+				+										
14. Beam									+		+				+										
15. Slab					+	+	+	+	+	+	+	+	+	+		+	+	+	+						
16. Column					+				+	+					+						+	+			
17. Column						+			+	+					+						+	+			
18. Column							+				+	+			+							+	+		
19. Column								+				+		+								+	+		
20. Beam																+			+		+		+	+	+
21. Beam																+	+			+		+		+	
22. Beam																+	+			+			+	+	+
23. Beam																		+	+	+		+			
24. Beam																				+		+			
25. Beam																				+		+			

Fig. 3 Interaction matrix of the experimental model

The script analyzes the data received from the interaction matrix and generates a matrix of functions according to the established rules. The matrix is shown in Fig. 4.

The influencing elements in the model are in the first line of the matrix, while the exposed ones are in the first column. Legend is used in the function matrix describing actions of influencing elements: C – compress; B – bend; B+T – bend and torque; B+C – bend and compress.

9.3   Functional Diagram Creation

According to the received data, the functional diagram (Fig. 5) is constructed in the drafting view in the software. The blocks of elements are created as Autodesk Revit® families. Arrows and notes are created using lines and text.

Influencing element (columns) / Exposed element (rows)

Exposed element	1. Foundation	2. Foundation	3. Foundation	4. Foundation	5. Columns	6. Columns	7. Columns	8. Columns	9. Beam	10. Beam	11. Beam	12. Beam	13. Beam	14. Beam	15. Slab	16. Columns	17. Columns	18. Columns	19. Columns	20. Beam	21. Beam	22. Beam	23. Beam	24. Beam	25. Beam
1. Foundation																									
2. Foundation																									
3. Foundation																									
4. Foundation																									
5. Columns	C																								
6. Columns		C																							
7. Columns			C																						
8. Columns				B+C																					
9. Beam					C			B+C		B		B	B	B											
10. Beam					C	C			B		B														
11. Beam						C	C			B		B	B	B											
12. Beam							C	B+C	B		B														
13. Beam									B+T		B														
14. Beam									B+T		B														
15. Slab									C	C	C	C	C	C											
16. Columns					C																				
17. Columns						C																			
18. Columns							C																		
19. Columns								B+C																	
20. Beam																C			B+C		B		B	B	B
21. Beam																C	C			B		B			
22. Beam																	C	C			B		B	B	B
23. Beam																		C	B+C	B		B			
24. Beam																				B+T		B			
25. Beam																				B+T					

Fig. 4 Function matrix of the experimental model

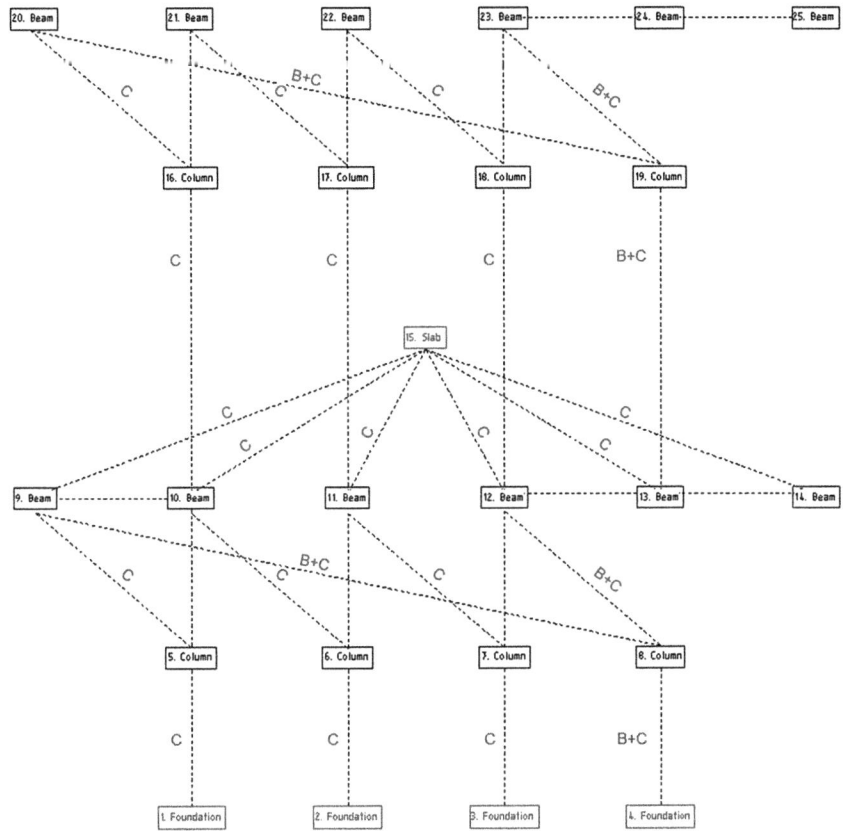

Fig. 5 Functional diagram of the experimental model

9.4   Ranking

According to ranking rules, the elements of the model are assigned a ranking factor and a rank. We have introduced the following rules: (1) Elements fall into three categories: "A"—high importance of elements, "B"—average significance of elements, and "C"—low significance; (2) So-called harmful functions are assigned a certain number of negative points: Compress—(−1), Bend—(−2), and Torque—(−3); and (3) So-called positive functions are assigned a certain number of positive points: Stretch—1 and Hold—2 (Table 1).

Table 1 Ranking results

Element's name	Closeness to the target function	Presence of negative functions	Presence of positive functions	Ranking factor	Rank
1 Foundation	A	–	–	A	1
2 Foundation	A	–	–	A	1
3 Foundation	A	–	–	A	1
4 Foundation	A	–	–	A	1
5 Column	B	−1	–	−1B	3
6 Column	B	−1	–	−1B	3
7 Column	B	−1	–	−1B	3
8 Column	B	−3	–	−3B	4
9 Beam	B	−12	–	−12B	9
10 Beam	B	−6	–	−6B	5
11 Beam	B	−10	–	−10B	8
12 Beam	B	−8	–	−8B	7
13 Beam	B	−7	–	−7B	6
14 Beam	B	−7	–	−7B	6
15 Slab	A	−6	–	−6A	2
16 Column	C	−1	–	−1C	10
17 Column	C	−1	–	−1C	10
18 Column	C	−1	–	−1C	10
19 Column	C	−3	–	−3C	11
20 Beam	C	−12	–	−12C	17
21 Beam	C	−6	–	−6C	13
22 Beam	C	−10	–	−10C	16
23 Beam	C	−8	–	−8C	15
24 Beam	C	−7	–	−7C	14
25 Beam	C	−5	–	−5C	12

9.5   Trimming

According to results of ranking, the program identifies elements that are subjects for trimming and elements for which design, geometric dimensions, material, and the way of installation need to be considered in more detail (Fig. 6).

Results of trimming:

1. Removal of elements occurs only if they fall into category "C";
2. Elements with harmful functions are highlighted in the model (yellow for elements in the "C" category, red for elements with harmful functions, and grey for elements in category "A" or elements without harmful functions).

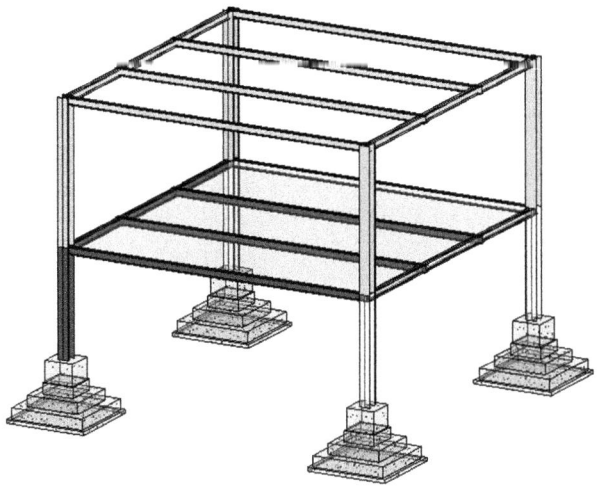

Fig. 6 Highlighting elements after trimming

As described previously, a list is formed and divided into two graphs. Elements in the list are in order of decreasing rank from the less functional elements to the more functional ones. Thus, the higher the element in the list, the higher probability for it to be excluded from the model or change its function (Table 2).

9.6   Result of "Contradiction Analysis" Technique

The analysis is carried out by selecting an element and choosing both improving and worsening parameters in the windows form. Once this is done, a result often consisting of several tips for resolving technical contradictions for the category that the element belongs to appears.

9.7   Case Study Conclusion

The presented case study shows that TRIZ tools such as Function Analysis and Trimming can be applied in framed structural systems design. Modern design software allows engineers to perform such analysis directly in the BIM environment. However, the need for some stages of functional analysis realization in construction software can be discussed. For

Table 2 Trimming report

Elements for trimming	Elements requiring attention
20. Beam (17)	9. Beam (9)
22. Beam (16)	11. Beam (8)
23. Beam (15)	12. Beam (7)
24. Beam (14)	13. Beam (6)
21. Beam (13)	14. Beam (6)
25. Beam (12)	10. Beam (5)
19. Column (11)	8. Column (4)
16. Column (10)	5. Column (3)
17. Column (10)	6. Column (3)
18. Column (10)	7. Column (3)

instance, the interaction matrix can be excluded from the functional analysis since it is positioned as a sub-step to construction of the functional matrix. Furthermore, the same algorithm of actions was repeated for both stages, so the stage of constructing the interaction matrix can be embedded into the algorithm of functional matrix construction. The same can be said about construction of a functional diagram. The functional diagram is quite clear and easy to use for manual analysis of small structures; however, with software implementation on large projects, the need for its construction can be discussed. The program accurately excludes the possibility of errors during analysis of the model and the interaction matrix and functional diagram can be created and shown optionally.

With regard to further research, the development of ranking rules for a wide range of structural schemes and assessment of additional groups of elements would make functional analysis in the construction field more accurate and reliable.

10   Conclusion and Discussions

In this research, we obtained a result that allows designers to automatically analyze and exclude non-functional elements from the model and propose a solution to prevent unfavorable functions of its elements. The key advantage is that this analysis is being done in the early design stage before deep structural analysis, which is time- and cost-consuming. After the Function Analysis is completed, we propose analyzing the trimmed model using another TRIZ tool called Contradiction Analysis. This part of the analysis is the final part of the conceptual design phase. Contradiction Analysis includes 40 techniques to eliminate technical contradictions. A technical contradiction in TRIZ is a situation where an attempt to improve one characteristic of a technical system causes worsening of another. For more effective organization of use of techniques, a special table has been developed. The table includes the characteristics of technical systems that need to be improved and those characteristics that are worsened. At the intersection of the table graphs, the numbers of solutions are indicated, which help to eliminate the arisen technical contradiction. For the construction field, the revision of all proposed technical characteristics was carried out in order to identify the methods most suitable for use. Our goal is to implement these tools in the design process. To achieve this goal, we decided to link the possibilities of this tool with categories of the model's elements. The software for this tool should be implemented as follows:

1. Select the model element;
2. Select the worsening parameter;
3. Select the improving parameter; and
4. Obtain a number of solutions to the technical contradiction for the category of selected element.

Perform Steps 2 and 3 using the Windows Form in the software program. Use the Dynamo visual programming tool to obtain information about the element, analyze the input data, and output the result.

For further research, we plan to validate the obtained results by giving the developed design scripts and user instructions to professional designers in order to independently test the algorithm and collect feedback. We are developing a questionnaire for this purpose.

Acknowledgments   The author would like to acknowledge Leonid S. Chechurin (Professor of Operations Management & Systems Engineering, LUT School of Engineering Science) for his supervision and fruitful discussions on the research topic. The author would also like to acknowledge the Finnish National Agency for Education for the positive decision on the financial support of this project (decision TM-17-10703 dated 08.02.2018).

References

Abrishami, S., Goulding, J. S., Rahimian, F. P., & Ganah, A. (2014). Integration of BIM and generative design to exploit AEC conceptual design innovation. *Journal of Information Technology in Construction, 19*(August 2013), 350–359.

Altshuller, G. S. (1999). The innovation algorithm: TRIZ, systematic innovation and technical creativity. *Innovations.* Retrieved from http://www.amazon.com/dp/0964074044

Coşkun K., Altun M. C. (2011). *Applicability of TRIZ to in-situ construction techniques* (p. 17). 2011 2nd International Conference on Construction and Project Management IPEDR vol.15 (2011) © (2011) IACSIT Press, Singapore.

Bakker, H. M., Chechurin, L. S., & Wits, W. W. (2011). Integrating TRIZ function modeling in CAD software. In *Proceedings of TRIZfest-2011* (p. 18). Издательство Политехнического университета. Retrieved from http://doc.utwente.nl/93359/1/paper_HB_LCH_WW_final.pdf

Chiu, R. S., & Cheng, S. T. (2012). The improvement of heat insulation for roof steel plates by triz application. *Journal of Marine Science and Technology, 20*(2), 122–131. Retrieved from http://jmst.ntou.edu.tw/marine/20-2/122-131.pdf

Conall Ó Catháin, Mann, D. (2009). Construction innovation using TRIZ. In *Global innovation in construction conference 2009* (pp. 296–306). Retrieved from http://www.irbnet.de/daten/iconda/CIB15607.pdf

G2 CROWD PUBLISHES WINTER 2015 RANKINGS OF THE BEST BUILDING DESIGN AND BIM SOFTWARE. (2015). Retrieved June 20, 2016, from https://www.g2crowd.com/blog/best-building-design-bim-software

Ikovenko, S. (2004). TRIZ and computer aided inventing. In *Building the information society* (Vol. 156, pp. 475–485). https://doi.org/10.1007/978-1-4020-8157-6_42.

Lee, D., & Shin, S. (2014). Advanced high strength steel tube diagrid using TRIZ and nonlinear pushover analysis. *Journal of Constructional Steel Research, 96*, 151–158. https://doi.org/10.1016/j.jcsr.2014.01.005.

León-Rovira, N. (2001). A proposal to integrate TRIZ and CAD (computer aided TRIZ-based design). TRIZCON2001, Philadelphia: The Third Annual Altshuller Institute for TRIZ Studies Conference.

Lin, Y. H. (2005). Applying TRIZ to the construction industry. In *Proceedings of the 10th international conference on civil, structural and environmental engineering computing*. Stirlingshire: Civil-Comp Press.

Mohamed, Y., & AbouRizk, S. (2003). Triz for structuring construction innovation. In *Canadian society for civil engineering – 31st annual conference: 2003 building our civilization* (pp. 241–249). Moncton. Retrieved from http://www.scopus.com/record/display.url?eid=2-s2.0-33645726845&origin=inward&txGid=99EC6DDA36CBBDAEE28EB03F865BC780.fM4vPBipdL1BpirDq5Cw%3a22

Mohamed, Y., & AbouRizk, S. (2005). Application of the theory of inventive problem solving in tunnel construction. *Journal of Construction Engineering and Management, 131*(10), 1099–1108. Retrieved from http://cedb.asce.org/cgi/WWWdisplay.cgi?148955

NIMBS Committe. (2007). National Building Information Modeling Standard. *Nbim*, 180.

Open source graphical programming for design. (n.d.). Retrieved from http://dynamobim.org

Renev, I. A., & Chechurin, L. S. (2016). Application of TRIZ in building industry: Study of current situation. *Procedia CIRP, 39*, 209–215. https://doi.org/10.1016/j.procir.2016.01.190.

Renev, I., Chechurin, L., & Perlova, E. (2017). Early design stage automation in architecture-engineering-construction (AEC) projects. In *Proceedings of the 35th eCAADe conference* (pp. 373–382). Rome: Sapienza University of Rome.

Salamatov, Y. (2005). *TRIZ: The right solution at the right time: A guide to innovative problem solving* (2nd ed.). Retrieved from http://www.xtriz.com/publications/Chapter2SalamatovBook.pdf

Teplitskiy, A. (2005a). Application of 40 inventive principles in construction. *The TRIZ Journal*, 1–23. Retrieved from http://www.triz-journal.com/application-40-inventive-principles-construction/

Teplitskiy, A. (2005b). Solving technical contradictions in construction with examples. *The TRIZ Journal*. Retrieved from http://www.metodolog.ru/triz-journal/archives/2005/07/04.pdf

Optimized Morphological Analysis in Decision-Making

Shqipe Buzuku and Andrzej Kraslawski

1 Introduction

Creativity has emerged as a driving force to innovation processes and systems' design for the industry of current and future generations. Many assessment methods for innovation are already developed and can be identified in the literature. One of these methods is the Theory of Inventive Problem-Solving, also known by its Russian acronym TRIZ. Developed by Altshuller (1984), it is a remarkable instrument to boost creativity in engineering.

TRIZ focuses on the creative idea-generation process in a company's environment (Chechurin 2016). Another promising method widely used

S. Buzuku (✉)
Lappeenranta University of Technology, Lappeenranta, Finland
e-mail: Shqipe.Buzuku@lut.fi

A. Kraslawski
Lappeenranta University of Technology, Lappeenranta, Finland

Faculty of Process and Environmental Engineering
Technical University of Lodz, Lodz, Poland

© The Author(s) 2019
L. Chechurin, M. Collan (eds.), *Advances in Systematic Creativity*,
https://doi.org/10.1007/978-3-319-78075-7_14

to support systematic creativity is morphological analysis (MA), developed by Zwicky (1969). MA is presented as a promising method for generating new concepts and finding the best solution to support decision-making. Both TRIZ and MA are well-known and commonly used methods for complex problem-solving, including techniques for idea generation and divergent thinking.

MA is a non-quantified modelling method for identifying, structuring, and studying complex problems. It is a suitable tool for addressing highly complex and ill-structured problems, also known as "wicked problems," which are becoming increasingly frequent in a globalized world. The term wicked problems, introduced by Rittel and Webber (1973), is used to describe complex social problems that "*do not have an enumerable set of potential solutions*" and are multi-dimensional and non-quantifiable, with inherent technical, social, institutional, and political difficulties of addressing them (Ney 2012). Such problems are found, for instance, in strategy development, product innovation, policy design, and so on. Hence, MA is used to structure and investigate the behavior of wicked problems and discover solutions (Ritchey 2011).

Moreover, wicked problems may deal with long-term policy design, planning and complexity management, and involve many individuals and organizations with overlapping roles. Tackling wicked problems in decision-making for industrial mega-project management can happen, for example, while seeking optimization of time, cost and quality, and reduction or management of risk. This type of problem is usually formulated with many contradictory requirements, and involves many system's elements from various functional areas. In such cases, the application of traditional multi-objective decision-making approaches has limitations. Hence, this situation allows applying a morphological approach developed by Zwicky (1969) and Ritchey (2011). Moreover, today's industry is constantly under high pressure to generate competitive advantages and improve resource efficiency. For example, time reduction is a significant target in overall industrial project management. To address this problem, several methods and tools exist focused for optimization.

Therefore, in this chapter, we propose further improving MA by integrating it with sensitivity analysis (SA), which has never been done before. SA is an important task in multi-criteria decision-making. Moreover, SA deals with the investigation of potential changes and errors in variables and assumptions, and their impacts on the results of underlying models.

SA, applied post hoc to decision models, deals with uncertainties related to the decision outcomes and/or to the preferential judgements (i.e., value function and criterion weights). The objective is to find out how the options' ranking changes by any modification made on the decision models (Mysiak 2010). Applying SA shows great potential for time reduction over the MA iterative part of the process.

In order to elaborate on the topic, we organize the chapter as follows: Section 2 provides background and literature review of MA and SA. Section 3 explains the proposed solution for optimization of MA through SA. Section 4 tests the theory using a numerical illustration to demonstrate the proposed method and reports results, managerial implications, and recommendations. Section 5 presents the research conclusions, limitations, and future work.

2   Methodological Background

This section explains the overall process of the study by providing a brief explanation of each method used. An innovative approach of applying MA complemented with SA to this problem is explored with the aim of seeking optimal solutions. Moreover, the methodological foundation of the study is based on the iterative nature of the cross-consistency assessment (CCA)—an integral part of MA—as an opportunity for further optimizing the overall process of MA. A decision tree is generated based on the morphological space and CCA is conducted in order to present the benefits of the method.

2.1   Morphological Analysis

Zwicky (1969) introduced MA as a non-quantified modelling method for identifying, structuring, and studying complex problems. In principle, MA is based on the "divide and conquer" technique, which tackles a problem using two basic approaches: "analysis" and "synthesis"—the basic method for developing scientific mode (Ritchey 2018). This technique is a decomposition method that breaks down a system into subsystems with several attributes and selects the most valuable alternative (Yoon and Park 2005). In other words, MA systematically categorizes the possible combinations of subsystems. The strength of the technique lies in its ability to provide structured models for complex problems into simpler problems, rather than offering solutions (Pidd 2009). MA can be considered as a contemporary heuristic method that has clear links with traditional approaches in different fields of problem-solving (Arciszewski 2018). It has been the subject of academic research over many years and it has been applied to a wide variety of fields and contexts such as shown in Table 1.

MA has been widely used for identifying possible variable combinations in different disciplines. It enables problem representations using a number of dimensions, which are permitted to assume the number of conditions (Eriksson and Ritchey 2002). The identified conditions in

Table 1 Literature survey of morphological analysis applications

Application	References
Engineering and Product Design	Jimenez and Mavris (2010), Ostertagová et al. (2012), Ölvander et al. (2009), Sholeh et al. (2018)
Design Theory and Architecture	Chen and Lai (2010), Ritchey (2011), Zeiler (2018), Kannengiesser et al. (2013)
Future Studies and Scenario Development	Ritchey (2011), Lopes Correia da Silva (2011), Duczynski (2017), Johansen (2018), Voros (2018), Haydo (2018)
Technological Forecasting/ Technology Foresight	Yoon (2008), Feng and Fuhai (2012), Takane et al. (2009), Yoon et al. (2014)
Management Science, Policy Analysis and Organizational Design, Strategic Planning	Ajith and Ganesh (2009), Pidd (2009), Plauché et al. (2010), Kuriakose et al. (2010), Storbacka and Nenonen (2012), Im and Cho (2013), Seidenstricker et al. (2014), Buzuku and Kraslawski (2015), Buzuku et al. (2015), Teles and Freire de Sousa (2017, 2018), Duczynski (2018)
Security Safety and Design Studies	Louise et al. (2009)
Creativity, Innovation and Knowledge Management	Seidenstricker and Linder (2014), Frow et al. (2015), Geum et al. (2016), Jeong et al. (2016)
Modeling Theory, OR Methods and GMA	Ritchey and de Waal (2007), Plauché et al. (2010), Ritchey (2012, 2014, 2018), Williams and Bowden (2013)

each dimension can be combined to derive all the possible alternatives that can solve the problem. MA requires a multidisciplinary group of five to seven experienced experts (Yoon and Park 2005), representing different aspects of the issues involved. In particular, MA provides a strong possibility for identifying unexpected combinations of new concepts. The basic procedure of the MA consists of the following stages:

- Problem definition,
- Analysis,
- Synthesis, and
- Exploration of results

Problem Definition

In this stage, the problem definition is to determine suitable problem characteristics. The individual problem-solver or a facilitated group brainstorms to define problem characteristics, also referred to as parameters. In the specific context of the study, the problem is to establish a systematic integrative approach for problem-solving, in such a way that the integrated approach supports complex problem-solving for efficient industrial management.

The specific goal is to utilize the MA and optimization of CCA through SA as modelling methods supporting decision-making in case of wicked problems, such as strategy development, product innovation, or policy design.

Analysis

After problem definition, in the analysis stage, the main task in MA is generating a morphological field, which consists of identifying and specifying the essential parameters or variables of the problem and generating the design alternatives or values for each parameter or component (i.e., a morphological class). At this stage, creative and design thinking is desirable for generating a set of potential solutions as large as possible. Finally, the morphological box (field) is created with a large solution space (see Fig. 1).

Parameter A	Parameter B	Parameter C	Parameter D	Parameter E
Value A1	Value B1	Value C1	Value D1	Value E1
Value A2	Value B2	Value C2	Value D2	Value E2
Value A3	...	Value C3	...	Value E3
Value A4		...		...
...				

Fig. 1 A morphological matrix consisting of five parameters and their ranges of values. The matrix represents 144 (= 4 × 2 × 3 × 2 × 3) formal configurations such as (A2, B1, C2, D2, E1)

Synthesis

The synthesis stage presents the CCA proposed by Ritchey (2015). CCA involves evaluation of the feasibility of all combinations of all values and of all the variables. CCA assesses compatibility for each value via pairwise comparison (the comparison of a value from one column, or morphological class, against another value from another morphological class). The purpose of CCA matrix is to filter out infeasible configurations. The CCA is represented in a multidimensional matrix, which then is reduced to find an approximation for an optimal solution. The process is iterative in nature and time-consuming. Manual configuration evaluations in physical settings are based on a cross-consistency matrix. Therefore, optimization of the iteration process is a relevant thing to do.

Exploration

In this stage, the internally consistent combinations are synthesized, with the objective to reduce the total set of possible combinations. This step is termed assessment of consistent configurations. The CCA is computed to ensure the consistency of pairwise comparisons. This step is based upon the insight that there may be numerous mutually incompatible pairs of conditions in the morphological matrix. Zwicky calls MA "totally research," which enables in an unbiased way attempts to derive all the possible solutions of any given problem.

The main advantage of the MA method is visualization support. With the usage of CCA, MA facilitates the calculation of a large subset of consistent configurations, which is impossible to do manually. It may help to discover new relationship or configurations, which may not be so evident, or which might have been overlooked by other unstructured or less-structured methods. Importantly, it encourages the identification and investigation of boundary conditions, that is, the limits and extremes of different contexts and factors. In addition, the MA method provides various kinds of representations of the consistent configuration space. Such representations help groups to structure discussions and support decision-making.

2.2   Sensitivity Analysis

SA serves a wide range of applications. The SA can be applied in:

(i) Decision-making for identifying critical value/criterion, testing robustness, and riskiness of decision;
(ii) Communication for increasing credibility and confidence; and
(iii) Modelling process, for better understanding of input-output relationship and for understating the model needs and restrictions (Mysiak 2010).

SA deals with the investigation of potential changes and errors associated to the inputs and assumptions, and their impact on the results of underlying models. While the impact of uncertainties on the decision outcomes is mostly analyzed by statistical modelling and simulation, the preferential judgments are objects of uncertainty during the modelling of weights and value function (Mysiak 2010). In normal mathematical modelling and simulation, SA is used to evaluate how the changes of the variables or assumptions affect the results. When analyzing a CCA matrix, all values contained within it (result of the evaluation and weighting of values by a panel of experts) can be considered as inputs and each of these inputs has a specific impact in the solution space (the output) generated by the evaluation of the matrix. The use of SA helps highlighting the combinations with higher impact within the solution space and of a specific value.

3   Proposed Solution

In this section, we propose and present an innovative solution using a case example. One method capable of complementing and optimizing MA is SA. According to Ferretti, Saltelli, and Tarantola (2016), SA has been applied in a very diverse range of fields, typically to calculate and estimate uncertainty of models. However, to the best of our knowledge, it has never before been applied with MA. Furthermore, the proposed use of SA aims to find the most influential combinations of the CCA matrix to reduce the time of iteration. In this study, we integrate MA with SA as a concept with high potential for time reduction.

3.1   Optimization of Morphological Analysis Through Sensitivity Analysis

MA is an approach successfully used for solving the problems where many solutions and alternatives are possible (Zwicky 1969). It consists in, first, analyzing all possible combinations of input variables, and then identifying the most promising solution. The exhaustive list of permutations of the values then forms the complete morphological space (often referred to as a Zwicky box by Ritchey (2018)). However, complications arise with this method because it is not always feasible to manually test all (or even many) of the potential solutions identified within a morphological space (Eriksson and Ritchey 2002). Therefore, Ritchey (2011) proposed the computer implementation of this method. A binary or numerical CCA matrix, generated through a panel of experts' evaluation over the combinatory interactions of the parameters and their value pairs generates a set of solution space (Ritchey 2015). The process is iterative in nature and it should be repeated as many times as required, to achieve a manageable solution space. A manageable solution space should have enough combinations of solutions for decision-makers to have a clear pathway, but not too many that decision-making is complex within the solution space again (Ritchey 2018). SA has been used in a multitude of fields (Ferretti et al. 2016), often aimed at testing the uncertainty levels derived of variable assumptions in mathematical models.

However, the presented approach proposes yet a different use of SA as a concept. The suggested approach proposes the use of SA to evaluate the effect of the change of each value in the matrix in the overall set of combinations of a specific value. In other words, if a pairwise value is changed from value "A" to "B" (zero to one in a binary matrix, or different numerical values in a numerical matrix), it will decrease the overall amount of combinations of value "A" in the solution space. Since every iteration of value revaluation takes time to the panel of experts, knowing which cells most affect the overall number of combinations of a certain value can save a significant amount of time for revaluating a large CCA matrix.

4   Numerical Illustration

To test the theory, a CCA matrix with values from one to three and dimensions as shown in Fig. 2 was created in a loop of randomly generated values for one thousand iterations in MatLab, and the relative amount of combinations in each iteration is calculated according to Eq. 1:

$$RWx_{(a,b)} = \frac{Kx_{(a,b)}}{\sum_{i=1}^{R} \sum_{j=1}^{C} Kx_{(i,j)}} \tag{1}$$

Equation 1 shows the method used to calculate the relative weight (RW) of an element pair, in respect to the total combinations "K," which are unique combinations of value "x," for the position (a, b) (where "a" is the row and "b" is the column in the CCA matrix) calculated as indicated by Ritchey (2010).

	P2,V1	P2,V2	P2,V3	P2,V4	P2,V5	P3,V1	P3,V2	P3,V3	P3,V4	P3,V5	P4,V1	P4,V2	P4,V3	P5,V1	P5,V2	P5,V3	P5,V4	P5,V5	P5,V6
P1,V1	3	1	2	3	1	3	1	1	2	3	1	3	3	3	2	3	3	3	3
P1,V2	1	3	1	2	1	1	3	3	2	1	1	1	3	1	3	3	1	1	3
P1,V3	3	3	3	2	1	3	3	3	1	1	3	3	3	1	3	3	1	1	3
P1,V4	1	2	3	3	2	1	2	3	3	1	1	3	3	3	2	3	3	3	3
P1,V5	1	2	1	1	3	3	1	1	3	3	1	2	2	1	3	1	1	1	1
P1,V6	1	2	1	1	3	1	1	1	3	3	1	2	2	3	1	1	1	1	3
P2,V1						3	1	1	2	2	1	1	3	3	2	3	1	3	3
P2,V2						1	3	3	2	2	1	2	3	2	2	3	1	1	3
P2,V3						3	1	1	1	3	3	3	3	1	3	1	1	1	1
P2,V4						2	2	3	1	2	2	2	3	1	3	1	1	2	3
P2,V5						1	1	1	3	3	2	3	3	3	1	1	3	3	3
P3,V1											3	3	3	3	3	3	1	1	1
P3,V2											3	2	1	1	3	3	1	1	1
P3,V3											1	2	3	1	1	3	1	1	1
P3,V4											1	3	3	3	2	2	2	2	3
P3,V5											1	3	3	3	2	2	2	2	3
P4,V1														2	1	2	3	2	1
P4,V2														2	2	2	3	3	2
P4,V3														3	3	3	2	2	2

Fig. 2 Example of a cross-consistency assessment (CCA) matrix revaluation layout highlighting the cells of value "3" and marking in red the cells generating 75% of the total combinations of value "3"

"R" denotes the total of rows; "C" denotes the total of columns, and "i" and "j" are the counters to calculate the total of every cell.

Figures 3 and 4 show the premise behind the CCA-plus-SA concept. The value pair P1V3 and P2V1 generates a set of 27 unique combinations of value "3" (Fig. 3), while the total unique combinations of value "3" for the row of P1V3 is 81. In contrast, the value pair P1V2 and P2V2 (Fig. 4) produces only six unique combinations of value "3." When all value pairs are evaluated, 993 unique combinations of value "3" are produced from the entire matrix. Therefore, the cell P1V3, P2V1 hosts 2.7% of the total unique combinations of value "3," while the cell P1V2, P2V2 generates only 0.6%. Figure 2 shows an example of the benefits of conducting SA into the CCA matrix. The picture highlights all cells with value "3" (a total of 108). Considering a scenario where it is desirable to reduce the overall amount of combinations of value "3," the panel of experts in charge of evaluating the iteration may intuitively review only the combinations of value "3" (108 highlighted in Fig. 2, both red and yellow) saving about 56% of time required to evaluate the whole CCA matrix. Furthermore, after applying SA, it is possible, in the example case, to focus on the cells with a higher amount of overall unique combinations of value "3." Highlighted in red are the cells of value "3" that host more than 1% of the overall unique combinations of value "3." By revaluating only the cells highlighted in red, it is necessary to iterate only 31

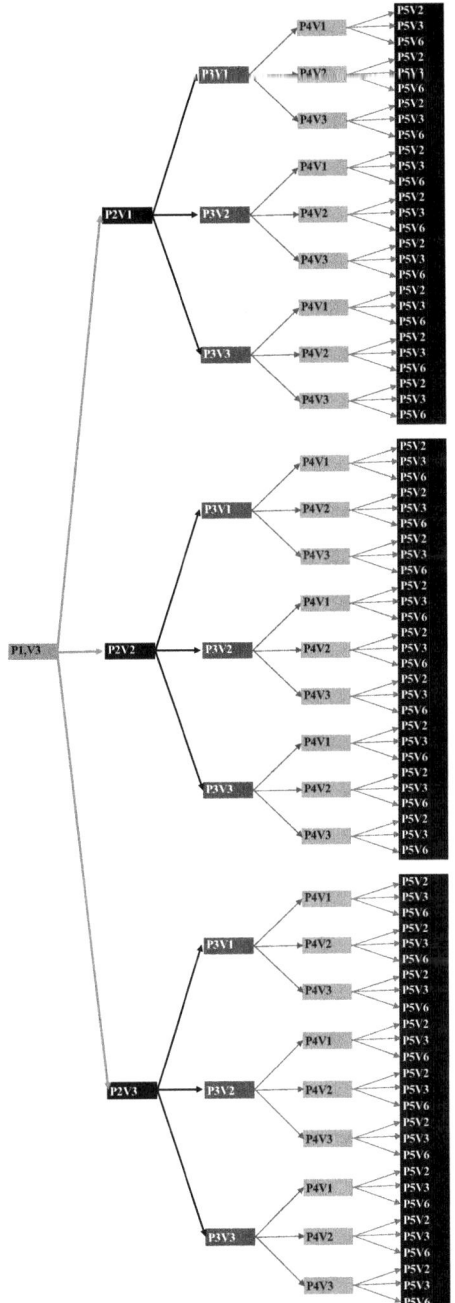

Fig. 3 Example of a large morphological decision tree for a value pair of high amount of unique combinations of value "3"

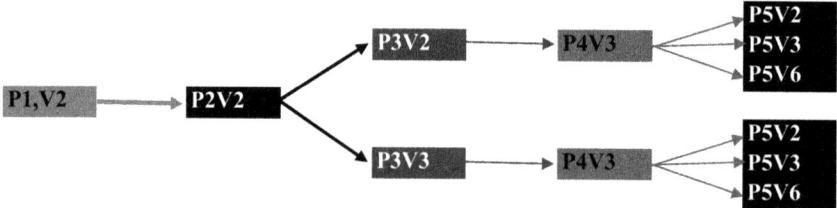

Fig. 4 Example of a small morphological decision tree for a value pair of low amount of unique combinations of value "3"

(highlighted in red) out of the 108 cells of value pairs marked with a "3." In this manner (assuming that the revaluation of each value takes the same amount of time), it is possible to save, in the example, 71% of the time of iteration when only the cells of value "3" are evaluated, or 84% of the time of re-evaluating all the 247 values in the CCA matrix. Furthermore, by evaluating the value of only the cells highlighted in red, it is possible to reduce up to 75% of the total amount of unique combinations of value "3" from the example given.

4.1   Results

Using a multi-paradigm numerical computing software, we created a loop of one thousand iterations to generate and evaluate through SA a CCA matrix of random values from "1" to "3" and of the dimensions shown in Fig. 2. In addition, a sensitivity limit of 1% was set, meaning that every cell of the CCA matrix containing more than 1% of the total amount of combinations of a certain value is then highlighted as target for revaluation when needed. From the analysis of the results, it was found that over the loop of one thousand iterations, it is possible to reduce the iteration time an average of 63.1% when only the combinations of a certain value are evaluated, with a maximum of 88.8% and minimum of 24.7%. When compared to the evaluation of all cells in the matrix, the average time saving was 87.8% with a maximum of 96.4% and a minimum of 77.7%. Furthermore, the relative amount of combinations contained within the SA highlighted cells averaged 73.9%, while the maximum registered was 91.8% and the minimum 53.1%.

In other words, over the one thousand-loop iterations test, an average of 73.9% of the combinations of any value from "1" to "3" can be reduced by evaluating only 36.9% of the cells of the CCA matrix with a specific value, or 12.2% of all the cells of the whole CCA matrix, hence saving 63.1% and 87.8% of the time respectively. The considered timesaving assumes an equal time required to evaluate every cell in the CCA matrix. The concept of optimization of MA through SA, as presented in this section, holds a great potential for time reduction of the MA iterative part of the process.

4.2   Managerial Implications

The results of this research reveal important implications for decision-makers and managers. Several managerial suggestions are from results analysis. Decision-makers are able to define which combinations within their industrial case require more attention and which combinations will be out of the box with less significance.

The proposed approach of integrating MA with SA as a concept with high potential for time reduction can effectively and efficiently be used for managing industrial complexity. The major contribution of the research is to facilitate the decision-making process in industry. The optimization of a CCA-based model helps to facilitate decision-makers and managers to determine the relative importance of the combination sets in the CCA.

The results of the method approach could encourage managers to implement the integrated approach of MA with SA with high effect on time reduction over the iteration process. In addition, SA results help managers determine the influence of the experts' opinion by evaluating the inputs against their impact.

4.3   Recommendations

The sensitivity limit should be chosen accordingly to the size of the CCA matrix. For a CCA matrix of, for example, 80 values, a sensitivity limit of 1% would be too low, and result in the majority of cells with the desired value being highlighted (and required to iterate), thus mostly neglecting the advantage of SA application. Alternatively, a 1% sensitivity limit into a CCA matrix of, for example, 500 values may exclude too many cells, therefore reducing the reach of combinations reduction by applying SA. As a general rule of thumb, the authors recommend a sensitivity limit of ~ $(1/n)$ ×2, where "n" stands for the number of cells of the CCA matrix. In the example case, $(1/247)$ ×2 = 0.8% ~ 1%. SA is advised to be used in order to reduce the total amount of combinations of the desired value only for a few iterations, but not to the length of generating the final solution. After the solution space has been significantly reduced, it is recommended that the experts evaluate a slightly extended pool of solutions. This is under the reasoning that the most optimal solution is not necessarily contained within a segregated group of optimal values, but may sometimes include compromises in some combinations.

5   Conclusions and Future Work

MA is a widely used method for decision-making in planning and management, and it is often complemented with other methods to tackle its limitations. Nevertheless, the authors are unaware of research in the existing literature on optimization of MA and CCA through SA and focus on time reduction in management of complex projects.

This chapter proposed a model for optimization of MA and CCA through SA to support decision-making in project management. The integration of MA and SA, under the test conditions presented, has shown promising potential for time reduction. MA optimization was achieved by reducing iteration time. The two morphology constituents—dimensions and values—show different characteristics. The construction of dimensions requires a significant level of expert judgments compared to that of a value, since the combination of dimensions should thoroughly reflect type of interaction. To succeed in developing new innovative concepts, dimensions should be mutually exclusive and collectively exhaustive. Furthermore, the iterative process of MA and CCA receives a significant improvement using SA.

SA as an optimization tool can be performed with any set of values, in every iteration, and target any specific value ("1," "2," or "3" from the example, but not limited to these). It is a powerful and adaptable tool capable of obtaining desirable solutions in a fraction of the time required otherwise. The analysis can also be automated in a spreadsheet.

In addition, the sensitivity limits can be tailored to adapt the method to CCA matrixes of any dimensions, adding another layer of flexibility. This work has some identified limitations. It is worth mentioning that the test conditions are randomly generated values and the time reduction is considering equal time requirements for evaluation of every value. Extensive application in real-world cases may show different distributions. Nevertheless, the integration of the methods should, in any case and to some extent, optimize the iteration and revaluation process with significant time reductions. Moreover, it could be argued that when focusing only on a highlighted set of cells due to their value, the possibility exists that a good or optimal solution contained within the non-highlighted cells could be missed. Nevertheless, by narrowing the combinations to iterate, a good solution (if not the optimal one) could be reached much faster, while in a complete iteration an optimal solution could still be lost among the many combinations to be reiterated. Furthermore, in order to prove or disprove this possibility, further research and application to real-life case studies is required. In a real-world situation such as that studied here, MA proved to be a successful approach for resolving complex problems and it can be applied with multidisciplinary group decision-makers. Future research can directly include integration of risk and uncertainty assessment in all modifications of MA decisions.

Acknowledgements This research was supported in part by The Research Foundation of Lappeenranta University of Technology [LUT Tukisäätiö grant number 122/16] and The Foundation for Economic Education, [Grant number 160039].

References

Ajith, K. J., & Ganesh, L. S. (2009). Research on knowledge transfer in organizations: A morphology. *Journal of Knowledge Management, 13*(4), 161–174. https://doi.org/10.1108/13673270910971905.

Altshuller, G. (1984). *Creativity as an exact science*. Pocket mathematical library. Boca Raton: CRC Press.

Arciszewski, T. (2018). Morphological analysis in inventive engineering. *Technological Forecasting and Social Change, 126*, 92–101. https://doi.org/10.1016/j.techfore.2017.10.013.

Buzuku, S., & Kraslawski, A. (2015). Engineering geology and health and safety. *Application of morphological analysis to policy formulation for wastewater treatment* (pp. 102–108). Notes of the Mining Institute 214. http://pmi.spmi.ru/index.php/pmi/article/view/138/159.

Buzuku, S., Kraslawski, A., & Harmaa, K. (2015, November 4–6). Supplementing morphological analysis with a design structure matrix for policy formulation in a wastewater treatment plant. In *Modeling and managing complex systems* (pp. 9–18). Fort Worth. https://doi.org/10.3139/9783446447264.002.

Chechurin, L. (2016). TRIZ in science. Reviewing indexed publications. *Procedia CIRP, 39*, 156–165. https://doi.org/10.1016/j.procir.2016.01.182.

Chen, J. H.-C., & Lai, C.-F. (2010). The theory of morphological analysis applied to western apparel-a case study of renaissance era. *International Journal of Computer Science and Network Security, 10*(4), 176–184.

Duczynski, G. (2017). Morphological analysis as an aid to organisational design and transformation. *Futures, 86*, 36–43. https://doi.org/10.1016/j.futures.2016.08.001.

Duczynski, G. (2018). Investigating traffic congestion: Targeting technological and social interdependencies through general morphological analysis. *Technological Forecasting and Social Change, 126*(February 2017), 161–167. https://doi.org/10.1016/j.techfore.2017.05.019.

Eriksson, T., & Ritchey, T. (2002). Scenario development using computerised morphological analysis. Papers presented at the *Cornwallis and Winchester international OR conference*, England. http://citeseerx.ist.psu.edu/viewdoc/download?doi=10.1.1.469.9096&rep=rep1&type=pdf. Accessed 1 June 2017.

Feng, X., & Fuhai, L. (2012). Patent text mining and informetric-based patent technology morphological analysis: An empirical study. *Technology Analysis and Strategic Management, 24*(5), 467–479. https://doi.org/10.1080/09537325.2012.674669.

Ferretti, F., Saltelli, A., & Tarantola, S. (2016). Trends in sensitivity analysis practice in the last decade. *Science of the Total Environment, 568*, 666–670. https://doi.org/10.1016/j.scitotenv.2016.02.133.

Frow, P., Nenonen, S., Payne, A., & Storbacka, K. (2015). Managing co-creation design: A strategic approach to innovation. *British Journal of Management, 26*(3), 463–483. https://doi.org/10.1111/1467-8551.12087.

Geum, Y., Jeon, H., & Lee, H. (2016). Developing new smart services using integrated morphological analysis: Integration of the market-pull and technology-push approach. *Service Business, 10*(3), 531–555. https://doi.org/10.1007/s11628-015-0281-2.

Haydo, P. A. (2018). From morphological analysis to optimizing complex industrial operation scenarios. *Technological Forecasting and Social Change, 126*, 147–160. https://doi.org/10.1016/j.techfore.2017.06.009.

Im, K., & Cho, H. (2013). A systematic approach for developing a new business model using morphological analysis and integrated fuzzy approach. *Expert Systems with Applications, 40*(11), 4463–4477. https://doi.org/10.1016/j.eswa.2013.01.042.

Jeong, S., Jeong, Y., Lee, K., Lee, S., & Yoon, B. (2016). Technology-based new service idea generation for smart spaces: Application of 5G mobile communication technology. *Sustainability, 8*(11), 1211. https://doi.org/10.3390/su8111211.

Jimenez, H., & Mavris, D. (2010). An evolution of morphological analysis applications in systems engineering. In *48th AIAA aerospace sciences meeting including the new horizons forum and aerospace exposition* (pp. 1–10). https://doi.org/10.2514/6.2010-972.

Johansen, I. (2018). Scenario modelling with morphological analysis. *Technological Forecasting and Social Change, 126*(February 2017), 116–125. https://doi.org/10.1016/j.techfore.2017.05.016.

Kannengiesser, U., Williams, C., & Gero, J. (2013). What do the concept generation techniques of Triz, morphological analysis and brainstorming have in common. *19th international conference on engineering design*, ICED 2013 7 DS75-07 (October 2015), 297–300.

Kuriakose, K. K., Raj, B., Murty, S. S. A. V., & Swaminathan, P. I. (2010). Knowledge management maturity models – A morphological analysis. *Journal of Knowledge Management Practice, 11*(3), 1–9.

Lopes Correia da Silva, L. (2011). Morphological analysis of electric vehicles introduction in urban traffic in Sao Paulo. *Future Studies Research Journal, 3*(1), 14–37.

Louise, L., Mapule, M., & le Roux, H. (2009). A model for peace support operations: An overview of the ICT and interoperability requirements. In *Proceedings of the 4th international conference on information warfare and security* (pp. 1–10).

Mysiak, J. (2010). *Decision methods*. 2010. http://www.netsymod.eu/mdss/mDSS_DECMETH.pdf

Ney, S. (2012). *Resolving messy policy problems: Handling conflict in environmental, transport, health and ageing policy, The earthscan science in society series.* Abingdon: Routledge.

Ölvander, J., Lundén, B., & Gavel, H. (2009). A computerized optimization framework for the morphological matrix applied to aircraft conceptual design. *Computer-Aided Design, 41*(3), 187–196. https://doi.org/10.1016/j.cad.2008.06.005.

Ostertagová, E., Kováč, J., Ostertag, O., & Malega, P. (2012). Application of morphological analysis in the design of production systems. *Procedia Engineering, 48*, 507–512. https://doi.org/10.1016/j.proeng.2012.09.546.

Pidd, M. (2009). *Tools for thinking: Modelling in management science.* Chichester: Wiley.

Plauché, M., de Waal, A., Grover, A. S., & Gumede, T. (2010). Morphological analysis: A method for selecting ICT applications in South African government service delivery. *Information Technologies & International Development, 6*(1), 1–20.

Ritchey, T. (2010). *Wicked problems, social messes: Decision support modelling with morphological analysis.* Stockholm: Ritchey Consulting.

Ritchey, T. (2011). *Wicked problems – Social messes decision support modelling with morphological analysis, Risk, governance and society.* Heidelberg: Springer. https://doi.org/10.1007/978-3-642-19653-9.

Ritchey, T. (2012). Outline for a morphology of modelling methods: Contribution to a general theory of modelling. *Acta Morphologica Generalis AMG, 1*(1), 1–20.

Ritchey, T. (2014). Four models about decision support modeling. *Acta Morphologica Generalis AMG, 3*(1), 1–15.

Ritchey, T. (2015). Principles of cross-consistency assessment in general morphological modelling. *Acta Morphologica Generalis, 4*(2), 1–20.

Ritchey, T. (2018). General morphological analysis as a basic scientific modelling method. *Technological Forecasting and Social Change, 126*(June 2017), 81–91. https://doi.org/10.1016/j.techfore.2017.05.027.

Ritchey, T., & de Waal, A. (2007). Combining morphological analysis and Bayesian networks for strategic decision support. *ORiON, 23*(2), 105–121. https://doi.org/10.5784/23-2-51.

Rittel, H. W. J., & Webber, M. M. (1973). Dilemmas in a general theory of planning. *Policy Sciences, 4*(2), 155–169. https://doi.org/10.1007/BF01405730.

Seidenstricker, S., & Linder, C. (2014). A morphological analysis-based creativity approach to identify and develop ideas for BMI: A case study of a high-tech manufacturing company. *International Journal of Entrepreneurship & Innovation Management, 18*(5/6), 409–424. https://doi.org/10.1504/IJEIM.2014.064716.

Seidenstricker, S., Scheuerle, S., & Linder, C. (2014). Business model prototyping – Using the morphological analysis to develop new business models. *Procedia – Social and Behavioral Sciences, 148*, 102–109. https://doi.org/10.1016/j.sbspro.2014.07.023.

Sholeh, M., Ghasemi, A., & Shahbazi, M. (2018). A new systematic approach in new product development through an integration of general morphological analysis and IPA. *Decision Science Letters, 7*, 181–196. https://doi.org/10.5267/j.dsl.2017.5.004.

Storbacka, K., & Nenonen, S. (2012). Competitive arena mapping: Market innovation using morphological analysis in business markets. *Journal of Business-to-Business Marketing, 19*(3), 183–215. https://doi.org/10.1080/10 51712X.2012.638464.

Takane, Y., Jung, S., Oshima-Takane, Y., Millsap, R. E., & Maydeu-Olivares, A. (2009). Multidimensional Scaling, *9*(3), 219–242.

Teles, M. d. F., & de Sousa, J. F. (2017). A general morphological analysis to support strategic management decisions in public transport companies. *Transportation Research Procedia, 22*, 509–518. https://doi.org/10.1016/j.trpro.2017.03.069.

Teles, M. d. F., & de Sousa, J. F. (2018). Linking fields with GMA: Sustainability, companies, people and operational research. *Technological Forecasting and Social Change, 126*, 138–146. https://doi.org/10.1016/j.techfore.2017.05.012.

Voros, J. (2018). On a morphology of contact scenario space. *Technological Forecasting and Social Change, 126*, 126–137. https://doi.org/10.1016/j.techfore.2017.05.007.

Williams, P. B., & Bowden, F. D. J. (2013). Dynamic morphological exploration. In *22nd national conference of the Australian society for operations research*, Adelaide, Australia, 1–6 December 2013 (pp. 232–238). www.asor.org.au/conferences/asor2013.

Yoon, B. (2008). On the development of a technology intelligence tool for identifying technology opportunity. *Expert Systems with Applications, 35*(1–2), 124–135. https://doi.org/10.1016/j.eswa.2007.06.022.

Yoon, B., & Park, Y. (2005). A systematic approach for identifying technology opportunities: Keyword-based morphology analysis. *Technological Forecasting and Social Change, 72*(2), 145–160. https://doi.org/10.1016/j.techfore.2004.08.011.

Yoon, B., Park, I., & Coh, B.-y. (2014). Exploring technological opportunities by linking technology and products: Application of morphology analysis and text mining. *Technological Forecasting and Social Change, 86,* 287–303. https://doi.org/10.1016/j.techfore.2013.10.013.

Zeiler, W. (2018). Morphology in conceptual building design. *Technological Forecasting and Social Change, 126,* 102–115. https://doi.org/10.1016/j.techfore.2017.06.012.

Zwicky, F. (1969). *Discovery, invention, research through the morphological approach.* New York: Macmillan.

Engineering Creativity: The Influence of General Knowledge and Thinking Heuristics

Iouri Belski, Anne Skiadopoulos, Guillermo Aranda-Mena, Gaetano Cascini, and Davide Russo

1 Introduction

1.1 The Need for Creative Solutions Beyond Profession

Rapid change in technology, growing availability of new materials as well as development of highly accurate models that underpin computer-based

I. Belski (✉) • G. Aranda-Mena
Royal Melbourne Institute of Technology, Melbourne, VIC, Australia
e-mail: iouri.belski@rmit.edu.au

A. Skiadopoulos
La Trobe University, Bundoora, VIC, Australia

G. Cascini
Polytechnic University of Milan, Milan, Italy

D. Russo
Department of Management, Information and Production Engineering, University of Bergamo, Bergamo, Italy
e-mail: davide.russo@unibg.it

© The Author(s) 2019
L. Chechurin, M. Collan (eds.), *Advances in Systematic Creativity*,
https://doi.org/10.1007/978-3-319-78075-7_15

simulators have resulted in shrinking lifetime of products and services created by engineers. These changes in resources available to engineering designers brought restrictions on product development time and raised a demand for creativity of engineering solutions. In order to win in this rapidly changing world, engineering companies need to offer novel products that are capable of outperforming those of their competitors. Moreover, these novel products need to have reasonably long life spans in order for companies to make profits and to prevent themselves from being outcompeted by new products of others. These constraints require engineering designers to gain skills in choosing winning designs from numerous suitable proposals that are based not only on diverse external appearances but also on different principles of operations.

Growing availability of advanced manufacturing technologies enables engineers to develop the same design functionality in more than one way (e.g. mechanically, chemically, electro-magnetically, deploying nanotechnology, etc.). Therefore, in order to succeed in today's rapidly changing world, engineers need to be abreast of novel technologies. This will help them to continuously deliver innovative products by choosing the most suitable technologies and principles of operation that suit the required product functionality.

Development of silicon wafer cleaning methods illustrates the changes in cleaning technologies that are based on different principles of operation. Over the last 60 years, cleaning methods evolved from traditional mechanical and chemical cleaning to vapor-phase cleaning, plasma stripping and cleaning as well as cryogenic aerosol/supercritical fluid cleaning (Kern 2008). Furthermore, the evolution of silicon wafer cleaning technologies on novel principles of operation is still continuing with new cleaning methods developed continuously (e.g. foam/bubble cleaning, laser cleaning, nanoprobe cleaning) (Reinhardt 2008).

In essence, changes in technology resources have strengthened the need for engineering designers to propose divergent ides for product implementation. Engineers of the twenty-first century need skills in choosing the most 'progressive' and 'winning' designs for development and fabrication. More and more often, these choices can be found outside of a knowledge base of a single engineering profession.

1.2   Definition of Engineering Creativity

Recent studies advocated that creativity is domain specific (Baer 2015). More and more scholars agree that being creative in one domain does not make an individual creative in another knowledge domain (Weisberg 2006; Baer 2012). This view is further supported by negligible transfer of the creativity training gains from one knowledge area to another (Baer 2016). Therefore this chapter will specifically consider engineering creativity and use the definition of engineering creativity developed by Belski (2017):

> Engineering creativity is the **ability** to generate novel solution ideas for open-ended problems, ideas that are not obvious to experts in a particular engineering discipline and that are considered by them as potentially useful. (Belski 2017, p. 327)

This definition is based on analysis of legal criteria of patentability and patent authorship. It implies that the knowledge outside of the discipline is essential for engineering creativity. In order for an idea to not be obvious to experts from a particular discipline, it needs to be somewhat unusual for this discipline. In other words, to be considered as creative in engineering an idea has to use information that is not well known to professionals from this discipline (i.e. located 'outside' of their discipline knowledge).

1.3   Sources of Engineering Creativity

In her pioneering work on creativity that is still considered as appropriately describing creative performance, Amabile (1983) suggested that creative performance depends on three main components: (1) domain-relevant skills, (2) creativity-relevant skills and (3) task motivation. Domain-relevant skills include domain knowledge and domain-relevant 'talent' and depend on innate cognitive abilities as well as on formal and informal education of an individual. Creativity-relevant skills comprise appropriate cognitive style and knowledge of ideation heuristics and depend on training and experience. Task motivation incorporates problem solver's attitudes towards the task and depends on his or her intrinsic motivation (Amabile 1983).

It appears that perpetual increase in supply of resources that engineering designers can use, which has been occurring over the last 20–30 years, advocates for the extension of the above-mentioned creativity components proposed by Amabile. More and more evidence that originated from industrial corporations demonstrate that the knowledge outside of profession is becoming essential for engineering creativity.

This importance of general knowledge, for example, is supported by the findings made by Belski, Adunka and Mayer (2016a), who reported on the outcomes of surveying 46 engineering experts from the most innovative world corporations. Belski et al. established that, although discipline knowledge, years of practice (i.e. experience) and proficiency with creativity techniques are still considered by the experts important for achieving creative designs, general knowledge outside of their profession has become more essential for creative performance. The paired-samples t-test of the means of survey responses showed statistically significant differences between the need of general knowledge versus discipline knowledge ($t = 4.3$, $p < 0.001$) as well as between the need of general knowledge versus years of practice ($t = 3.8$, $p < 0.001$) for creative engineering work (Belski et al. 2016a). Such positive view of engineering experts on the role of general knowledge in creativity is well expressed by the above-mentioned definition of engineering creativity. In order for an idea to be not obvious to experts in a particular engineering domain, it needs to integrate knowledge outside of this domain. Otherwise, the idea will be 'obvious' to the domain experts.

Accordingly, in order to adequately model creative performance of engineers of the twenty-first century, knowledge from outside of the profession needs to be incorporated as an additional module into the Amabile's components.

1.4   Expert Schemas Versus Memory Search by Novices

The difference in problem-solving strategies between experts and novices has been extensively studied over the last 60 years (e.g. Gick 1986; Simon 1996; Harlim and Belski 2013, 2017; Cross 2004; Weisberg 2006). It has been concluded that one of the key differences in problem solving between experts and novices is associated with expert knowledge schemas. Over the years of practice, experts develop effective schemas that arm them with excellent task-recognition skills and allow them to propose sound solutions in their professional domain quickly (Belski and Belski 2013; Simon 1996). Novices, on the other hand, have not had sufficient practical experience to build effective knowledge schemas and, therefore, require searching a problem space in order to find solutions (Belski and Belski 2008; Gick 1986). Although expert knowledge schemas have been considered as advantageous in problem solving, it was posited that they may hinder experts' creativity (Belski and Belski 2013). Moreover, it was suggested that one of the effective ways for enhancing creativity of experts lies in engaging them in deliberate search of their knowledge repository for solution ideas that lie beyond their area of expertise (Belski and Belski 2013).

The suggestion that deliberate search of own general knowledge can bring more novel ideas to experts was confirmed by the study of Dobrusskin, Belski, and Belski (2014). They surveyed a team of 13 experts from Philips who were involved in developing solutions for a technical problem that had been thoroughly protected by intellectual property by other companies. In order to search their knowledge, these experts used the Theory of Inventive Problem Solving (TRIZ) heuristic of systematised Substance-Field Analysis (Su-Field) (Belski 2007). Although prior to utilisation of Su-Field they spent two weeks on situation and function analyses and were involved in industry scouting, the team did not find many suitable solutions. Su-Field, together with the eight fields of Mechanical, Acoustic, Thermal, Chemical, Electric, Magnetic, Intermolecular and Biological (MATCEMIB) that it uses for idea search, was much more helpful for generating novel ideas. Most of the survey participants believed that '*the use of the Su-Field procedure generated ideas that would have been overlooked otherwise*' (p. 126). Furthermore, the survey participant agreed that '*the eight fields of MATCEMIB have helped [them] to thoroughly search [their] knowledge for solution ideas*' (p. 126).

1.5   Enhanced Idea Generation with Su-Field

Recent studies reported on the positive influence of systematised Su-Field on students' ability to generate diverse ideas for a knowledge-rich, open-ended engineering problem (how to clean lime deposits from inside water pipes) (Belski et al. 2014, 2015, 2016b). It was discovered that the first-year students from universities in Australia, Czech Republic, Finland and Russia who were simply exposed to the names of the eight fields of MATCEMIB during idea generation produced at least two times more ideas than the students from a control group that were not shown any prompts. Also, students that were exposed to the names of the eight fields of MATCEMIB proposed solution ideas that covered a significantly broader range of engineering principles of operation (Belski et al. 2015). The experiment conducted with university students in Germany, which involved undergraduate and postgraduate engineering students, supported the findings of the above-mentioned experiments with the first-year students. It also recorded a boost in both a number of distinct ideas and breadth of these ideas for the students who were shown the names of the eight fields of MATCEMIB (Belski et al. 2016b).

1.6   General Knowledge and Differences in Performance

The results of the above-mentioned idea-generation experiments showed significant differences in the numbers and the breadth of ideas generated by the control groups of students from all five countries. Table 1 depicts the information on the composition of the control groups, semester of study at university, an average number (mean) of independent ideas generated by students of a particular group and the breadth of these ideas (Belski et al. 2015, 2016b).

Table 1 A number (mean) and breadth of distinct ideas generated by students from control groups

Country	Students	Breadth	Mean
Australia	21 (s1)	2.05	2.00
Czech Republic	18 (s1)	2.53	3.56
Russia	21 (s1)	2.57	4.32
Finland	8 (s1)	2.75	5.81
Germany	37 (s3)	2.30	3.90

The breadth of ideas was calculated as a sum of eight terms, each equal to a fraction of students from the control group that proposed ideas that were assigned by the assessors to each field of MATCEMIB (Belski et al. 2015). For example, the following is the spread of the ideas proposed by the students from Australia: 95% of students proposed Mechanical ideas; 5% – Acoustic; 14% – Thermal; 86% – Chemical; 0% – Electric; 0% – Magnetic; 0% – Intermolecular; 5% – Biological. Therefore, the breadth of ideas proposed by the control group from Australia was equal to:

$$\text{Breadth} = 0.95 + 0.05 + 0.14 + 0.86 + 0 + 0 + 0 + 0.05 = 2.05 \quad (1)$$

Belski and Belski (2016) analysed the results of students from Australia, Czech Republic, Russian Federation and Finland presented in Table 1 and concluded that the difference in the numbers and the breadth of ideas generated by students from different countries can be explained by dissimilarities in the following: (a) the depth of their knowledge in science that has been acquired during secondary school study and (b) entry prerequisites to degrees at the universities that participated in the study. Students from countries with better educational systems, who were required to demonstrate their sound knowledge in science prior to entering engineering study at university, proposed more independent ideas that also covered a wider set of principles of operation (Belski and Belski 2016).

The conclusion that science knowledge influences the breadth and the number of the ideas proposed by individual students has recently been supported by a study by Buskes and Belski (2017) that repeated the above-mentioned lime deposit cleaning experiment with the students from the University of Melbourne. It was discovered that both the breadth and the number of the ideas generated by a student from the control group at the University of Melbourne were moderately and statistically significantly correlated with the number of science subjects the student completed at secondary school. Moreover, the number of science subjects studied by the student explained 17% variation in the number and the breadth of the ideas proposed (Buskes and Belski 2017). These results further support the inclusion of the component of 'general knowledge outside the professional domain' into the creativity model developed by Amabile.

Almost all students from Australia, Czech Republic, Russian Federation and Finland entered university directly from secondary school. Therefore, they have not acquired substantial discipline knowledge and did not have enough practical experience for the expert schemas to form in their engineering disciplines. This means that whilst generating ideas for cleaning pipes of lime deposit they searched their database of general knowledge in a way expected from novices. Table 1 (Breadth) and the Breadth formula (1) indicate that this search for solutions was not very efficient. Most of the ideas proposed by the students from all control groups recommended solutions based on two principles of operation: either cleaning the lime deposit from pipes mechanically or by using some chemical substance to remove it.

Students from the German control group were in their second year of study. The first year of engineering study at university is usually devoted to introduction to the profession and to expansion of general knowledge. Therefore, it was anticipated that these students gained more knowledge outside of their profession than the students from the other four control groups who had just entered university did. Hence, the students from Germany were expected to perform better than the students that had just entered universities. As shown in Table 1, this did not occur. Moreover, although the results of the experimental groups from Germany that combined students of different study years, which are discussed in Belski et al. (2016b), were aligned with the results for the experimental groups from the other four countries as presented in Belski et al. (2015), the influence of general knowledge versus discipline knowledge on idea generation was not fully clear. In order to establish the influence of experience and knowledge on creative performance more accurately, it was necessary to analyse idea-generation performance of users with practical experience and education that significantly exceeded that of the first-year students from universities in Australia, Czech Republic, Finland and Russia.

This chapter presents the results of the study that repeated the experiments conducted in the five countries at two universities in Italy. This time, all 64 students that participated in the study were enrolled in the engineering master's programs. The study planned to establish correctness of two hypotheses that have been discussed by Belski et al. (2016a): (1) general knowledge is more important than discipline knowledge for attaining creative solution ideas in engineering and (2) engaging a user in searching the person's knowledge repository by prompting a user with the eight fields of MATCEMIB of Su-Field accelerates idea generation more effectively than additional discipline knowledge and years of practical experience.

2   Methodology

Sixty-four master's students from the University of Bergamo and the Polytechnic University of Milan, who were in their seventh or eighth semester of university, participated in this study that repeated the experiment conducted by Belski et al. (2014). The following is a short record of activities that the participants were involved in.

At each university, students were divided into four tutorial groups: one control and three experimental. All students were given 16 minutes of tutorial time to individually generate as many ideas as possible for the same problem (to remove the lime build-up in water pipes). Initially, the same PowerPoint slide that contained the problem statement translated into Italian and a photo of a cross-section of a pipe, half of which was covered with lime deposit, was presented to the students for two minutes by their tutors. Figure 1a depicts the English version of the problem statement that was presented to students from all groups. After two minutes of problem introduction that covered only the information presented in Fig. 1a, all students were asked to work individually and to record as many ideas to clean the lime build-up from the pipes as possible (ideas were recorded by students in Italian). The form to record ideas was distributed to the students just before the problem was presented. The form was the same for the students of all four groups. It was a copy of the Australian form that was translated into Italian.

Students from the control group were not influenced by any ideation methodology. After two minutes of problem introduction, they were allowed to think of solution ideas and to record them for 16 minutes. The slide shown in Fig. 1a was presented to the students from the control group for the whole duration of the idea-generation session.

Calcium carbonate, or lime, is a hard deposit found in kettles, the inner surface of pipes and other surfaces.
How to Remove the Lime Build Up in Pipes?

Write down as many ideas as you can   a

Calcium carbonate, or lime, is a hard deposit found in kettles, the inner surface of pipes and other surfaces.
How to Remove the Lime Build Up in Pipes?

Lotus eater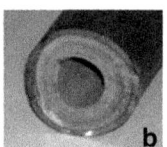

Write down as many ideas as you can   b

Calcium carbonate, or lime, is a hard deposit found in kettles, the inner surface of pipes and other surfaces.
How to Remove the Lime Build Up in Pipes?

Biological

Write down as many ideas as you can   c

Calcium carbonate, or lime, is a hard deposit found in kettles, the inner surface of pipes and other surfaces.
How to Remove the Lime Build Up in Pipes?

Biological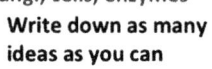
Microbes, bacteria, living organisms, plants, fungi, cells, enzymes

Write down as many ideas as you can   d

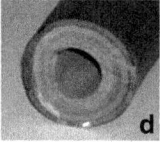

Fig. 1 The English version of the PowerPoint slides presented to students in Italian: (**a**) task introductory and the Control Group; (**b**) Random Word group; (**c**) MATCEMIB group; (**d**) MATCEMIB+ group. (Belski et al. 2015)

After two minutes of problem presentation, students from the experimental groups were told that during their idea-generation session they would be shown some words. No explanation of what these words were and what to do with them were given. Students from the Random Word groups were offered the Italian translation of the eight random words that were used in all other experiments. Students from the MATCEMIB group were offered the translations of the names of the eight fields of MATCEMIB in Italian. The MATCEMIB+ group students were shown the names of the eight fields (in big font) as well as some words (in small font) that illustrated the interactions of the particular field (e.g. friction, direct contact, collision, wind, etc. for the Mechanical field). The name of each field as well as each random word was shown to the students from the experimental groups for two minutes. Every two minutes a tutor changed the word on the screen and read the new word aloud. When a tutor of the MATCEMIB+ group changed slides every two minutes, he read aloud only the name of the field of MATCEMIB that was displayed, but did not read the words that illustrated field interactions that were displayed together with the field's name. Altogether, the students from all groups were generating and recording ideas for 16 minutes.

Figure 1 depicts the English version of one of the eight PowerPoint slides that were shown to the Italian students: Fig. 1a – the Control groups; Fig. 1b – the Random Word groups; Fig. 1c – the MATCEMIB groups; Fig. 1d – the MATCEMIB+ groups.

3   Results

3.1   Italian Study Only

Two independent assessors that used the same criteria as assessors from all other counties evaluated student ideas at each university. These criteria were developed for the original Australian study (Belski et al. 2014). Among other items, assessors counted the number of distinct (independent) ideas proposed by each student. In order to judge how broad or 'divergent' these independent ideas were, each idea was assigned to a field of MATCEMIB that most corresponded to the proposed principle of operation. The inter-rater reliability of assessment by independent assessors was evaluated for each university separately with SPSS software by establishing the Cronbach's Alpha for the number of independent ideas proposed by each individual student. Cronbach's Alphas for both the University of Bergamo and the Polytechnic University of Milan were very close to 0.9. The Cronbach's Alpha coefficient of 0.9 indicates excellent internal consistency. Thus, the assessment of students from both universities was evaluated as very reliable. As expected, due to similar students' backgrounds, the results of students from the same groups at both universities were very similar. Therefore, for further analysis the number of independent ideas proposed by each individual student made by the assessors was averaged and the results of students from both universities were combined.

Table 2 A number (Mean) and the Breadth of distinct ideas generated by students from Italy

Group	Students	Breadth	Mean
Control	16 (s7,8)	2.8	4.4
Random Word	15 (s7,8)	3.0	6.5
MATCEMIB	18 (s7,8)	4.7	6.4
MATCEMIB+	15 (s7,8)	6.1	8.1

Table 2 presents the results for the average number of independent ideas proposed by the Italian students in each group (mean) and the breadth of these ideas. It also contains information on the group sizes and their study semester.

The number of distinct solution ideas proposed by students was distributed normally in all groups, therefore one-way ANOVA with post-hoc Bonferrony tests were conducted. ANOVA showed significant differences between the groups ($F = 8.545$, $p < 0.001$). Bonferrony tests revealed statistically significant differences between the control group and each of the three experimental groups. Differences in the numbers of independent ideas between the experimental groups were not statistically significant.

The distribution of breath of ideas was not normal in some of the groups; therefore, the Kruskal-Wallis test of independent samples was conducted. It showed statistically significant difference in breadth between the groups ($p < 0.001$). The Matt-Whitney U tests were conducted to establish statistical differences in breadth of ideas between the groups. They showed statistically significant difference in breadth of ideas between the control group and the MATCEMIB group ($Z = -3.896$, $p < 0.001$) as well as the MATCEMIB+ group ($Z = -4.068$, $p < 0.001$). The tests also revealed statistically significant difference in breadth of ideas between the Random Word group and the MATCEMIB group ($Z = -3.625$, $p < 0.001$) as well as the MATCEMIB+ group ($Z = -3.918$, $p < 0.001$). Differences between all other groups were not statistically significant.

3.2   Control Groups: Italians and Students from Other Counties

The number of solution ideas in all control groups was distributed normally, therefore one-way ANOVA with post-hoc Bonferrony tests were conducted to establish differences in performance of the control groups of the first-year students with that of the master's students. ANOVA displayed significant statistical differences between the control groups for the number of ideas ($F = 7.050$, $p < 0.001$). Bonferrony tests showed that the Italian control group statistically significantly outperformed only the control group from Australia. Differences in the numbers of independent ideas between the control groups from Italy and the other four countries were not statistically significant.

The distribution of breath of ideas was not normal in some control groups; therefore, the Kruskal-Wallis test of independent samples was conducted. The test showed that the distribution of the ideas' breadth between the control groups was not statistically significant.

4   Discussion

4.1   General Knowledge Versus Discipline Knowledge

Comparison of performance of the control group from Italy with that from the other five countries only partly supported the first hypothesis (general knowledge is more important than discipline knowledge for attaining creative solution ideas). This conclusion was made on the basis of the following analysis.

Secondary schooling of students from Italy, which is a part of the European Union, as well as university entry requirements for engineering degrees in Italy, had more similarity to that of Czech Republic, Finland, Germany and even the Russian Federation than to Australia. Therefore, likewise to the conclusions of Belski and Belski (2016), statistically significant difference in the number of ideas proposed by the control groups from Italy and Australia can be explained by significant differences in science (general) knowledge of students from these two groups. This explanation supports the first hypothesis.

The absence of statistically significant differences in the number and breadth of solution ideas proposed by the students from the control groups from Czech Republic, Finland, Russian Federation, Germany and

Italy suggests that extension of discipline knowledge and practical experience that the Italian students attained over at least four years of extra study caused minimal influence on their creative performance. Assuming that the Italian students have not gained much general knowledge over the four years, this conclusion also supports the first hypothesis.

Indeed, significant expansion of general knowledge by the Italian students seems very unlikely. Engineering curricula are overloaded with discipline-related subjects and are focused on specialisation of students in their professions. Only very few subjects taught to engineering students over the three years of bachelor's degree programs in the European Union are devoted to expansion of their general knowledge. This means that the Italian students had significantly extended professional knowledge to that of the students from Czech Republic, Finland, Germany and the Russian Federation. At the same time, the gain in general knowledge by the Italian students over the additional four years of university study as well as due to extra practical experience might have been minimal and insufficient for enhancing their creative performance.

Alternatively, it is possible that over the four years of study students from Italy also substantially expanded their general knowledge (that is outside of their profession). Hence, if the first hypothesis is true, and additional general knowledge enhances creativity, Italian control group students were expected to outperform students from the control groups from all other countries, not only the students from Australia. Actually, Italian students performed somewhat better than students from the control groups of similar size. They did slightly better than their counterparts from Germany, Czech Republic and the Russian Federation in both the number of ideas (4.40 versus 3.90; 3.56 and 4.32 respectively) and the breadth of ideas (2.80 versus 2.30; 2.53 and 2.57). The fact that the difference in performance of Italian students was not statistically significantly higher than that of their peers from Germany, Czech Republic and the Russian Federation may imply that the gain in general knowledge by Italian students over the additional four years of study was not big enough to result in statistical significant difference in idea-generation performance.

Still, the assumption that Italian students gained substantial general knowledge over the four years cannot be ruled out. If it is the case, poor improvement in creative performance by the Italian control group compared to their first-year counterparts could be explained by reframing the interpretation of the opinions of experts surveyed by Belski, Adunka, and Mayer (2016a) on superiority of general knowledge over discipline knowledge for attaining creative solutions in engineering. The study of Buskes and Belski (2017) established that the variation of the number and the breadth of generated ideas that is explained by differences in science knowledge is quite moderate (17%). This means that extra general knowledge may not effectively facilitate improvement in creative performance on its own. Such conjecture does not contradict with the conclusions of Belski and Belski (2016) and Buskes and Belski (2017) on the influence of extensive prior knowledge in science on creative performance of engineering students. It just advocates reconsidering modelling the way general knowledge influences creativity. Most likely, students from the Italian control group were unable to use their additional general knowledge effectively, so their performance did not significantly differ statistically from that of the first-year students from the control groups.

It can be posited that extra general knowledge does not enhance creative performance on its own unless it has been transformed into appropriate problem-solving schemas. And in order for the general knowledge that has not been 'schematised' to lift creativity, it requires a 'catalyst' that helps a user to utilise his or her general knowledge effectively. The need for a catalyst is also supported by the opinions of the experts from the Belski et al. (2016a) study. The study engaged engineering experts from the most innovative world companies, who had good knowledge of ideation techniques, and also practiced them regularly. It is possible that regular practice in creativity techniques helped the experts to utilise their general knowledge and generate novel solution ideas effectively. This might explain why the experts' agreement with the statement 'creativity techniques that I have learnt over the years have significantly improved my ability to solve engineering problems creatively' (7.74/10) was second highest after the importance of general knowledge (8.41/10) and exceeded in value the discipline knowledge (7.00/10) and practical experience (7.21/10) (Belski et al. 2016a).

Thus, in order for the first hypothesis to be fully supported by the experimental data that is available so far, it needs to be reformulated as:

'general knowledge is more essential for creative performance than discipline knowledge if a practitioner is capable of searching his/her knowledge data base for general knowledge effectively'.

4.2   Idea Generation with Su-Field: Influence of MATCEMIB

The outcomes of the experiment conducted in Italy with the master's degree students were similar to the outcomes of the experiments that engaged the first-year bachelor's degree students from four other countries (Belski et al. 2015). The influence of the eight fields of MATCEMIB on both the number of ideas generated and on the breadth of these ideas had been fully replicated. Italian students from the MATCEMIB and the MATCEMIB+ groups proposed statistically significantly more ideas than their counterparts from the control group. The differences in breadth of solution ideas between each of the MATCEMIB groups and the control group were also statistically significant.

In essence, the results of idea generation of students from the Italian control group, those who have completed their bachelor's degrees, those may have had practical experience in industry and those who have returned back to university to get master's degrees, matched that of the control groups of recent school leavers but was well below the performance of the recent school leavers from the MATCEMIB and MATCEMIB+ groups from the other five countries. This means that the additional knowledge that the Italian students gained over four years of studying engineering did not make as significant a positive influence on their ability to utilise their knowledge as the prompts of the eight fields of MATCEMIB that were shown to the first-year students from Australia, Czech Republic, Finland and the Russian Federation. It appears that the names of the fields of MATCEMIB acted as the catalyst. They engaged students in searching their knowledge repositories and accelerated idea generation much more effectively than significant additional discipline knowledge and years of practical experience the Italian students had gained over at least four years. These results support the second hypothesis (engaging a user in searching the person's knowledge repository by prompting a user with the eight fields of MATCEMIB of Su-Field accelerates idea generation more effectively than additional discipline knowledge and years of practical experience).

5   Conclusions

The results of this study support the need to incorporate the component of 'general knowledge' into the model of creativity developed by Amabile (1983) in order to make this model suitable for the engineering profession of the twenty-first century. Due to a rapid change in technologies and growing availability of new materials, expert engineers can achieve patentable solutions when their knowledge spans beyond their traditional domain knowledge. Moreover, it seems that in order for engineering experts to utilise this general knowledge effectively, they need to apply sound ideation heuristics, like that of the eight fields of MATCEMIB.

References

Amabile, T. M. (1983). The social psychology of creativity: A componential conceptualization. *Journal of Pelsonality and Social Psychology, 45*(2), 357–376.

Baer, J. (2012). Domain specificity and the limits of creativity theory. *The Journal of Creative Behavior, 46*(1), 16–29. https://doi.org/10.1002/jocb.002.

Baer, J. (2015). The importance of domain-specific expertise in creativity. *Roeper Review, 37*(3), 165.

Baer, J. (2016). Content matters: Why nurturing creativity is so different in different domains. In R. A. Beghetto & B. Sriraman (Eds.), *Creative contradictions in education* (pp. 129–142). Switzerland: Springer International Publishing.

Belski, I. (2007). *Improve your thinking: Substance-field analysis.* Melbourne: TRIZ4U.

Belski, I. (2017). Engineering creativity – How to measure it? In N. Huda, D. Inglis, N. Tse, & G. Town (Eds.), *Proceedings of the 28th annual conference of the Australasian Association for Engineering Education (AAEE 2017)* (pp. 321–328). Sydney: School of Engineering, Macquarie University.

Belski, I., & Belski, I. (2008). Cognitive foundations of TRIZ problem-solving tools. In T. Vaneker (Ed.), *Proceedings of the TRIZ-future conference 2008* (pp. 95–102). Enschede: University of Twente.

Belski, I., & Belski, I. (2013). Application of TRIZ in improving the creativity of engineering experts. In A. Aoussat, D. Cavallucci, M. Trela, & J. Duflou (Eds.), *Proceedings of TRIZ future conference 2013* (pp. 67–72). Paris: Arts Et Metiers ParisTech.

Belski, I., & Belski, R. (2016). Influence of prior knowledge on students' performance in idea generation: Reflection on university entry requirements. In S. T. Smith, Y. Y. Lim, A. Bahadori, N. Lake, R. V. Pagilla, A. Rose, & K. Doust (Eds.), *Proceedings of the 27th annual conference of the Australasian association for engineering education – AAEE2016* (pp. 1–9). Lismore: Southern Cross University.

Belski, I., Hourani, A., Valentine, A., & Belski, A. (2014). Can simple ideation techniques enhance idea generation? In A. Bainbridge-Smith, Z. T. Qi, & G. S. Gupta (Eds.), *Proceedings of the 25th annual conference of the Australasian association for engineering education* (pp. 1C, 1–9). Wellington: School of Engineering & Advanced Technology, Massey University.

Belski, I., Belski, A., Berdonosov, V., Busov, B., Bartlova, M., Malashevskaya, E., et al. (2015). Can simple ideation techniques influence idea generation: Comparing results from Australia, Czech Republic, Finland and Russian Federation. In A. Oo, A. Patel, T. Hilditch, & S. Chandran (Eds.), *Proceedings of the 26th annual conference of the Australasian association for engineering education – AAEE2015* (pp. 474–483). Geelong: School of Engineering, Deakin University, Victoria, Australia.

Belski, I., Adunka, R., & Mayer, O. (2016a). Educating a creative engineer: Learning from engineering professionals. *Procedia CIRP, 39,* 79–84. https://doi.org/10.1016/j.procir.2016.01.169.

Belski, I., Livotov, P., & Mayer, O. (2016b). Eight fields of MATCEMIB help students to generate more ideas. *Procedia CIRP, 39,* 85–90. https://doi.org/10.1016/j.procir.2016.01.170.

Buskes, G., & Belski, I. (2017). Prior knowledge and student performance in idea generation. In N. Huda, D. Inglis, N. Tse, & G. Town (Eds.), *Proceedings of the 28th annual conference of the Australasian association for engineering education (AAEE 2017)* (pp. 354–361). Sydney: School of Engineering, Macquarie University.

Cross, N. (2004). Expertise in design: An overview. *Design Studies, 25*(5), 427–441.

Dobrusskin, C., Belski, A., & Belski, I. (2014). On the effectiveness of systematized substance-field analysis for idea generation. In C. Tucci, T. Vaneker, & T. Nagel (Eds.), *Proceedings of the TRIZ future conference: Global innovation convention (TFC 2014)* (pp. 123–127). Freiburg: The European TRIZ Association.

Gick, M. L. (1986). Problem-solving strategies. *Educational Psychologist,* *21*(1/2), 99–120.

Harlim, J., & Belski, I. (2013). Long-term innovative problem solving skills: Redefining problem solving. *International Journal of Engineering Education,* *29*(2), 280–290.

Harlim, J., & Belski, I. (2017). Stages of engineering problem solving: Learning from the experts. In N. Huda, D. Inglis, N. Tse, & G. Town (Eds.), *Proceedings of the 28th annual conference of the Australasian association for engineering education (AAEE 2017)* (pp. 295–302). Sydney: School of Engineering, Macquarie University.

Kern, W. (2008). 1 – Overview and evolution of silicon wafer cleaning technology. In *Handbook of silicon wafer cleaning technology* (2nd ed., pp. 3–92). Norwich: William Andrew Publishing.

Reinhardt, K. A. (2008). 11 – New cleaning and surface conditioning techniques and technologies. In *Handbook of silicon wafer cleaning technology* (2nd ed., pp. 661–688). Norwich: William Andrew Publishing.

Simon, H. A. (1996). *The sciences of the artificial* (3rd ed.). Cambridge, MA: MIT.

Weisberg, R. W. (2006). *Creativity: Understanding innovation in problem solving, science, invention, and the arts.* Hoboken: Wiley.

Part III

Advances in Managing Innovations and the Innovation Process

The four chapters in Part III describe novel ways and approaches for managing innovations and the innovation process as well as discuss evaluating and commercializing innovations.

Levelized Function Cost: Economic Consideration for Design Concept Evaluation

Mariia Kozlova, Leonid Chechurin, and Nikolai Efimov-Soini

1 Introduction

The ability to produce innovations systematically is rewarding. The design of a tangible, especially technology-driven innovation is mostly the design of a product. The latter stage of the innovation process has deserved a special title in modern communications, namely new product design (NPD). A typical NPD roadmap encompasses the following stages or milestones: Marketing (analysis of customer needs) → Specification (technical requirements for engineering) → Conceptual design (idea or the general concept for a new product) → Detailed design → Manufacture → Selling (Pugh and Clausing 1996). Sometimes marketing and selling are not needed or are performed in a reduced form (for example, when we develop a product or technology for our own needs).

Conceptual design is an important but vulnerable stage of the life cycle of a new product/technology. It is important because the price of a good

M. Kozlova (✉) • L. Chechurin • N. Efimov-Soini

School of Business and Management, Lappeenranta University of Technology, Lappeenranta, Finland

e-mail: mariia.kozlova@lut.fi; leonid.chechurin@lut.fi

© The Author(s) 2019

L. Chechurin, M. Collan (eds.), *Advances in Systematic Creativity*,
https://doi.org/10.1007/978-3-319-78075-7_16

idea can be very high. Once the concept has been selected, detailed design will start. The closer the design to its end, the higher the price of the changes, especially at the conceptual level.

On one hand, ideation is very important for NPD, and on the other hand, this design process stage is the least documented, standardized, and supported by methods. There are tools to stimulate brain activity in general (lateral thinking, brainstorming, synectics, etc.) and tools that provide systematic design problem analysis and synthesis or concept generation (e.g. morphological analysis, axiomatic design [AD], design for manufacturing and assembly, and theory of inventive problem solving [TRIZ]). The latter group can also be tagged as artificial intelligence methods because they teach algorithms to support new concept generation.

In fact, idea generation cannot be separated from idea evaluation. Separating the ideation phase from the evaluation phase will result in big losses: every design idea that does not make it through the evaluation gate wastes both efforts. The most typical evaluation gates for engineering are the following: Does the design contradict physical laws? Is the design manufacturable (especially for mass production)? Is the design economically viable? (Wallace and Burgess 1995, 429–446). We will not discuss patent and licensing issues here. The drama is that in big technology-driven business these questions are answered by different people. Although engineering designers are aware of the basic costs of the design, the detailed estimation of the profits is performed by economists. Ideally, the concept-generation process goes hand in hand with evaluation and decision-making.

Interestingly, some systematic creativity methods assist solely idea generation (morphological box), some provide certain, although non-economic evaluation criteria to tell good design apart from bad design (axiomatic design), and some toolkits, like TRIZ, contain both instruments.

TRIZ seems to be exceptionally general and powerful compared to other theories. The first publication on the method appeared in the mid-1950s in the USSR (Altshuller and Shapiro 1956, 37–49). It was developed by Altshuller and his followers into a set of tools for situation modelling (function modelling, substance-field models, contradictions), formal rules of model transformation (trimming, inventive standards, inventive principles), and the design evaluation axioms (Ideal Final Result [IFR], trends for engineering system evolution). There is evidence of wide applications of TRIZ in design practice reported by many innovation leaders (General Electric, Samsung group, Intel, Procter and Gamble etc.) and design engineers (Moehrle 2005, 285–296; Ilevbare et al. 2013, 30–37). The approach became the subject of scientific publications much later. However, about 1400 research papers in the SCOPUS database have been devoted to TRIZ since 2000 (Chechurin and Borgianni 2016, 119–134). One fundamental difference of TRIZ from other design techniques is the focus shift from design (material objects, components, links, etc.) to functions. Thus, an ideal system is believed to be a system that does not exist (and therefore does not cost anything) but performs the required function.

The ideality-driven idea evaluation in TRIZ leads to minimalistic design concepts with a minimal amount of components that perform the required function only when necessary and where necessary. The focus on the function provokes the idea that the economic evaluation of the design is to be performed on the function level as well.

Let us consider a concept design stage for a very simple system as an example. We need to design a holder for a whiteboard eraser. An immediate idea would be a plate at the bottom of the board, a kind of a shelf. Indeed, a narrow board would be a simple concept with small material, manufacturing, and maintenance costs. However, to perform the required function we would need to introduce an element (a board). Alternatively, the issue can be addressed with the concept of the IFR, which states that the "ideal" system is no system at all, but rather the ensured function. Even more, the function is ensured only when it is necessary (when the eraser exists and we do not use it) and where it is necessary (where we realize that the eraser is not needed any more). To embody this model we can teach the eraser to hold itself somewhere on the board. It could be realized with the help of several physical phenomena: (i) magnetizing the eraser or a part of it, (ii) giving the eraser a static charge, or (iii) making the surface of the eraser porous or sticky, and so on. In this concept, the "shelf" is ideal (according to the TRIZ concept); it (almost) does not exist and it requires zero material, but the eraser can be held.

When alternative conceptual designs are ready, we need to select which one to develop further. All three trade-offs represent the IFR (in terms of TRIZ) from the engineering point of view. However, their expected economic viability can differ a lot and must be considered at an early stage of NPD. Any cost-benefit analysis would stumble on the estimation of revenue from holding the eraser. Such attempts would complicate evaluation and open the door for manipulation of the results. This problem requires an indicator that is simple, clear, and not too demanding for calculation that focuses on the function performed in order to provide engineers with a tool to compare and choose between the alternative concepts from the economic point of view.

Some researchers have revealed possible psychological biases in the decision-making within the design selection process (Dong et al. 2015, 37–58; Nikander et al. 2014, 473–499; Toh and Miller 2015, 111–138). Although a broad set of methods exists to assist the selection process (Okudan and Tauhid 2008, 243–277), only a few incorporate economic viability.

The classical technique to estimate the profitability of any project, investment analysis (referred to as the capital budgeting in the finance literature) (Ryan and Ryan 2002, 355–364), implies estimating future benefits and costs, and offers some profitability indicators. However, such analysis requires more or less detailed estimates of future cash flows, which can be intricate regarding the benefit side of some design functions, such as, for example, metering. Moreover, investment analysis is a demanding process that often requires expertise of specialists from different fields. These facts impede the application of investment analysis in design evaluation. A modified measure for investment profitability, equivalent annual annuity or cost, allows comparing alternatives with different lifetimes, but does not solve the other abovementioned disadvantages connected with investment analysis and estimating future project cash flows.

A wide literature review conducted by Okudan and Tauhid (2008, 243–277) classifies a variety of concept-selection methods. Among those classes are multi-criteria decision-making (MCDM) methods decision matrices, optimization-based methods, and so on. The only method profoundly accounting for economic assessment is the utility theory-based model. However, manual assigning of "utility scores" to design attributes would not necessarily reflect the actual profitability of a design.

In practice, the Value Engineering (VE) approach (Miles and Boehm 1967), which is embedded into some international (e.g. ISO 9000) and corporate standards (SAVE International 2015), is often used to manage the cost-benefit ratio of a design, However, it is a complex approach and represents a management system rather than a valuation technique. It consists of a set of practical principles for quality management, but does not provide any particular solution for economic assessment.

Some researchers have incorporated market factor consideration into design valuation to reflect possible adoption success of a new product (Besharati et al. 2006, 333–350; Park and Park 2004, 387–394; Malen 1996, 105–122), however, required assumptions can be difficult to make during the early stages of design development.

Thus, the conceptual design stage, where the decisions are the most critical and costly, is left without a reliable economic indicator that would navigate the selection of ideas that are successful not only technically but economically as well.

Inspired by the practice of using a single simple indicator in the energy industry, we suggest paying attention to the levelized cost concept. It is widely used in the energy sector, where it is known as the levelized cost of electricity (*LCOE*). *LCOE* is a relative indicator that aggregates all project-related costs per unit of electricity produced (by a particular project over its lifetime). The different technology types in the energy sector may differ substantially by capital and operating costs, lifetime, electricity production performance, and so on. *LCOE* reflects all these features in one figure, and thus enables comparison of different technologies. In addition, it is convenient to display trends of technology development in time and analyze its overall learning curve. *LCOE* has become a common indicator in energy industry analysis, used actively by both business (see, e.g., US Energy Information Administration 2015; World Energy Council 2013; Bloomberg New Energy Finance 2015), and academia (see, e.g., Branker et al. 2011, 4470–4482; Campbell et al. 2008; Breyer and Gerlach 2013, 121–136; Ouyang and Lin 2014, 64–73; Hernández-Moro and Martínez-Duart 2013, 119–132).

Being essential in the power sector, the levelized cost concept has not conquered other industries yet, but examples of limited applications have been presented (Ogden et al. 1996, 115–130; Khastagir and Jayasuriya 2011, 3769–3784). However, its applicability in the valuation of any products or services has been shown by Reichelstein and Rohlfing (2014). In particular, they show that levelized cost as a single aggregated indicator is a feasible economic measure of planned investment.

Essential for the power-generation sector, the function-orientation makes *LCOE* convenient for design evaluation with a focus on its function. Therefore, in this chapter we generalize levelized cost of energy to levelized function cost (*LFC*).

Similar to the *LCOE* concept idea, life-cycle cost analysis is implemented in some industries to support the decision-making (Fuller and Petersen 1995), for example, the British Standard BS ISO 15686-5:2008 for building and construction assets, as well as to complement VE (Younker 2003). However, life-cycle cost only accounts for project costs, but not for its function, leaving such parameters as its productivity and lifetime beyond the evaluation scope.

We claim that the *LFC* concept can be utilized to assist decision-making in the engineering design choice. We continue with overview and comparison of several potential indicators for economic appraisal of design concepts, showing relative advantages of *LFC*. Before switching to a detailed explanation of the *LFC* equation, we first present the definition of function. Several generic examples serve to demonstrate the usefulness of the *LFC* estimate for different types of problems. With a numerical example of flow-meter design types, we illustrate how the levelized cost concept can provide insights into expected economic success of different design concepts.

2   Overview of Selected Indicators for Economic Appraisal of Design Concepts

A traditional, well-known, and broadly used profitability indicator for economic appraisal of any investment is Net Present Value (*NPV*) (Ryan and Ryan 2002, 355–364; Graham and Harvey 2001, 187–243). Its rise dates to times of Karl Marx (1894) and Irving Fisher (1907). It represents the sum of all project-related cash flows, discounted to properly account for the time value of money. (There are also such indicators as internal rate of return and profitability index, but similar to *NPV* they are based on the discounted cash flow notion, so we leave them outside of the scope of this comparison.) *NPV* is often used to assess investment projects, though for smaller capital investments like separate technical devices the Equivalent Annual Annuity (*EAA*) approach is shown as more suitable (Jones and Smith 1982, 103–110). It converts *NPV* to equivalent annuity payments. This allows for comparing projects with different lifespans that is not possible with *NPV*. However, when considering such technical devices embedded into a bigger system, estimation of revenues can be complicated and leaves room for manipulation of results. Perhaps because of that, the VE concept adapts the life-cycle cost (*LCC*) estimate, an *NPV* analog, but without revenue consideration (Younker 2003). It represents all design-related costs, including its initial cost (e.g. manufacturing and installation costs), operating and maintenance cost incurred over the whole lifetime, and disposal cost if any. Nevertheless, excluding cash inflows from the calculation, a new drawback arises, namely, inability of *LCC* to capture possibly different production profiles of estimated alternatives. Taking advantage of this exclusion from the estimation revenues, this drawback can be solved by introducing costs weighted per production unit, which is reflected in the levelized cost of energy indicator, widely used in the power production field (Short et al. 2005). The formulation of each indicator and the summary of their advantages and disadvantages are given in Table 1.

The table contains the general equation of the indicator, its simplified form, when annual cash flows (and production) are constant over time, and analysis of such advantages as comparability for different lifetime, absence of necessity to estimate revenues, and comparability for different productivity.

One can see that *LCOE* is the only indicator that possesses all mentioned features that is able to tackle design concepts with different lifetime and productivity simultaneously while not requiring estimating revenues. However, whereas electricity production is an unambiguous measurable output, definition of design function requires some further clarification.

Table 1 Comparison of selected indicators for design evaluation

	NPV Net present value	EAA Equivalent annual annuity	LCC Life-cycle cost	LCOE Levelized cost of energy
General equation where n is time period, N is total number of time periods in lifetime, I is initial cost, $Rev.$ is revenues, $O\&M$ is operating and maintenance costs, d is discount rate, and CRF is a capital recovery factor. (I and $O\&M$ are negative and $Rev.$ is positive)	$NPV = \sum_{n=0}^{N} \dfrac{I + (Rev + O\&M)_n}{(1+d)^t}$	$EAA = NPV^*CRF$	$LCC = \sum_{n=0}^{N} \dfrac{I + O\&M_n}{(1+d)^n}$	$LCOE = \dfrac{\sum_{n=0}^{N} \dfrac{(I + O\&M)_n}{(1+d)^n}}{\sum_{n=0}^{N} \dfrac{Q_n}{(1+d)^n}}$
Equation (equal CF over time)	$NPV = I + \dfrac{(Rev - O\&M)_a}{CRF}$	$EAA = I^*CRF + (Rev - O\&M)_a$	$LCC = I^*CRF + O\&M_a$	$LCOE = \dfrac{I^*CRF + O\&M_a}{Q_a}$
Comparable for different lifetime	NO	YES	NO	YES
No need to estimate revenues	NO	NO	YES	YES
Comparable for different productivity	YES	YES	NO	YES

3 Definition of Function

The primary goal of any engineering design efforts (in contrast to artistic design in general) is to ensure certain demand for function. Something does not happen itself, naturally, therefore we order engineering of an artificial device that performs the required function. Having ensured the specified function of the system, we can think of other important features of the product: cost, quality, sustainability, and so on. There are special tools or design management approaches to address all the requirements: Product Cost Analysis or Design-to-Cost (DTC), design for manufacturing and assembly (DFMA), design for quality (DFQ), and many others (Magrab et al. 2009).

Thus, the design for function is the first and inevitable stage of new system design. Therefore, how the function is described and how the quantitative specification for function is given are very important for success. Based on nothing but engineering folklore, the following story provides an understandable illustration. A team of space engineers worked on a search lamp bulb shell design for Lunar vehicles somewhere in the USSR in the 1960s. The bulb had to withstand substantial landing accelerations. The main function of the bulb shell was defined as "to protect the tungsten filament" that sits inside the bulb and emits the light. Certain design efforts were spent to develop such a glass shell and bind it reliably to the metal bulb base. When someone reformulated the function requirements for the shell "to stop the oxygen" or "to maintain the vacuum," suddenly the designers realized that there is no need for the shell at all: there is no oxygen on the moon or, in other words, there is a perfect vacuum there.

This significant simplification of design originated from careful function definition was given a systematic treatment in the TRIZ school of thought (Altshuller and Shapiro 1956, 37–49). The first principles of "correct" function formulation can be found in the concept of IFR and the model of an engineering system in classical TRIZ. A bit later, these principles took form of an extended roadmap for the inventive design-based function analysis called function-cost analysis (first published in Litvin et al. 1991, in Russian). Function analysis and tools for modification of design based on it form the modern TRIZ approach. In essence, the approach helps to get rid of function chimeras and declarative functions in new system specification that can substantially simplify the design. Another remarkable tool for formal function-based analysis of an engineering system is the method standardized by the US National Institute of Standards and Technology (NIST) (Hirtz et al. 2001). The approach suggests a list of standard functions that are typically required for an engineering system. A popular function-based design tool is axiomatic design (AD) (Suh 1990), which decomposes the function requirements and links them to design parameters (see Magrab et al. 2009).

The main difference in the TRIZ approach and the NIST and AD approaches is how the function is defined. In NIST and AD, the set of functions is large and can include field-specific or declarative functions expressed in almost natural (common-sense) language. The TRIZ-based approach requires careful analysis before the function formulation becomes formally legitimate. In fact, the list of legitimate functions rarely exceeds five verbs. Thus, functions like "to protect" or "to measure" can be easily met in AD or NIST but never in TRIZ-based function models.

Thus, the function analysis of TRIZ suggests a simple language or set of functions for the description of an engineering system. It is so simple that it can be called modelling. A complex product can be decomposed into a set of units in the same way the main function (or a function for which the product has been designed for) can be decomposed into the elementary functions from a set of legitimate ones. If multiple function requirements are exposed to the design, they should be considered one by one or in the same function model. (A kettle is to heat the water and to hold the water. Thus we should either build two function models for functions "heat" and "hold" or one function model where the flow of functions shows how the kettle's components contribute to perform the two function requirements). It is important to acknowledge another benefit of TRIZ-based function modelling: these functions are very simple and measurable by definition.

The function is legitimate if

- there are two material objects (or fields): function carrier and function recipient;

- they interact directly; and
- there is a parameter of function recipient that has been changed or maintained due to the function.

   A reader can experiment with these definitions, revealing that common-sense models such as "casing protects the chip" or "the thermometer measures the temperature" are not legitimate any more (Gerasimov et al. 1991, 40).

4   Levelized Function Cost for Design Valuation

This section introduces the concept of *LFC*. The rationale behind it is simple and straightforward: *LFC* is all design- (or project-) related costs per a unit of the total function output, properly adjusted for the time value of money. We start with the derivation of *LFC* arriving in its definition. Further, we show how the different features of a design influence the resulting *LFC* estimate, and based on it, we reveal when using *LFC* is relevant. The section ends with several generic examples illustrating the applicability of the *LFC* estimation.

4.1   Definition and Derivation of Levelized Function Cost

Levelized cost (of electricity or any other function unit) can be defined as a price level that covers exactly all project or design costs

$$TLCC = \sum_{n=0}^{N} \frac{Q_n^* LFC}{(1+d)^{n'}} \tag{1}$$

where

TLCC is the total life-cycle cost,
Q_n is the (electricity) output or productivity,

d is discount rate,

n is period index, and

revenue $Q*LFC$ is discounted to account for the time value of money.

The total life-cycle cost is simply the sum of all discounted project costs

$$TLCC = \sum_{n=0}^{N} \frac{C_n}{(1+d)^{n'}} \qquad (2)$$

where

C_n is annual costs, including initial investment and operating expenses.

Combining Eqs. (1) and (2) gives (3),

$$\sum_{n=0}^{N} \frac{Q_n{}^*LFC}{(1+d)^n} = \sum_{n=1}^{N} \frac{C_n}{(1+d)^{n'}} \qquad (3)$$

from which we arrive at the general definition of LFC (7).

Definition 1

LFC is a single indicator that shows the price level of a unit of function produced that would cover all costs related to the usage of a design, including
initial manufacturing and installation, as well as operation and maintenance costs, properly accounted for a time value of money

$$LFC = \frac{\sum_{n=0}^{N} \dfrac{C_n}{(1+d)^n}}{\sum_{n=0}^{N} \dfrac{Q_n}{(1+d)^n}}. \qquad (4)$$

If the initial costs occur within the base period $n=0$, then (4) can be rewritten as follows

$$LFC = \frac{I + PV(O\,\&\,M)}{\sum_{n=0}^{N} \dfrac{Q_n}{(1+d)^n}}, \tag{5}$$

where

I is the total initial cost of a design (device), including its manufacturing and installation, and
$PV(O\&M)$ are discounted operation and maintenance costs through the service life.

Equation (5) simplifies the situation to a "no taxes" environment that is suitable when assessing small elements of the total business. For other options, see details in Short et al. (2005).

LFC calculation can be simplified if the annual output Q and/or $O\&M$ costs are constant over time:

- If Q is equal over time,

$$LFC = \frac{I + PV(O\,\&\,M)}{Q_a / CRF} = \frac{I + PV(O\,\&\,M)^*}{Q_a}\,CRF, \tag{6}$$

where

Q_a is an annual function output, and

CRF is the capital recovery factor $\dfrac{d(1+d)^N}{(1+d)^N - 1}$ defined by discount rate d

and the total number of periods (years) of the design/project service life N.

If Q and O&M costs are equal over time,

$$LFC = \frac{I + \dfrac{O\&M_a}{CRF}^*}{Q_a} \quad CRF = \frac{I^*CRF + O\&M_a}{Q_a}, \tag{7}$$

where

$O\&M_a$ is the annual operation and maintenance costs.

Definition 2

If the function output and O&M costs are equal over time, levelized cost can be defined as the sum of annualized initial costs and annual O&M costs over the annual function output

$$LFC = \frac{I^*CRF + O\&M_a}{Q_a}. \tag{8}$$

For the purpose of design valuation, we recommend using the risk-free rate as the discount rate in the *LFC* equation, or its national equivalent, if the scope of design applicability is limited to one country.

It is important to treat inflation correctly in the calculation differentiating between real and nominal terms. Equation (8) implies *O&M* costs expressed by one figure in real terms (without inflation effect), so the real discount rate should be used to calculate *CRF*. Since the risk-free rate, as any other interest rates, is usually reported in nominal terms, it should be converted into the real one. For further details on calculating real and nominal *LFC*, see Short et al. (2005). Disposal costs must be included in *LFC* as well, if any occur for the estimated concepts.

Concisely, the *LFC* represents all the life-cycle, concept-related costs properly accounted for the time value of money per a unit of function output expected to be produced by the concept. Thus, a higher *LFC* would signify a more expensive solution to performing a particular function and vice versa. Therefore, *LFC* can be used to compare alternative design concepts.

4.2   Factors Influencing Levelized Cost

The simplified equation of levelized cost (8) shows the influence of the contributing factors on the *LFC* estimate explicitly. It allows performing analytical sensitivity analysis by evaluating partial derivatives.

The sensitivity of *LFC* to initial and *O&M* costs is linear and proportional to annual productivity (9 and 10)

$$\frac{\partial LFC}{\partial I} = \frac{CRF}{Q_a}, \tag{9}$$

$$\frac{\partial LFC}{\partial O\,\&\,M_a} = \frac{1}{Q_a}. \tag{10}$$

Levelized function cost from annual productivity is a nonlinearly decreasing (11) and convex (12) function, implying that the more productivity there is, the less marginal *LFC* there is

$$\frac{\partial LFC}{\partial Q_a} = -\frac{I^*CRF + O\,\&\,M_a}{Q_a^2}, \tag{11}$$

$$\frac{\partial^2 LFC}{\partial Q_a^2} = \frac{2\left(I^*CRF + O\,\&\,M_a\right)}{Q_a^3}. \tag{12}$$

The influence of the device lifetime and the discount rate on *LFC* is also nonlinear (13 and 14)

$$\frac{\partial LFC}{\partial N} = \frac{\dfrac{d(d+1)^N \log(d+1)I}{(d+1)^N - 1} - \dfrac{d(d+1)^{2N} \log(d+1)I}{\left((d+1)^N - 1\right)^2}}{Q_a}, \tag{13}$$

$$\frac{\partial LFC}{\partial d} = \frac{\dfrac{d(d+1)^{N-1} IN}{(d+1)^N - 1} - \dfrac{d(d+1)^{2N-1} IN}{\left((d+1)^N - 1\right)^2} + \dfrac{(d+1)^N I}{(d+1)^N - 1}}{Q_a}, \tag{14}$$

Table 2 The effects of variables on *LFC*

Variable	Component in the *LFC* equation (8)	Type of function
Service life	*CRF*	Decreasing, nonlinear, convex
Productivity	*Q*	
Initial cost	*I*	Increasing, linear
O&M costs	*O&M*	
Discount rate	*CRF*	Increasing, nonlinear, convex

The *LFC* derivative with respect to *N* is negative, showing the decrease of *LFC* with increasing lifetime, while the opposite occurs with respect to the discount rate. The latter reflects the cost of financing the investment and contributes essentially to *LFC* rise. The second derivatives for both *N* and *d* are positive.

Table 2 summarizes the sensitivity of *LFC* to the variables. Increasing the service life or productivity of a design *ceteris paribus* would decrease its levelized function cost. The convex character of the function implies that the maximum effect can be achieved with the initial improvement of service life or productivity of the design, but further improvement decreases the marginal benefit from it. Essentially, higher initial or *O&M* costs or discount rate would increase the *LFC*.

Overall, the type and sign of *LFC* change with different parameter variations are known. However, sensitivity analysis for each particular case is relevant because of possibly different ranges of values, as well as units of input variables for different problems.

4.3 Range of Problems That Can Be Tackled with Levelized Cost

We have recognized two major types of questions in engineering concept selection that can be addressed with *LFC*:

1. Which design is cheaper for a particular task?
2. Which design offers more cost reduction in the market?

The first question is relevant in the context of a particular operation that requires a particular function to be performed. In this case, the costs of different design options should be compared directly with each other.

The second question deals with a general NPD selection problem, focusing on predicting the success of different designs in the market. In this case, the potential cost savings of an invented design is defined over the current market technology (the difference in the *LFC* of the existing and the new design) and the one that offers more cost reduction is selected.

Levelized cost allows comparing solutions designed for different sets of application objects performing the same function. To illustrate, let us assume that design D_1 performs the function for three objects $\{a,b,c\}$, D_2 for $\{a,b\}$ and D_3 for $\{c\}$. Then, to define the least-cost solution (question I), the *LFC* of D_1 should be compared with the sum of the *LFC* of D_2 and D_3 weighted by the production volume.

$$
\begin{aligned}
&D_1 \to \{a,b,c\}, D_2 \to \{a,b\}, D_3 \to \{c\}, \text{then} \\
&LC(D_1) \perp LC(D_2)w_2 + LC(D_3)w_3,
\end{aligned}
\tag{15}
$$

where

$\to$ stands for to which objects the design performs the function, $\perp$ is the comparability sign, and w_2 and w_3 are the weights of the respective design production in the total volume.

When comparing the new design cost with the market benchmark, either a design with the same set of objects should be found or a logic similar to (15) should be applied, if the same set of objects can be constituted with several existing solutions.

The *LFC* concept becomes relevant when it is not clear whether a more simple solution from the engineering point of view would deliver more value for the customers. Simplification often limits the applicability of a design to a specific object. Alternatively, a modest increase in complexity can offer some extension of the applicability. *LFC* provides a rough and quick estimation of such a choice from the economic viability perspective.

Considering the parameters participating in the *LFC* calculation (see Table 2), it can offer comparison of designs with:

1. Different ratio of *O&M* expenses to the initial cost (e.g. the *O&M* cost of a design can be reduced to nothing, but it would increase the manufacturing cost);
2. Different service life (e.g. increase in the service life can be achieved only with some increase in manufacturing cost and/or *O&M* costs);
3. Different productivity (that would also cause increase in the initial and/or *O&M* costs);
4. Different sets of objects of a function (e.g. expanding object coverage with some increase in the cost or loss in service life duration); or
5. Any combination of the abovementioned.

The listed scenarios challenge the choice and require an economic viability assessment for solid decision-making.

4.4   Generic Examples

This section presents a simplified case of design comparison to demonstrate levelized cost in use. We consider all the scenarios presented above, except the last one, which is introduced in a real case illustration in the next section.

1. Different ratio of *O&M* expenses to the initial cost

Let us take as a basis an imaginary design D_1 that performs a function on object A_1. Let its initial cost be equal to 10 currency units (*c.u.*), the annual productivity 2 function units at 1 *c.u.* of *O&M* costs, and its lifetime 50 years.

Now let us assume that the same function on the same object can be realized by another design, D_2, which eliminates *O&M* costs totally, but its initial cost would be 50% higher, other things being equal. The question arises, which of them is cheaper overall?

LFC provides a straightforward answer to this question. Assuming the real risk-free rate at 5% in (8), we get the following result:

$$LFC(D_1) = \frac{10^* \dfrac{5\%(1+5\%)^{50}}{(1+5\%)^{50}-1}+1}{2},$$

$$LFC(D_2) = \frac{15^* \dfrac{5\%(1+5\%)^{50}}{(1+5\%)^{50}-1}+0}{2}, \tag{16}$$

$$LFC(D_1) = 0.77 > LFC(D_2) = 0.41.$$

With the given assumptions, the second design can provide the same function at almost half the cost of the first one. However, with other inputs, the result can be the opposite (e.g. if the initial cost is 100 instead of 10).

2. Different service life

Another engineering idea can offer a design D_3 (for the same function and object) with doubled service life, but it can be achieved only with a 20% increase in the initial cost.

$$LFC(D_1) = \frac{10^* \dfrac{5\%(1+5\%)^{50}}{(1+5\%)^{50}-1}+1}{2},$$

$$LFC(D_3) = \frac{12^* \dfrac{5\%(1+5\%)^{100}}{(1+5\%)^{100}-1}+1}{2}, \tag{17}$$

$$LFC(D_1) = 0.77 < LFC(D_3) = 0.80.$$

Although the idea may seem appealing, the levelized cost estimate shows that it is not economically viable.

3. Different productivity

Another NPD direction can target the increase in productivity. Alternative design D_4 can double the productivity while increasing the $O\&M$ costs by 50%.

$$LFC(D_1) = \frac{10^* \dfrac{5\%(1+5\%)^{50}}{(1+5\%)^{50}-1} + 1}{2},$$

$$LFC(D_4) = \frac{10^* \dfrac{5\%(1+5\%)^{50}}{(1+5\%)^{50}-1} + 1.5}{4}, \tag{18}$$

$$LFC(D_1) = 0.77 < LFC(D_4) = 0.51.$$

The *LFC* estimate confirms the expediency of such a solution.

4. Different sets of objects of a function

Often, one function must be applied to different function recipients (objects). If the issue is to find the lowest cost solution of performing this function within a particular operation or business, it is reasonable to compare whether two different designs specialized on distinct objects are cheaper than the one more sophisticated that can handle the whole set of required objects.

Let us assume that in addition to object a_1, the same function is required for a_2. While object a_1 can be served by D_1 and a_2 by D_5 (with all the same features, except 20% less initial cost and 50% higher $O\&M$

costs), both objects can be treated by a more expensive D_6 with initial cost equal to 10 c.u., the rest being equal.

$$\frac{\left(LFC(D_1)+LFC(D_5)\right)}{2}=$$

$$\frac{10^*\dfrac{5\%(1+5\%)^{50}}{(1+5\%)^{50}-1}+1}{2}+\dfrac{8^*\dfrac{5\%(1+5\%)^{50}}{(1+5\%)^{50}-1}+1,5}{2}}{2},$$

(19)

$$LFC(D_6)=\frac{12^*\dfrac{5\%(1+5\%)^{50}}{(1+5\%)^{50}-1}+1}{2},$$

$$\frac{\left(LFC(D_1)+LFC(D_5)\right)}{2}=0.87>LFC(D_6)=0.83.$$

Since D_1 and D_5 have the same productivity, we simply take the average of their LFC to obtain the function cost for both objects (see 15). The LFC estimate suggests that with the given assumptions, using the more sophisticated design is cheaper.

Different designs can also be compared with respect to potential success in the market. If D_1 and D_6 represent the NPD choice, they should be compared to their market analogs that perform the same function for the same set of objects. Assuming the D_1 analog costs 30 c.u. and serves for 35 years, and the D_6 analog costs 20 c.u. and serves for 25 years, we can estimate which design would deliver more cost reduction to the market. The chosen values indicate that D_1 offers a greater cut in the initial cost but less increase in the service life in comparison with D_6, but the overall effect is unclear, however.

$$1 - \frac{LFC(D_1)}{LFC(D_{1a})}, 1 - \frac{LFC(D_6)}{LFC(D_{6a})},$$

$$1 - \frac{\dfrac{1}{2}\left(10^* \dfrac{5\%(1+5\%)^{50}}{(1+5\%)^{50}-1} + 1\right)}{\dfrac{1}{2}\left(30^* \dfrac{5\%(1+5\%)^{35}}{(1+5\%)^{35}-1} + 1\right)},$$

(20)

$$1 - \frac{\dfrac{1}{2}\left(12^* \dfrac{5\%(1+5\%)^{50}}{(1+5\%)^{50}-1} + 1\right)}{\dfrac{1}{2}\left(20^* \dfrac{5\%(1+5\%)^{25}}{(1+5\%)^{25}-1} + 1\right)},$$

$$45\% > 31\%.$$

According to this example, the first design D_1 offers more relative cost reduction per unit of function than D_6, implying the decision to be in favor of D_1. In spite of the fact that these designs have different function recipients, they become comparable with the *LFC* estimate.

Overall, the levelized cost estimate can be used to evaluate and compare different designs with different features against each other or against market benchmarks. As it represents a relative indicator of the overall costs per unit of performed function, it can be applied to any conceptual product or service with defined functionality.

5   Case Illustration

Below, we illustrate the problem of design concept choice and the application of *LFC* for making an economically-wise decision with a case of flowmeters. The necessity of flow measurement of a liquid or a gas arises in a number of different fields, including water management, the energy sector, mining, food processing, agriculture, and so on. The different concepts of flowmeter design are based on different physical phenomena. We focus on three of them, namely, turbine, electromagnetic, and ultrasonic designs.

The turbine-type flowmeters simply convert the flow to the rotation of the rotor proportionally. The physical principle behind the electromagnetic flowmeter concept is electromagnetic induction. A magnetic field creates a potential difference that is proportional to the flow velocity and is sensed by electrodes. The ultrasonic flowmeter concept utilizes the Doppler Effect, registering the difference in frequencies of ultrasound waves emitted along and against the direction of the flow. These types of flowmeters are applicable to different objects due to their different physical principles, although providing the same function.

The choice between these three types of flowmeters, whose design is elaborated and requires economic assessment, serves as the case study here. As these three types are designed to operate in different liquids, we treat them as different products. Thus, we compare the three designs not directly to each other but to their existing analogs. Then the marginal benefit introduced by a new design is evaluated (in other words, how much cheaper the new design is). After these cost-reduction estimates, the three target designs are compared to each other. Thus, the objective of our evaluation is to estimate what design type offers higher cost reduction to the market.

For the purpose of illustration, we use six commercially available solutions. Three of them represent "invented designs in process," while the three others are assigned to be "existing market products" for comparison. Their specifications are presented in rows 1–10 of Table 3. All the presented designs differ in the set of objects (liquids), lifetime, repair cycle, productivity, electricity consumption, as well as manufacturing and installation costs. *LFC* is calculated per one cubic meter of measured liquid. We use (6), because productivity is constant over time, but not the *O&M* costs, for which we calculate the present value separately.

The *LFC* calculation is reflected in rows 11–23 of Table 3 and includes the following steps:

Table 3 Calculation of *LFC* for three designs and their benchmarks

Design specifications	Electromagnetic		Turbine		Ultrasonic	
1 Type	Electromagnetic		Turbine		Ultrasonic	
2 Liquids	Water, salt water, dirty water		Only clear water		All liquids	
3 Commercialization	New	Existing	New	Existing	New	Existing
4 Design	Piterflow RS50	PREM	Okhta T50	VSHN	Vzloyt MR	PortaFlow220
5 Lifetime, hours	80,000	80,000	100,000	100,000	75,000	13,250
6 Verification cycle, years	4	4	5	6	2	–
7 Flow, m³/hour	36	36	30	50	35	800
8 Electricity consumption, V*A	6	5	0	0	12	19
9 Cost of meter, c.u.	16,150	16,530	4240	10,875	34,800	418,000
10 Cost of installation, c.u.	3000	3000	1500	1500	1500	1500
LFC calculation						
11 Lifetime, years	9.1	9.1	11.4	11.4	8.6	1.5
12 Measured flow per year, m³	315,360	315,360	262,800	438,000	306,600	7,008,000
13 Electricity consumption, kWh/year	53	44	0	0	105	168
14 Electricity price, c.u./kWh	2					
15 Electricity cost, c.u./year	105	88	0	0	210	337
16 Present value (PV) of electricity cost	611	509	0	0	1173	452
17 PV of maintenance cost	230	230	201	113	508	0
18 O&M costs (PV), c.u.	841	739	201	113	1681	452
19 Initial costs, c.u.	19,150	19,530	5740	12,375	36,300	419,500
20 Discount rate (real)	10%					
21 Capital recovery factor	17%	17%	15%	15%	18%	74%
22 Levelized function cost (*LFC*), c.u./m³	$10.9*10^{-3}$	$11.1*10^{-3}$	$3.4*10^{-3}$	$4.3*10^{-3}$	$22.2*10^{-3}$	$44.6*10^{-3}$
23 Cost reduction	1%		21%		50%	

1. Input estimation:

   • Calculating annual productivity as the number of measured cubic meters per annum (row 12);
   • Estimation of the O&M costs and discounting them (row 18). The O&M costs consist of electricity cost (with assumed electricity price 2 c.u./kWh) and repairing costs that occur in accordance with the repairing cycle for each device. All costs are discounted at a 10% rate.
   • Calculating the initial cost (row 19) as a sum of manufacturing and installation costs. No discounting is needed, as this expense occurs in the initial period.
   • Calculating the capital recovery factor, taking technology lifetime into account (row 21);

2. Calculating LFC in accordance with (6) (row 22);
3. Defining cost reduction as a difference in LFCs of the existing analog and the new design, divided by the LFC of the analog to get the percentage value (row 23).

The obtained levelized cost estimate in c.u./cubic meter shows that turbine flowmeters are the cheapest alternative if only clean water is to be measured, followed by the electromagnetic flowmeters that serve a broader set of liquids, concluding with ultrasonic ones that can be applied to any liquid. The overall picture looks reasonable, as the more objects are to be treated by the function, the more expensive the solution is. However, the fact that different designs can be applied to different sets of liquids makes the direct cost comparison biased. Therefore, we have compared each solution to the existing analog, checking how much value the new design delivers.

The results demonstrate that the ultrasonic flowmeter design offers the highest cost reduction (50%), followed by the turbine design (21%), and ending with a minor achievement by the electromagnetic design (1%). With such estimations, the decision should be made in favor of the ultrasonic design, as it represents the highest cost-reduction potential to the existing analog and thus the most profitable trade-off.

For further analytics and assistance in further design development, sensitivity analysis of its LFC to different parameters can be recommended.

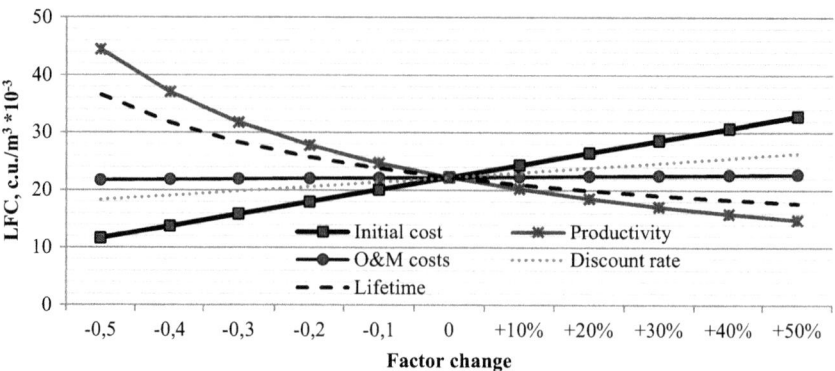

Fig. 1 *LFC* sensitivity analysis for ultrasonic design

Here, we present sensitivity analysis for the *LFC* of the "winning" ultrasonic flowmeter. Figure 1 shows how levelized cost changes (*y* axis) with one-by-one change in the design parameters (*x* axis) from −50% to 50% with 10% steps.

As it can be seen in the graph, the initial cost cut has the biggest potential for the *LFC* reduction. The *O&M* costs being relatively low compared to the initial cost do not affect *LFC* much. The discount rate is of minor influence as well. Sensitivity to lifetime and productivity is similar, though with more amplitude for productivity, and it offers some potential for *LFC* reduction. However, it is worth mentioning that the bigger these parameters, the smaller the marginal saving in *LFC*. Such sensitivity analysis provides an economic basis for the decision of which direction the further design development should take.

To sum up, the *LFC* indicator can enhance decision-making in engineering with economic considerations, and it can be used in particular for choosing the most economically viable trade-off and to support the roadmap of further design elaboration.

6   Discussion and Conclusion

Design evaluation from the economic point of view is crucial in the early stages of new product development. The earlier an engineer is able to evaluate the design from the economic perspective, the less effort is wasted on perfecting the ideas that are technologically attractive but not economically viable.

This chapter presents a method for express economic evaluation of the design concept based on the levelized cost approach adopted from the power-generation sector. *LFC* represents a single indicator of all life-cycle, concept-related costs per a unit of performed function, reflecting also design productivity and lifetime. Thus, it is easy to use and interpret for engineers.

The proposed approach suits essentially the trend of focusing on function or service rather than on a device, material, or object in new product design. The approach is illustrated by a case study of evaluating three flowmeter designs.

A known drawback of the levelized cost indicator in its original application area is possible variations driven by location differences (Borenstein 2011, 67–92). This includes local labor costs, and access to fuel transportation and electricity transmission, which contribute to operational and initial costs, not to mention climate conditions for renewable energy technologies, which affect the electricity output strongly. However, this disadvantage is hardly inherent to the design assessment, except in technology cases with similar dependence on the location. Nevertheless, a new limitation of the approach arises from its application to design evaluation. Since the conceptual design stage lacks details on the concept even from the engineering side, assumptions made regarding its costs are essentially vague and can bias the resulting estimate. For this reason, cost assumptions should be made carefully. To improve the reliability of the estimation, the effect of assumption variation should be captured with, for example, sensitivity analysis as illustrated in this chapter, or with soft computing-based techniques.

LFC in the context of MCDM methods also deserves discussion. As shown in numerous reviews, often chosen criteria for MCDM methods constitute quality (failure rate), production performance or efficiency, lifetime, maintenance, and so on (Ho et al. 2010, 16–24). In cases when all these criteria can be reflected in the *LFC* equation, we recommend directly assessing cost of function delivered to a customer based on the known relationship between criteria and the outcome, instead of conducting valuation with MCDM methods, where set weights or estimate aggregation algorithms may not necessarily reflect the reality in the best way. However, when criteria that is important to consider cannot be captured by the *LFC* estimate, MCDM methods are the better choice. Nevertheless, instead of multiple cost criteria, one aggregated *LFC* can clarify and simplify the decision problem. In other words, *LFC* can be used as a standalone indicator or can also be integrated in MCDM approaches and existing evaluation roadmaps.

Further research should consider the integration of *LFC* into systematic approaches to design development, like value engineering and quality management systems. Enhancing the *LFC* estimation with fuzzy set theory-based techniques will increase the reliability of the method by capturing possible imprecision in target design specifications and expected performance. It is also supposed to integrate *LFC* into design software to provide engineers with a practical tool for economic evaluation of new ideas.

Acknowledgements The authors would like to acknowledge the support by Fortum Foundation (grant No. 201700063), the Finnish Strategic Research Council project "Manufacturing 4.0" (grant No. 313396), and TEKES, the Finnish Funding Agency for Innovation, and its program FiDiPro. We would also like to acknowledge the input of Prof. Mikael Collan from LUT School of Business and Management. He provoked and inspired the authors to bridge design and evaluation methods.

References

Altshuller, G. S., & Shapiro, R. B. (1956). Psychology of inventive creativity. *Issues of Psychology, 6*, 37–49.

Besharati, B., Azarm, S., & Kannan, P. K. (2006). A decision support system for product design selection: A generalized purchase modeling approach. *Decision Support Systems, 42*(1), 333–350.

Bloomberg New Energy Finance. (2015, October 12). *The cost landscape of solar and wind.* Available from http://www.senate.mn/committees/2015-2016/3058_Committee_on_Environment_and_Energy/2015-1_BNEF_presentation_final.pdf

Borenstein, S. (2011). The private and public economics of renewable electricity generation. *Journal of Economic Perspectives, 26*(1), 67–92.

Branker, K., Pathak, M. J. M., & Pearce, J. M. (2011). A review of solar photo-voltaic levelized cost of electricity. *Renewable and Sustainable Energy Reviews, 15*(9), 4470–4482.

Breyer, C., & Gerlach, A. (2013). Global overview on grid-parity. *Progress in Photovoltaics: Research and Applications, 21*(1), 121–136.

Campbell, M., Aschenbrenner, P., Blunden, J., Smeloff, E., & Wright, S. (2008). The drivers of the levelized cost of electricity for utility-scale photovoltaics. In *White Paper: SunPower Corporation.*

Chechurin, L., & Borgianni, Y. (2016). Understanding TRIZ through the review of top cited publications. *Computers in Industry, 82*, 119–134.

Dong, A., Lovallo, D., & Mounarath, R. (2015). The effect of abductive reasoning on concept selection decisions. *Design Studies, 37*, 37–58.

Fisher, I. (1907). *The rate of interest: Its nature, determination and relation to economic phenomena.* New York: Macmillan.

Fuller, S. K., & Petersen, S. R. (1995). *Life-cycle costing manual for the federal energy management program.* [cited December 3 2015]. Available from http://www.nist.gov/customcf/get_pdf.cfm?pub_id=907459

Gerasimov, V., Kalish, V., Kuzmin, A., & Litvin, S. S. (1991). Basics of function-cost analysis approach. *Guidlines (in Russian).* Moscow: Moscow, *Inform-FSA*: 40.

Graham, J. R., & Harvey, C. R. (2001). The theory and practice of corporate finance: Evidence from the field. *Journal of Financial Economics, 60*(2–3) (5), 187–243.

Hernández-Moro, J., & Martínez-Duart, J. M. (2013). Analytical model for solar PV and CSP electricity costs: Present LCOE values and their future evolution. *Renewable and Sustainable Energy Reviews, 20*, 119–132.

Hirtz, J. M., Stone, R. B., Szykman, S., McAdams, D. A., & Wood, K. L. (2001). *Evolving a functional basis for engineering design.* Paper presented at proceedings of the ASME design engineering technical conference: DETC2001, Pittsburgh.

Ho, W., Xiaowei, X., & Dey, P. K. (2010). Multi-criteria decision making approaches for supplier evaluation and selection: A literature review. *European Journal of Operational Research, 202*(1), 16–24.

Ilevbare, I. M., Probert, D., & Phaal, R. (2013). A review of TRIZ, and its benefits and challenges in practice. *Technovation, 33*(2–3) (0), 30–7.

Jones, T. W., & Smith, J. D. (1982). An historical perspective of net present value and equivalent annual cost. *The Accounting Historians Journal, 9*, 103–110.

Khastagir, A., & Jayasuriya, N. (2011). Investment evaluation of rainwater tanks. *Water Resources Management, 25*(14), 3769–3784.

Litvin, N. D., Gerasimov, V. M., Kalish, V. S., Karpunin, M. G., & Kuzmin, A. M. (1991). *Basic methodology for functional-cost analysis.* Moscow: InformFCA. (in Russian).

Magrab, E. B., Gupta, S. K., Patrick McCluskey, F., & Sandborn, P. (2009). *Integrated product and process design and development: The product realization process.* Boca Raton: CRC Press.

Malen, D. E. (1996). Decision making in preliminary product design: Combining economic and quality considerations. *The Engineering Economist, 41*(2), 105–122.

Marx, K. (1894). *Capital: A critique of political economy, vol. III. the process of capitalist production as a whole* [Das Kapital, Kritik der politischen Ökonomie] (trans: Untermann, E., ed. F. Engels). Chicago: Charles H. Kerr and Co.

Miles, L. D., & Boehm, H. H. (1967). *Value engineering.* Landsberg: Verlag Moderne Industrie.

Moehrle, M. G. (2005). How combinations of TRIZ tools are used in companies–results of a cluster analysis. *R&D Management, 35*(3), 285–296.

Nikander, J. B., Liikkanen, L. A., & Laakso, M. (2014). The preference effect in design concept evaluation. *Design Studies, 35*(5), 473–499.

Ogden, K. L., Ogden, G. E., Hanners, J. L., & Unkefer, P. J. (1996). Remediation of low-level mixed waste: Cellulose-based materials and plutonium. *Journal of Hazardous Materials, 51*(1), 115–130.

Okudan, G. E., & Tauhid, S. (2008). Concept selection methods-a literature review from 1980 to 2008. *International Journal of Design Engineering, 1*(3), 243–277.

Ouyang, X., & Lin, B. (2014). Levelized cost of electricity (LCOE) of renewable energies and required subsidies in China. *Energy Policy, 70*, 64–73.

Park, Y., & Park, G. (2004). A new method for technology valuation in monetary value: Procedure and application. *Technovation, 24*(5), 387–394.

Pugh, S., & Clausing, D. (1996). *Creating innovative products using total design: The living legacy of Stuart Pugh.* London: Addison-Wesley Longman Publishing.

Reichelstein, S., & Rohlfing, A. (2014). Levelized product cost: Concept and decision relevance. *The Accounting Review 90*(4), 1653–1682.

Ryan, P. A., & Ryan, G. P. (2002). Capital budgeting practices of the fortune 1000: How have things changed. *Journal of Business and Management, 8*(4), 355–364.

SAVE International. (2015, December 3). *Value methodology standard.* Available from http://www.value-eng.org/pdf_docs/monographs/vmstd.pdf

Short, W., Packey, D. J., & Holt, T. (2005). *A manual for the economic evaluation of energy efficiency and renewable energy technologies.* Honolulu: University Press of the Pacific.

Suh, N. P. (1990). *The principles of design.* New York: Oxford University Press.

Toh, C. A., & Miller, S. R. (2015). How engineering teams select design concepts: A view through the lens of creativity. *Design Studies, 38,* 111–138.

US Energy Information Administration. (2015, October 12). *Levelized cost and levelized avoided cost of new generation resources in the annual energy outlook 2015.* Available from http://www.eia.gov/forecasts/aeo/pdf/electricity_generation.pdf

Wallace, K., & Burgess, S. (1995). Methods and tools for decision making in engineering design. *Design Studies, 16*(4) (10), 429–46.

World Energy Council. (2013). *World energy perspective.* cost of energy technologies. [cited October 12 2015]. Available from https://www.worldenergy.org/wp-content/uploads/2013/09/WEC_J1143_CostofTECHNOLOGIES_021013_WEB_Final.pdf

Younker, D. (2003). *Value engineering: Analysis and methodology* (Vol. 30). New York: CRC Press.

Reflecting Emotional Aspects and Uncertainty in Multi-expert Evaluation: One Step Closer to a Soft Design-Alternative Evaluation Methodology

Jan Stoklasa, Tomáš Talášek, and Jana Stoklasová

1 Introduction

Multiple-expert evaluation is based on the assessment of the given object (e.g. a decision, project or design alternative) by several experts. The aim frequently is to obtain an objective assessment and to consider as wide a

J. Stoklasa (✉)
School of Business and Management, Lappeenranta University of Technology, Lappeenranta, Finland

Department of Applied Economics, Faculty of Arts, Palacký University Olomouc, Olomouc, Czech Republic
e-mail: jan.stoklasa@lut.fi

T. Talášek
Department of Applied Economics, Faculty of Arts, Palacký University Olomouc, Olomouc, Czech Republic
e-mail: tomas.talasek@upol.cz

J. Stoklasová
Marital and Family Counseling Centre Prostějov, Prostějov, Czech Republic
e-mail: stoklasova.jana@ssp-ol.cz

© The Author(s) 2019
L. Chechurin, M. Collan (eds.), *Advances in Systematic Creativity*,
https://doi.org/10.1007/978-3-319-78075-7_17

range of viewpoints as possible, so that no drawback of the alternative is overlooked. This is understandable and well in line with the basic ideas of operations research, mainly with the requirement of multi-disciplinarity. Obviously the amount of expertise, the relevance of the experts for the given purpose, their "decision power" and so on can be reflected in the process (i.e. by specifying the weights of their opinions/evaluations). The more diverse the set of experts is, the more comprehensive the overall evaluation obtained from them can be. Alternatively, this diversity introduces several issues concerning the aggregation of the evaluations provided by these individuals. Even if we consider criteria that are measurable (or at least quantifiable), we need to make sure that the same measuring instrument is used, the same scales are applied and that all the experts have access to all the relevant information. Even when this is achieved and the weights of the experts (representing the value of their opinion in the particular situation) are determined, the confidence of the experts' answers can be variable rendering the overall evaluation difficult to interpret.

When we consider less tangible criteria, the situation becomes even more challenging. With a decreasing ability to measure the values of the criterion, the need for a qualitative approach to its assessment increases. Linguistic scales (Zadeh 1975), Likert-type scales (Likert 1932; Stoklasa et al. 2017) or semantic-differential-type scales (Osgood et al. 1957, 1964) anchored with linguistic values are used. Unfortunately, the use of linguistic values introduces another "degree of freedom" in the evaluation process. The words (linguistic expressions) used to anchor the scales can be understood differently (or in some cases even not understood at all) and even if there was some level of consensus concerning the denotative meaning of these linguistic terms, their connotations will most probably vary from person to person. A selection of the same linguistic value by two different evaluators thus no longer guarantees that the same evaluation was expressed by each of them. Unifying the understanding of the meaning of the linguistic terms can be a tedious task (Stoklasa 2014; Stoklasa and Talášek 2015). The uncertainty inherent in the use of linguistic labels and linguistic variables has to be reflected appropriately (see e.g. Stoklasa 2014; Stoklasa et al. 2014; Talašová et al. 2014) and in many cases this is done by the use of fuzzy modelling (Fiss 2011; Stoklasa and

Talášek 2016; Stoklasa et al. 2011, 2014, 2018a; Talašová et al. 2014), interval valued modelling (Stoklasa et al. 2016, 2018b, c) or by finding alternative lossless representations of the set of evaluations instead of their direct aggregation (Stoklasa et al. 2017).

There is one more (and a rather interesting) perspective we can take on multi-expert evaluation. In practice we frequently need to obtain assessment of alternatives, products and projects that are not only "objective" but also reflect the "gut feeling" of the evaluators and their emotions triggered by the alternative. This is important, since a "bad feeling" or "fear" of the suggested alternative can indicate that some criteria (potentially relevant only for a subset of the evaluators) might not have been considered, or even might not be consciously known to the evaluators. This is not a new finding in the field of operations research. Brill, Chang and Hopkins (1982) proposed the "modelling to generate alternatives" (MGA) approach and since then it has been frequently applied in various fields (see e.g. Yeomans (2011), Yeomans and Gunalay (2011) for some recent municipal waste management and environmental management applications). The main idea behind MGA is to replace the search for the best solution by searching for a set of sufficiently good solutions that are comparably good but that differ as much as possible from each other in their characteristics. This is supposed to provide solutions to the "gut unacceptability" of the best solution by providing comparably good alternatives to it that are sufficiently different in terms of their characteristics but not in terms of their outcome. This showcases that "hidden," unknown or forgotten criteria do exist and their identification can prove to be crucial to a successful multi-expert evaluation. It thus seems that emotions can be a relevant factor in the evaluation—and should be reflected in the evaluation models. The emotional component of evaluation has been stressed in the context of Kansei engineering by Jindo, Hirasago, and Nagamachi (1995), Nagamachi (1995) and Kobayashi and Kinumura (2017) and in the field of design (see e.g. Huang et al. 2012) and marketing, where the connections of emotions and products is very relevant.

Mainly alternatives that are good enough in terms of the measurable criteria and that do not trigger a defensive emotional reaction in the decision-makers responsible for the final choice, when suggested as solutions, have the potential to be accepted. It is thus reasonable to reflect the emotional component when needed and to use the information concerning the prevailing emotional tone (and its consistency among the experts) as an additional resource in final decision-making. This way, a soft emotion-oriented evaluation can be considered either as an alternative to standard multi-expert evaluation methods using measurable criteria and well-defined aggregation function, or as an additional approach to the quantitative one providing a qualitative insight, information on less tangible aspects of the evaluation situation and on the consistency of the understanding of (or feeling about) the linguistic values used in the more qualitative context.

In this chapter, we focus on this softer component of evaluation and show how the uncertainty stemming from a lower understanding of the linguistic labels, their perceived irrelevance or lower confidence concerning the final answers can be combined with the information concerning variable emotional responses of the evaluators to the labels (the effect of the connotative aspects of their meaning) in a multi-expert evaluation methodology. We suggest substituting crisp (real-number) values with interval values when the uncertainty is present. We also need to keep in mind that as the scales for measurable criteria need to be of the same type and of the same ranges to be meaningfully aggregated, so do the uncertainties—not all the types of uncertainty can be combined into a single overall uncertainty without the loss of meaning. We propose how to deal with these different types of uncertainty.

To have a clear application framework for such a soft multi-expert evaluation methodology, we choose the area of product design, where emotions not only play a crucial role but also where the stimulation of a specific emotion in the user of the product can even be one of the goals. In this area, the emotional design and Kansei engineering (Nagamachi 1995) approaches, introduced to reflect the consumers' needs in the design process, have already justified the focus on the emotional aspects of the evaluation. More specifically, we are proposing a generalization of the product classification method in emotional design proposed by Huang et al. (2012). The original method uses Kansei adjectives and semantic differential scales (in their standard, real-numbered version) and introduces an inter-expert "emotional connotation" variability check through the assessment of Kansei tags in terms of their emotional connotation. This method is summarized in the following section.

In the third section, we propose a generalization of the data-gathering procedure for the method that reflects the perceived irrelevance of the Kansei tags for the evaluation of a given alternative by introducing an uncertainty into the evaluation—converting the real-number evaluation into an interval one. This step is inspired by Stoklasa, Talášek, and Stoklasová (2018b, c). Also the confidence of the evaluators' answers concerning the emotional connotation of the Kansei tags is recorded and reflected analogously. A new measure of emotional dissensus on the emotional-loading of the Kansei tags is proposed and its use in product evaluation and classification is discussed. The conclusions section follows.

Key Concepts Summary Box

Modelling to generate alternatives (MGA)—an approach to optimization aiming at providing not one but more feasible solutions as different from each other as possible, all with the values of the objective function close to the optimal value (see e.g. Brill et al. 1982 or Yeomans 2011).

Kansei engineering—a consumer-oriented approach to product design based on the reflection of less tangible aspects such as feelings concerning the product in the design process. The aim is to inspire specific feelings by the features of the design alternative (see Nagamachi 1995; Jindo et al. 1995; or Kobayashi and Kinumura 2017).

Kansei adjectives—Kansei words in the form of adjectives, that is, words describing customers' or consumers' needs, feelings and perceptions concerning the product (see e.g. Jiao, Zhang, and Helander (2006) for a Kansei mining system).

Kansei tag—a group or cluster of Kansei adjectives corresponding to the same concept or basic emotion (Xu and Wunsch (2009) provide an example of a clustering algorithm suitable for the creation of Kansei-adjectives clusters, i.e. Kansei tags).

Likert scale—a psychometric measurement instrument popularized by Likert (see e.g. Likert 1932) frequently used in questionnaires. Likert scales are discrete scales with linguistic labels on the agree-disagree or similar continuums, which are supposed to be symmetrical with respect to the middle point (either present in the scale itself, or theoretical; e.g. strongly agree, agree, undecided, disagree, strongly disagree) of the scale. Usually the equidistance of the scale values is assumed.

Semantic differential—a method proposed by Osgood, Suci, and Tannenbaum (1957) for the measurement of attitudes. The method utilizes discrete bipolar-adjective scales to get input information and uses factor analysis to define the semantic space and represents the attitude towards a concept (or its connotative meaning) as a point in this n-dimensional semantic space.

2   Basic-Emotion Based Semantic Differential Method for Product Classification

As suggested by Jiao, Zhang, and Helander (2006), the Kansei adjectives can be used to facilitate the expressing of consumers' needs, emotional states and feelings in connection with the product that is being evaluated. The use of Kansei adjectives (or their clusters represented by Kansei tags; some clustering algorithms suitable for this purpose can be found e.g. in Xu and Wunsch (2009)) is well compatible with semantic-differential-type scales (or Likert-type scales) and as such presents a simple enough combination of tools for obtaining inputs for the evaluation process. It is therefore suggested also in the emotional design semantic-differential method based on basic emotions introduced in Huang et al. (2012).

Let us now consider p evaluators need to evaluate n alternatives with respect to m criteria (represented here by Kansei tags). We also consider q basic emotions, which will be used to assess the variance in understanding the Kansei tags by different evaluators (the number and list of these basic emotions is dependent on the underlying theory we choose for the purpose). Huang et al. (2012) propose a seven-step procedure consisting of the following steps (here we present just a brief description with comments, see Huang et al. (2012, pp. 571–575) for more details):

1. *Selection of the Kansei adjectives* for the purpose of the evaluation and their grouping into *Kansei tags*. This step also involves specifying the

set of alternatives to be evaluated. It is an initial step, which in general terms requires the criteria (here represented by clusters of Kansei adjectives grouped under a unifying Kansei tag) and alternatives to be specified. In addition, the *set of basic emotions* should be specified in this step. We will consider all the Kansei tags to be represented by continuous universes $[-r, r]$, where $r > 0$ (i.e. by intervals of the length $2r$). Note, that any other interval of the same length can be used without any loss of information (just a linear transformation of the values of the interval would be required; e.g. Huang et al. (2012, pp. 573) use intervals $[1, 7]$). The basic emotions will also be represented by continuous universes of a length possibly different from the length of the Kansei-tag universe, denoted by $[-d, d]$, $d > 0$, that is, by intervals of the length $2d$ (Huang et al. (2012, p. 573) consider intervals $[0, 10]$ for this purpose).

2. *Selection of the survey participants.* In other terms, this step requires the selection of evaluators, that is, experts. Different groups of evaluators can be considered (e.g. product users and designers). All the necessary points of view should be represented and the number of the evaluators needs to be reasonable. If needed, weights of the evaluators (i.e. the value of their opinion for the given purpose) can be specified.

3. *Evaluation of the alternatives with respect to the Kansei tags.* A schematic representation of a questionnaire that could be used for this purpose is summarized in the top part of Fig. 1. The evaluation of the alternative a_i with respect to the Kansei tag KT_j by the evaluator k is represented by $x_{K_{ijk}} \in [-r, r]$ in further calculations; $i = 1, \ldots, n, j = 1, \ldots, m$ and $k = 1, \ldots, p$.

In essence, the use of Kansei adjectives as anchors for the poles of Likert-type or semantic-differential-type scales is an example of simple linguistic modelling. As such, it requires a uniform understanding of these adjectives (or Kansei tags) if the information has to be aggregated across the experts/evaluators.

Also the points of view and hence the evaluations of and attitudes towards the object can differ significantly in different subgroups of experts

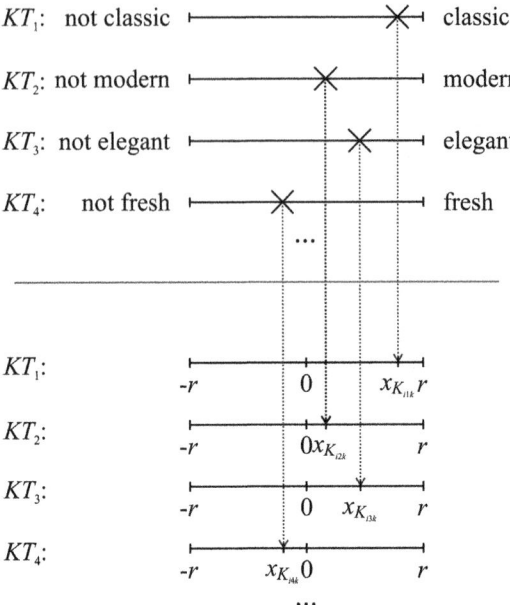

How would you describe a_i:

Fig. 1 Evaluation form for the alternative a_i by the evaluator k with respect to the given Kansei tags, $i = 1, ..., n$ and $k = 1, ..., p$. The upper part represents the tool as seen and used by the evaluator, the lower part represents the conversion of the inputs into model variables' values $x_{K_{ijk}} \in [-r, r]$, where KT_j, $j = 1, ..., m$, represents the j-th Kansei tag

(as confirmed e.g. by Hsu, Chuang and Chang 2000), the importance of criteria can be seen differently and the emotional connotation of the evaluation can be different. It thus makes sense to at least investigate how consistent the group of evaluators, or its subgroups, are in their interpretation of the criteria or linguistic labels used to represent them. Hence, the connotation of the Kansei tags is checked in terms of their association with basic emotions in the next step.

4. *Assessment of the Kansei tags in terms of their emotional associations.* The upper part of Fig. 2 presents the questionnaire used for this purpose and its lower part the conversion of the answers into the values

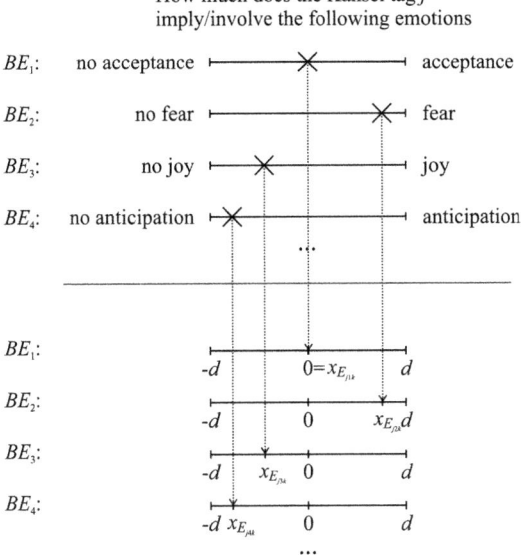

Fig. 2 Assessment form for the Kansei tag j by the evaluator k with respect to the pre-specified basic emotions, $j = 1, ..., m$ and $k = 1, ..., p$. The upper part represents the tool as seen and used by the evaluator, the lower part represents the conversion of the inputs into model variables' values $x_{E_{jlk}} \in [-d, d]$, where BE_l, $l = 1, ..., q$, represents the l-th basic emotion

$x_{E_{jlk}} \in [-d, d]$, that is, numerical values representing the assessment of the Kansei tag KT_j by the evaluator k with respect to the basic emotion BE_l; $j = 1, ..., m$, $k = 1, ..., p$ and $l = 1, ..., q$.

5. *Calculation of the mean value of each Kansei tag*, which is done by (1). This value $\alpha_{K_{ij}}$ is supposed to represent the overall group evaluation of the alternative i with respect to the Kansei tag KT_j.

$$\mu_{K_{ij}} = \frac{\sum_{k=1}^{p} x_{K_{ijk}}}{p} \tag{1}$$

Since the Kansei tags can be interpreted differently by the evaluators, Huang et al. (2012) suggest investigating the "semantic meaning" of the Kansei tags in terms of basic emotions. The idea behind this being that if the perception of the Kansei tag KT_j is very different among the evaluators (i.e. the variability of its evaluation in terms of the basic emotions is too high), then the aggregated value $\mu_{K_{ij}}$ has very difficult interpretation and needs to be modified to account for this large variability. First, the *mean basic-emotion value* $\alpha_{E_{jl}}$ is computed for a Kansei tag KT_j using (2) and then the respective variance V_{jl} is computed using (3).

$$\mu_{E_{jl}} = \frac{\sum_{k=1}^{p} x_{E_{jlk}}}{p} \tag{2}$$

$$V_{jl} = \sum_{k=1}^{p} \frac{\left(x_{E_{jlk}} - \mu_{E_{jl}}\right)^2}{p} \tag{3}$$

Finally, a measure of the total variability V_{KT_j} for each Kansei tag KT_j is calculated using (4). Note, that Huang et al. (2012) compute the total variance as a square root of our V_{KT_j}.

$$V_{KT_j} = \sum_{l=1}^{q} V_{jl} \tag{4}$$

Once the total variability of each Kansei tag is known, it can be interpreted and used in several ways. Generally the higher the value of V_{KT_j}, the larger the inconsistency of understanding and interpreting the j-th Kansei tag is among the evaluators (or to be more specific the more variable emotional associations are triggered by the Kansei tag in the evaluators). One possible way of using these total variance values would be to discard those Kansei tags with the total variability larger than a given threshold, since their aggregated value is almost impossible to interpret correctly. Huang et al. (2012), however, suggest modifying the values of $\alpha_{K_{ij}}$ based on the values of V_{KT_j} as described in the following step.

6. *Calculating the adjusted mean values of the Kansei tags* using (5), where F is a linear or nonlinear mapping function. Huang et al. (2012, pp. 574–575) suggest several possible mapping functions, yet their rationale is not very clear (note that (5) actually moves the average based on the variability).

$$\mu_{K_{ij}}^{adj} = \mu_{K_{ij}} - F\left(V_{KT_j}\right) \tag{5}$$

In fact, the evaluator (i.e. the decision-maker responsible for the final decision) might not know how to choose one of these functions, since no good practices or lists of "mapping functions of choice for particular problems" exist. Even though the authors claim that the actual choice of the mapping function does not have an effect on the final outcome, we consider this step to be a questionable one and as such it is not supported or further commented by us in this chapter. We, however, acknowledge the value of the information carried in V_{KT_j}.

7. *Presenting the results and drawing conclusions*—final classification or evaluation of the alternatives. Let us for now consider that the adjustment represented by (5) is done in a reasonable and meaningful way. Then either threshold values can be specified to see whether an object (alternative) should be classified under a specific Kansei tag, or simply a profile of Kansei mean values can be provided for each alternative. In the latter case an "ideal" or "desired" evaluation in terms of Kansei values can be specified and the alternative closest to this ideal can be chosen.

Although the method suggested by Huang et al. (2012) provides means for the assessment of consistency of understanding (or feeling about) the Kansei tags by the group of evaluators, it can still be further developed. First, the modification of mean Kansei values based on their variance is not well justified and might not even be necessary. Second, the scale relevance issue (i.e. the possibility that some evaluators might consider some of the Kansei tags less than fully appropriate for the evaluation purposes; or that the emotional assessment might be difficult for the evaluators because they might not be entirely confident about their answers in this step, see e.g. Heise 1969) as well as the unclear interpretation of values close to the middle one (Kulas and Stachowski 2009) are not dealt with. Hence, there are still several possible sources of uncertainty that are not accounted for.

In the next section we therefore suggest a modified data-collection procedure in line with Stoklasa, Talášek, and Stoklasová (2016, 2018b, c), which can reflect the lower perceived scale relevance and also lower confidence of the evaluators with their answer. We then adopt the emotional-assessment variance perspective and suggest a measure of inconsistency of the perceptions of Kansei tags and its possible use in the evaluation process.

3   Interval-Valued Generalization of the Basic-Emotion Based Semantic Differential Method

Semantic differential scales are a popular tool for data acquisition, mainly due to their simplicity. Unfortunately, this simplicity comes with a price. During the 60 years since the introduction of semantic differential by Osgood et al. (1957), there have been several studies published concerning the problems possibly associated with the use of the bipolar-semantic differential scales (both discrete and continuous). The main objections were directed towards

- the inability of the original method to reflect lower scale relevance (i.e. the impossibility of expressing perceived partial or complete irrelevance of the scale for the purpose of evaluation by the evaluators);
- concept-scale interactions (Heise 1969)—the need of tailoring the semantic differential scale for each purpose/study;
- the impossibility of expressing ambivalent attitudes (Kaplan 1972)—note that a single value is required from the evaluator on each scale in the standard version of the semantic differential method; and

- the problematic interpretability of middle answers as stressed by Kulas and Stachowski (2009)—it is virtually impossible to know, whether a middle value of the scale provided by the evaluator should be interpreted as a "neutral answer," an answer indicating the irrelevance of the scale for the given purpose or the fact that the evaluator does not understand the anchoring linguistic labels well enough in the given context.

Recently, a solution to many of these issues has been suggested in Stoklasa et al. (2016, 2018b, c) by the enrichment of the data-gathering procedure and by a transition to interval-valued answers. This way, the uncertainty stemming from lower perceived relevance of the scales and lower confidence of the answers no longer remains hidden, but is directly transformed into a multi-valued answer. The difference in the data-gathering procedure with respect to the original semantic differential method lies in the administration of a second scale with each semantic differential one. This scale is represented by a [0%, 100%] universe and is used to obtain the information of the relevance of the scale for the given purpose as perceived by the decision-maker (or it can also be framed as confidence with the answer, etc.). The expressed decrease in relevance or confidence is then proportionally transformed into an interval on the original bipolar-adjective semantic differential scale. Figure 3 summarizes the generalized data-gathering procedure that would in this case replace the one discussed in step 4 of the method by Huang et al. (2012) and depicted in Fig. 1. Note that the perceived Kansei-tag relevance $y_{K_{ijk}}$ expressed by the evaluator on the relevance scales r_j is transformed into the values $w_{K_{ijk}} \in [0, 2r]$ using (6), $i = 1, \ldots, n, j = 1, \ldots, m$ and $k = 1, \ldots, p$. These values represent a part of the Kansei-tag universe proportional in size to the perceived irrelevance of the Kansei tag for the purpose of the evaluation.

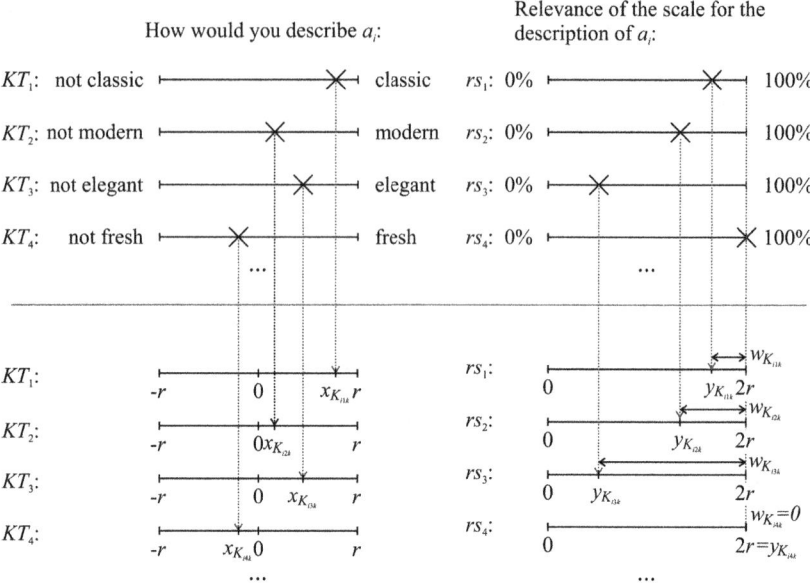

Fig. 3 Evaluation form for the alternative a_i by the evaluator k with respect to the given Kansei tags, $i = 1, \ldots, n$ and $k = 1, \ldots, p$ – extended version inspired by Stoklasa, Talášek, and Stoklasová (2016, 2018c). The upper part of the figure represents the tool as seen and used by the evaluator, the lower part represents the conversion of the inputs on the Kansei tag scales into model variables' values $x_{K_{ijk}} \in [-r, r]$, where KT_j, $j = 1, \ldots, m$, represents the j-th Kansei tag, and of the perceived scale relevance into uncertainty regions of the width $w_{K_{ij}}$. The right part of the figure (titled "Relevance of the scale for the description of a_i:") denotes the addition with respect to the original semantic differential method

$$w_{K_{ijk}} = 2r - y_{K_{ijk}} \tag{6}$$

Based on these values, the resulting interval-valued evaluation $\left[x^L_{K_{ijk}}, x^R_{K_{ijk}}\right]$ is computed. The procedure first checks if an uncertainty interval of the width $w_{K_{ijk}}$ can be defined symmetrically around $x_{K_{ijk}}$ and still fit into the $[-r, r]$ interval (i.e. if $\left[x_{K_{ijk}} - \frac{w_{K_{ijk}}}{2}, x_{K_{ijk}} + \frac{w_{K_{ijk}}}{2}\right] \subseteq [-r, r]$). If this is possible, then $\left[x^L_{K_{ijk}}, x^R_{K_{ijk}}\right]$ is defined symmetrically around $x_{K_{ijk}}$, otherwise the uncertainty interval is shifted in such a way that it remains a subset of $[-r, r]$ and retains its width $w_{K_{ijk}}$. This is summarized in formula (7).

$$
\left[x_{K_{ijk}}^L, x_{K_{ijk}}^R \right] = \begin{cases} \left[-r, -r + w_{K_{ijk}} \right] & \text{for} \left(x_{K_{ijk}} - \dfrac{w_{K_{ijk}}}{2} \right) < -r, \\[3mm] \left[r - w_{K_{ijk}}, r \right] & \text{for} \left(x_{K_{ijk}} + \dfrac{w_{K_{ijk}}}{2} \right) > r, \text{ and} \\[3mm] \left[x_{K_{ijk}} - \dfrac{w_{K_{ijk}}}{2}, x_{K_{ijk}} + \dfrac{w_{K_{ijk}}}{2} \right] & \text{otherwise} \end{cases} \tag{7}
$$

Analogously, the interval-valued assessments of the Kansei tags with respect to the basic emotions $\left[x_{E_{jlk}}^L, x_{E_{jlk}}^R \right]$ can be computed using formula (8), where $w_{E_{jlk}}$ is computed using (9), and the uncertainty stems from a lower confidence of the answer concerning a basic emotion BE_l expressed on the scale ca_l represented by the [0%, 100%] interval. An example of the data input form along with the necessary notation is summarized in Fig. 4.

$$
\left[x_{E_{jlk}}^L, x_{E_{jlk}}^R \right] = \begin{cases} \left[-d, -d + w_{E_{jlk}} \right] & \text{for} \left(x_{E_{jlk}} - \dfrac{w_{E_{jlk}}}{2} \right) < -d, \\[3mm] \left[d - w_{E_{jlk}}, d \right] & \text{for} \left(x_{E_{jlk}} + \dfrac{w_{E_{jlk}}}{2} \right) > d, \text{ and} \\[3mm] \left[x_{E_{jlk}} - \dfrac{w_{E_{jlk}}}{2}, x_{E_{jlk}} + \dfrac{w_{E_{jlk}}}{2} \right] & \text{otherwise} \end{cases} \tag{8}
$$

$$
w_{E_{jkl}} = 2d - y_{E_{jlk}} \tag{9}
$$

Let us now consider applying the extended data-gathering procedure as summarized in Figs. 3 and 4, that is, that we obtain the interval values $\left[x_{K_{ijk}}^L, x_{K_{ijk}}^R \right]$ instead of $x_{K_{ijk}}$ and $\left[x_{E_{jlk}}^L, x_{E_{jlk}}^R \right]$ instead of $x_{E_{jlk}}$ for all $i = 1$, ..., n, $j = 1$, ..., m, $k = 1$, ..., p and $l = 1$, ..., q. Note that in any case the

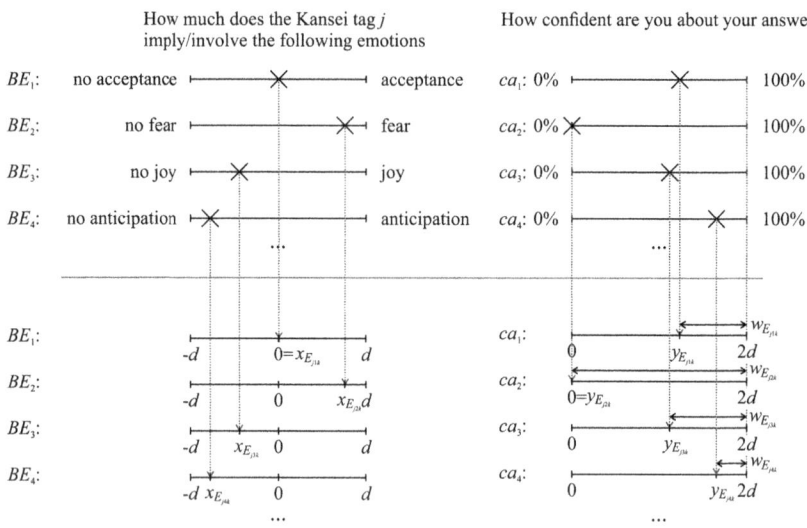

Fig. 4 Assessment form for the Kansei tag j by the evaluator k with respect to the pre-specified basic emotions, $j = 1, \ldots, m$ and $k = 1, \ldots, p$ – extended version inspired by Stoklasa, Talášek, and Stoklasová (2016). The upper part of the figure represents the tool as seen and used by the evaluator, the lower part represents the conversion of the inputs on the Kansei tag scales into model variables' values $x_{E_{jlk}} \in [-d, d]$, where $BE_l, l = 1, \ldots, q$, represents the l-th basic emotion, and of the perceived confidence of the answer into uncertainty regions of the width $w_{E_{jlk}}$. The right part of the figure (titled "How confident are you with your answer:") denotes the addition with respect to the original semantic differential method

original crisp values expressed on the Kansei-tag scales always lie in the uncertainty intervals (i.e. $x_{K_{ijk}} \in \left[x_{K_{ijk}}^L, x_{K_{ijk}}^R \right]$). The same holds analogously for the basic-emotion assessment of the Kansei tags (i.e. $x_{E_{jlk}} \in \left[x_{E_{jlk}}^L, x_{E_{jlk}}^R \right]$). The proposed modification is so far a direct generalization of the original method, since if there is no uncertainty, that is, when $w_{K_{ijk}} = 0$ for some i, j and k, we get the interval-valued evaluation computed using (7) in the form $\left[x_{K_{ijk}}^L, x_{K_{ijk}}^R \right] = \left[x_{K_{ijk}}, x_{K_{ijk}} \right]$, which is in fact nothing else than an interval representation of the real number $x_{K_{ijk}}$. The same holds for (8) and the assessment of Kansei tags with respect to the basic emotions. The *interval-valued basic-emotion based semantic differential method* can now be summarized in the following steps:

1. *Selection of the Kansei adjectives* for the purpose of the evaluation and their grouping into *m Kansei tags*, specification the *set of q basic emotions*. There is no difference in this step with respect to the method proposed in Huang et al. (2012). We will again consider all the Kansei tags to be represented by continuous universes $[-r, r]$, where $r > 0$ and the basic emotions to be represented by continuous universes of a possibly different length, denoted by $[-d, d]$, $d > 0$.

2. *Selection of the survey participants.* Again, no change with respect to Huang et al. (2012).

3. *Evaluation of the alternatives with respect to the Kansei tags.* The enhanced questionnaires depicted in Fig. 3 will be used. Although these questionnaires double the number of answers needed from the evaluators, the information concerning the relevance of the Kansei tag is not a difficult one to provide. In fact, a 100% relevance can be considered a default value and only if the perceived relevance is lower, would an input from the evaluator specifying how low it is be required. The evaluations of the alternative a_i with respect to the Kansei tag KT_j by the evaluator k are now represented by $\left[x^L_{K_{ijk}}, x^R_{K_{ijk}} \right] \subseteq [-r, r]$, $i = 1$, ..., n, $j = 1$, ..., m and $k = 1$, ..., p.

4. *Assessment of the Kansei tags in terms of their emotional associations.* Again, an enhanced questionnaire (see Fig. 4) will be used to obtain inputs for this purpose. The intervals representing the assessment of the Kansei tag KT_j by the evaluator k with respect to the basic emotion BE_l are thus obtained in the form of $\left[x^L_{E_{jlk}}, x^R_{E_{jlk}} \right] \subseteq [-d, d]$, $j = 1$, ..., m, $k = 1$, ..., p and $l = 1$, ..., q.

5. *Calculation step – determination of the Kansei-tag means and assessment of the consistence of understanding of the Kansei tag by the evaluators.* The Kansei tag mean values, represented again by intervals, are now calculated by (10).

$$\mu^I_{K_{ij}} = \frac{\sum_{k=1}^{p} \left[x^L_{K_{ijk}}, x^R_{K_{ijk}} \right]}{p} = \left[\sum_{k=1}^{p} \frac{x^L_{K_{ijk}}}{p}, \sum_{k=1}^{p} \frac{x^R_{K_{ijk}}}{p} \right] \quad (10)$$

This value $\alpha^I_{K_{ij}}$ is supposed to represent the overall group evaluation of the alternative i with respect to the Kansei tag KT_j. It can, however, be properly interpreted only if the understanding of the Kansei tags was identical (or very similar) for all the evaluators. If some of the evaluators have different emotional associations with the Kansei tags than the others, the aggregated value $\alpha^I_{K_{ij}}$ might be difficult to interpret. In line with Huang et al. (2012), we will therefore investigate the consistency of the emotional associations triggered by the Kansei tags in the group of evaluators. First, we calculate the *mean basic-emotion value* $\alpha^I_{E_{jl}}$ for each Kansei tag KT_j and each basic emotion BE_l using (11).

$$\mu^I_{E_{jl}} = \frac{\sum_{k=1}^{p}\left[x^L_{E_{jlk}}, x^R_{E_{jlk}}\right]}{p} = \left[\sum_{k=1}^{p}\frac{x^L_{E_{jlk}}}{p}, \sum_{k=1}^{p}\frac{x^R_{E_{jlk}}}{p}\right] \qquad (11)$$

Now we need to assess to what extent the intervals $\alpha^I_{E_{jl}}$ differ among the evaluators. To do so, we will now apply the concept of strong consensus in the BE_{jl} dimension as introduced in Stoklasa, Talášek, and Stoklasová (2018b). A set of p interval-valued evaluations $\{I_1, \ldots, I_p\}$ with their associated crisp values $x_{I_k} \in I_k$, for all $k = 1, \ldots, p$, is considered to represent a strong consensus in the given evaluation dimension if and only if $I_1 \cap \ldots \cap I_p \neq \varnothing$ and $x_{I_k} \in \left(I_1 \cap \ldots \cap I_p\right)$ for all $k = 1, \ldots, p$. A strong consensus is thus present if there exists an interval of values on which all the evaluators agree and that comprises all the crisp evaluations. Having introduced this concept, we now define a measure of inconsistency (variability) of the emotional assessment of the Kansei tag KT_j with respect to the basic emotion BE_l as V^I_{jl}, which is computed using (12), where

$$CI_{jl} = \bigcap_{k=1}^{p}\left[x^L_{E_{jlk}}, x^R_{E_{jlk}}\right] = \left[CI^L_{jl}, CI^R_{jl}\right], \quad MI_{jl} = \left[\min_k x^R_{E_{jlk}}, \max_k x^L_{E_{jlk}}\right] \text{ and}$$

$d\left(x_{K_{ijk}}, CI_{jl}\right)$ is defined by (13).

$$V_{jl}^I = \begin{cases} 0 & \text{if } CI_{jl} \neq \varnothing \text{ and } x_{E_{jlk}} \in CI_{jl} \forall k, \\[2ex] \dfrac{\sum_{k=1}^{p} d\left(x_{E_{jlk}}, CI_{jl}\right)}{p} & \text{if } CI_{jl} \neq \varnothing \text{ and } x_{E_{jlk}} \notin CI_{jl} \text{ for some } k, \\[3ex] \left|MI_{jl}\right| + \dfrac{\sum_{k=1}^{p} d\left(x_{E_{jlk}}, MI_{jl}\right)}{p} & \text{if } CI_{jl} = \varnothing \end{cases} \tag{12}$$

$$d\left(x_{E_{jlk}}, CI_{jl}\right) = \begin{cases} \left[\left[x_{E_{jlk}}, CI_{jl}^L\right]\right] & \text{if } x_{E_{jlk}} < CI_{jl}^L, \\[2ex] \left[\left[CI_{jl}^R, x_{E_{jlk}}\right]\right] & \text{if } CI_{jl}^R < x_{E_{jlk}}, \\[2ex] 0 & \text{otherwise} \end{cases} \tag{13}$$

The idea behind (12) is, that if a strong consensus of the interval evaluations of the Kansei tag KT_j with respect to the basic emotion BE_l exists, then it is possible to find a consensual evaluation and as such the evaluations can be considered consistent (hence the zero value of V_{jl}^I in this case). If there is no strong consensus, but if the interval CI_{jl} is nonempty (i.e. if at least *weak consensus* exists), the variability is calculated as the average of the distances of the crisp evaluations from this interval defined by (13). If there is even no weak consensus, we define the smallest interval that would represent a weak consensus, MI_{jl}, calculate the average distance of the crisp evaluations from this interval (again by (13)) and add it to the length of MI_{jl} to obtain the value of the variability.

Now that we have a measure of the variability of the assessment of each Kansei tag with respect to a single basic emotion (expressed in fact as the measure of dissensus of the evaluations), we need to define an overall variability measure for the Kansei tag across all the basic emotions. Since only basic emotions are considered (and not complex ones derived or composed from them), we can consider them to be independent and to constitute q dimensions of an emotional-assessment Cartesian space. In this space, we can define the up-to-q dimensional area of variability VA_j as an up-to-q dimensional box with edges of the lengths V_{jl}^I by (14).

$$VA_j = \left[0,V_{j1}^I\right] \times \ldots \times \left[0,V_{jq}^I\right] \tag{14}$$

The larger the area VA_j is, the more substantial change of evaluations would have to take place for a strong consensus to be reached. Note, that VA_j represents an up-to-q dimensional block in the basic-emotion space. A measure of its size could therefore be an applicable measure of the overall variability of the emotional assessment of KT_j by the evaluators. We suggest the length of the body diagonal as the overall variability measure $V_{KT_j}^I$. More specifically we use a normalized body diagonal length computed by (15), that is, $V_{KT_j}^I \in [0,1]$.

$$V_{KT_j}^I = \frac{\sqrt{\left(V_{j1}^I\right)^2 + \left(V_{j2}^I\right)^2 + \cdots + \left(V_{jq}^I\right)^2}}{\sqrt{q(2d)^2}} \tag{15}$$

6. *Reflection of the variability of understanding of Kansei tags in terms of basic emotions.* The variability of the emotional interpretation of the Kansei tags and the uncertainty thus introduced in the evaluation model stems from a different source than the uncertainty defining the intervals $\left[x_{K_{ijk}}^L, x_{K_{ijk}}^R\right]$. It is therefore difficult to combine these two different pieces of information into one and to modify the Kansei-tag mean values based on their variability. There are, however, several methodologically safer ways of using the information concerning the variability of understanding of Kansei tags by the evaluators:

- the simplest way being just presenting the variability of Kansei tags along with the interval-valued Kansei tag means in the evaluation profile. This way no information is lost or distorted. However, it requires more competencies and skills from the decision-maker responsible for the final decision.

- or a threshold for acceptable inconsistency can be specified and those Kansei tags that do not meet this minimum consistency requirement might be discarded from the evaluation. This way, however, we are potentially loosing information and it may render a significant part of the data we have gathered useless, if many Kansei tags have higher overall variability.
- since $V_{KT_j}^I \in [0,1]$, these values can be used as weights of the Kansei tags in further aggregation of the results, or even as weights of fuzzy rules using the Kansei-tag mean values for classification and/or interpretation purposes.

This way we have covered step 7 of the original method as well.

The interval-valued version of the basic-emotion based semantic differential method suggested in this chapter can be used as a soft counterpart to standard multi-expert, multiple-criteria decision-making methods, offering both means for the assessment of less tangible criteria and tools for consistency checking of the connotative meanings of the linguistic labels used in the model. Although the basic-emotion perspective might not be applicable in all problems, it constitutes a blueprint for analogous assessments of the connotative component of the meaning of linguistic terms used as anchors in the semantic differential method.

4   Conclusion

This chapter suggests a soft design-alternative evaluation methodology using the basic-emotion based semantic differential method by Huang et al. (2012) as its basis and utilizing the interval-valued extension of the semantic differential proposed by Stoklasa, Talášek, and Stoklasová (2018c) and the concepts of strong and weak consensus introduced by Stoklasa, Talášek, and Stoklasová (2018b). The combination of these approaches introduces two new possible sources of uncertainty in the method originally proposed by Huang et al. (2012) and offers means for dealing with the low-scale relevance issue as well as with some other commonly identified drawbacks of semantic differential scales. It presents a generalization of the method by Huang et al., but does not perform the Kansei-tag-mean modification step. Instead, alternative uses of the variability of Kansei tags are suggested. The proposed emotion-based, linguistic, multi-expert evaluation method constitutes a tool for the evaluation of less tangible (and difficult to measure) aspects of the alternatives in multi-expert evaluation problems not restricted to the area of design (or consumer product) evaluation, but generally to every problem where qualitative criteria need to be reflected by a group of experts. As such, it might be an interesting source of inspiration also for social sciences and humanities, that is, areas of research dealing with difficult-to-measure concepts.

Acknowledgments This research was partially supported by the grant IGA_FF_2018_002 of the internal grant agency of Palacký University Olomouc.

References

Brill, E. D., Chang, S.-Y., & Hopkins, L. D. (1982). Modeling to generate alternatives: The HSJ approach and an illustration using a problem in land use planning. *Management Science, 28*(3), 221–235. https://doi.org/10.1287/mnsc.28.3.221.

Fiss, P. C. (2011). Building better causal theories: A fuzzy set approach to typologies in organization research. *Academy of Management Journal, 54*(2), 393–420. https://doi.org/10.5465/AMJ.2011.60263120.

Heise, D. R. (1969). Some methodological issues in semantic differential research. *Psychological Bulletin, 72*(6), 406–422. https://doi.org/10.1037/h0028448.

Hsu, S. H., Chuang, M. C., & Chang, C. C. (2000). A semantic differential study of designers' and users' product form perception. *International Journal of Industrial Ergonomics, 25*(4), 375–391. https://doi.org/10.1016/S0169-8141(99)00026-8.

Huang, Y., Chen, C. H., & Khoo, L. P. (2012). Products classification in emotional design using a basic-emotion based semantic differential method. *International Journal of Industrial Ergonomics, 42*(6), 569–580. Elsevier Ltd. https://doi.org/10.1016/j.ergon.2012.09.002.

Jiao, J., Zhang, Y., & Helander, M. (2006). A Kansei mining system for affective design. *Expert Systems with Applications, 30*(4), 658–673. https://doi.org/10.1016/j.eswa.2005.07.020.

Jindo, T., Hirasago, K., & Nagamachi, M. (1995). Ergonomics development of a design support system for office chairs using 3-D graphics. *International Journal of Industrial Ergonomics, 15*, 49–62.

Kaplan, K. J. (1972). On the ambivalence-indifference problem in attitude theory and measurement: A suggested modification of the semantic differential technique. *Psychological Bulletin, 77*(5), 361–372. https://doi.org/10.1037/h0032590.

Kobayashi, M., & Kinumura, T. (2017). A method of gathering, selecting and hierarchizing Kansei words for a hierarchized Kansei model. *Computer-Aided Design and Applications, 14*(4), 464–471. Taylor & Francis. https://doi.org/10.1080/16864360.2016.1257188.

Kulas, J. T., & Stachowski, A. A. (2009). Middle category endorsement in odd-numbered Likert response scales: Associated item characteristics, cognitive demands, and preferred meanings. *Journal of Research in Personality, 43*(3), 489–493. Elsevier Inc. https://doi.org/10.1016/j.jrp.2008.12.005.

Likert, R. (1932). A technique for the measurement of attitudes. *Archives of Psychology, 22*(140), 5–55.

Nagamachi, M. (1995). Kansei engineering: A new consumer-oriented technology for product development. *International Journal of Industrial Ergonomics, 15*, 3–11.

Osgood, C. E. (1964). Semantic differential technique in the comparative study of cultures. *American Anthropologist, 66*(3), 171–200. https://doi.org/10.1515/9783110215687.109.

Osgood, C. E., Suci, G. J., & Tannenbaum, P. H. (1957). *The measurement of meaning*. Chicago: University of Illinois Press.

Stoklasa, J. (2014). *Linguistic models for decision support*. Lappeenranta: Lappeenranta University of Technology.

Stoklasa, J., & Talášek, T. (2015). Linguistic modelling in economics and management practice – Some open issues. In *Proceedings of the international scientific conference knowledge for market use* (pp. 959–969). Olomouc: Societas Scientiarum Olomucensis II.

Stoklasa, J., & Talášek, T. (2016). On the use of linguistic labels in AHP: Calibration, consistency and related issues. In *Proceedings of the 34th international conference on mathematical methods in economics* (pp. 785–790). Liberec: Technical University of Liberec.

Stoklasa, J., Talašová, J., & Holeček, P. (2011). Academic staff performance evaluation – Variants of models. *Acta Polytechnica Hungarica, 8*(3), 91–111.

Stoklasa, J., Talášek, T., & Musilová, J. (2014). Fuzzy approach – A new chapter in the methodology of psychology? *Human Affairs, 24*(2), 189–203. https://doi.org/10.2478/s13374-014-0219-8.

Stoklasa, J., Talášek, T., & Stoklasová, J. (2016). Semantic differential and linguistic approximation – Identification of a possible common ground for research in social sciences. In *Proceedings of the international scientific conference knowledge for market use* (pp. 495–501). Olomouc: Societas Scientiarum Olomucensis II.

Stoklasa, J., Talášek, T., Kubátová, J., & Seitlová, K. (2017). Likert scales in group multiple-criteria evaluation. *Journal of Multiple-Valued Logic and Soft Computing, 29*(5), 425–440.

Stoklasa, J., Talášek, T., & Luukka, P. (2018a). Fuzzified Likert scales in group multiple-criteria evaluation. In M. Collan & J. Kacprzyk (Eds.), *Soft computing applications for group decision-making and consensus modeling* (Vol. 357, pp. 165–185). https://doi.org/10.1007/978-3-319-60207-3_11.

Stoklasa, J., Talášek, T., & Stoklasová, J. (2018b). *Attitude-based multi-expert evaluation of design.* (unpublished).

Stoklasa, J., Talášek, T., & Stoklasová, J. (2018c). *Semantic differential for the twenty-first century: Scale relevance and uncertainty entering the semantic space.* Quality & Quantity. https://doi.org/10.1007/s11135-018-0762-1.

Talašová, J., Stoklasa, J., & Holeček, P. (2014). HR management through linguistic fuzzy rule bases – A versatile and safe tool? In *Proceedings of the 32nd international conference on mathematical methods in economics* (pp. 1027–1032). Olomouc: Palacký University.

Xu, R., & Wunsch, D. C. (2009). *Clustering.* Hoboken: Wiley.

Yeomans, J. S. (2011). Efficient generation of alternative perspectives in public environmental policy formulation: Applying co-evolutionary simulation-optimization to municipal solid waste management. *Central European Journal of Operations Research, 19*(4), 391–413. https://doi.org/10.1007/s10100-011-0190-y.

Yeomans, J. S., & Gunalay, Y. (2011). An efficient modelling to generate alternatives approach for addressing unmodelled issues and objectives in public environmental planning. *Asian Journal of Information Technology, 10*(3), 122–128. https://doi.org/10.3923/ajit.2011.122.128.

Zadeh, L. A. (1975). The concept of a linguistic variable and its application to approximate reasoning-I. *Information Sciences, 8*(3), 199–249. https://doi.org/10.1016/0020-0255(75)90036-5.

Using Innovation Scorecards and Lossless Fuzzy Weighted Averaging in Multiple-criteria Multi-expert Innovation Evaluation

Mikael Collan and Pasi Luukka

1   Introduction

When evaluating innovations it is important to be smart about it and to consider the nature of such evaluations very carefully. Typically, evaluation in this context is "man-made," that is, the measurement tool, or the yardstick used, is a human expert, often supported by an evaluation structure, such as a list of the to-be-considered criteria. The more criteria are considered, the more complex the evaluation becomes. Furthermore, to gain a holistic view, the evaluations made on the multiple criteria must be somehow aggregated.

Estimates made about innovations are such that they typically try to capture within them events that will happen in the future. This means that the estimates are forward-looking and that innovation evaluation, as any evaluation for that matter, suffers from imprecision related to the measurement tool and caused by the complexity of the future. The

M. Collan (✉) • P. Luukka

School of Business and Management, Lappeenranta University of Technology, Lappeenranta, Finland

e-mail: mikael.collan@lut.fi; pasi.luukka@lut.fi

© The Author(s) 2019

L. Chechurin, M. Collan (eds.), *Advances in Systematic Creativity*,

https://doi.org/10.1007/978-3-319-78075-7_18

imprecision comes from the less-than-perfect ability of the experts to forecast how an innovation will fare in the future. If the experts are able to express the imprecision they experience within their estimates, it is quite intuitive to say that capturing this imprecision would make sense, as capturing it would mean not losing information about it.

In other words, to be able to perform evaluation of innovations, one needs an apparatus that is able to both consider and help in the estimation of multiple criteria in a way that the estimates can be expressed in a manner that captures the imprecision of the said estimates (as viewed by the estimators). In this chapter, we have chosen to use scorecards as a tool for the elicitation of multiple-criteria estimations of innovations, and we take use of fuzzy sets and fuzzy logic (Zadeh 1965, 1978), or as some like to call it "the logic of the imprecise", for capturing the estimation imprecision.

The origins of management use of scorecards can be said to go back at least fifty years; however, it seems to be rather difficult to know when exactly the first management scorecards were used. It is only via the "balanced scorecard" (Kaplan and Norton 1992, 1993) that scorecards rose to prominence in management. Scorecards are, however, a much older invention and "regular" use of scorecards to record scores and to aggregate results has been used in sports, for example, golf, for a long time. Scorecards are what can be called "weighted sum models" (WSM), where the weights the criteria in the model receive are uniform. WSM are simple decision analysis models for the treatment and aggregation of information given on multiple criteria.

In this chapter, we concentrate on the aggregation procedure used in aggregating scorecards, when the scorecards use imprecise estimates modeled with fuzzy logic (modeled as fuzzy numbers). The focus is especially on the idea that it would make sense to include all the information from the "start to the finish" of the procedure, without losing some of it in the process. This is why we have chosen to use a new aggregation operator, the lossless fuzzy weighted averaging (LFWA) operator (Luukka et al. 2018), and to test it in the context of innovation management, together with scorecards. To the best of our knowledge, this is the first time the LFWA is used in this context. The main contribution of this chapter is to show how the LFWA is able to preserve the information contained in expert estimates and how aggregation of innovation scorecards can be done by using LFWA.

We also show how the ranking of innovations may change on the basis of the aggregation procedure used and point out that this may have an effect on the possible investment choices made based on such rankings. The numerical example used to illustrate the methods uses a simple procedure of calculating the center of gravity (COG) to perform the ranking (alternatives are ranked based on the COG). We acknowledge that in order to keep all the relevant information "with us" for as long as possible, the choice of COG may not be the best one, however, in this chapter we focus on the notion of lossless aggregation and not on the notion of lossless "defuzzification." Furthermore, for the purposes of this research we concentrate our focus on situations, where we have "quite consensual" estimates, that is, we have experts who see things quite similarly. These are simplifying assumptions that help us frame the problem. The study of more problematic cases, such as extreme dissensus between evaluators' evaluations, falls outside the scope of this chapter and remains an issue for further research.

This chapter continues with an introduction of the main background concepts underlying this research and then with a numerical case illustration of how new innovation designs can be evaluated, the evaluations aggregated and ultimately ranked by way of using the LFWA method. The chapter closes with a discussion and some conclusions.

2　Background Concepts

As discussed above, the main mathematical framework underlying the research introduced in this chapter is fuzzy logic (Zadeh 1965); we are using fuzzy numbers and fuzzy sets in capturing the imprecision connected to expert estimates. Specifically, we use triangular fuzzy numbers. Fuzzy sets (A) in general can be understood as subsets of a universal set (X) $A \subseteq X$, where partial degree of membership to a set is allowed.

Definition 1 A fuzzy set $A \subseteq X$ is given by its membership function

$$\mu_A : X \rightarrow [0,1]$$

A membership function of A in X (μ_A) can also be denoted by $A(x)$. Nowadays, the latter is even more commonly used. Every element $x \in X$ has a membership degree $\mu_A(x) \in [0, 1]$. A set of pairs

$$A = \left\{ x, |\; \mu_A(x), |\; x \in X \right\}$$

defines a fuzzy set A.

Fuzzy numbers are special cases of fuzzy sets. Triangular fuzzy numbers are defined below.

Definition 2 A triangular fuzzy number A can be defined by a triplet $A = (a_1, a_2, a_3)$. The membership function $A(x)$ is defined as (Kaufmann and Gupta 1985):

$$A(x) = \begin{cases} \dfrac{x - a_1}{a_2 - a_1}, & a_1 \leq x \leq a_2 \\ \dfrac{a_3 - x}{a_3 - a_2}, & a_2 \leq x \leq a_3 \\ 0, & otherwise \end{cases} \tag{1}$$

The arithmetic operations for triangular fuzzy numbers have been covered in a number of previous works (see e.g. Kaufmann and Gupta 1985).

2.1   Creating Fuzzy Numbers from Expert Estimates

Expert estimates are collected by using scorecards, where the experts are asked to give, in addition to a best estimate, a minimum possible and a maximum possible estimate for the criterion estimated. These three estimates can be understood as three scenarios, where the minimum and the maximum possible estimates represent "all goes bad" and "all goes well" scenarios. Elicitation of information in this way is very similar to what is presented, for example, in Collan and Heikkilä (2011) and Collan et al. (2013).

From these three criteria-per-criteria estimates, we proceed in the following way:

(i) we observe that the best-guess scenario is the most likely one and we assign it "full degree of membership in the set of possible outcomes";

(ii) we observe that the minimum possible and the maximum possible estimates are the upper and the lower bounds of what is possible according to the expert—we assign a "limit to zero degree of membership in the set of possible outcomes."

(iii) we make the simplifying assumption that a linear relationship exists between the three points and that the three points thus define a triangular set that fulfills the definition of a triangular fuzzy number. These sets are treated as fuzzy numbers in any following operations.

The triangular fuzzy numbers constructed in this way are considered to be satisficing representations of the expert estimates that are suitable for the purposes of this research and more importantly, suitable for real-world management of innovations.

2.2   Lossless Fuzzy Weighted Averaging

The LFWA operator was first created for the purpose of aggregating multiple-expert estimates of one-off operational risks in the context of banking (Luukka et al. 2018). The idea of the operator is that the aggregation of the information obtained from expert estimates should be reflected in the aggregation end result in a way that does not suffer from information loss. The starting point for the creation of the operator was the observation that when the fuzzy weighted average (FWA) operator (Bojadziev and Bojadziev 2007; Collan and Luukka 2016) is used, relevant information about the minima and the maxima of the aggregated fuzzy numbers is being lost in the process. FWA is defined as:

Definition 3 The FWA operator of dimension n is a mapping FWA: X^n $\rightarrow X$ that has an associated weighting vector W, of dimension n, such that $\sum_{i=1}^{n} w_i = 1$ and $w_i \in [0, 1]$, then:

$$FWA\left(A_1,, A_2,,\ldots A_n\right) = \sum_{i=1}^{n} w_i A_i \qquad (2)$$

where $(A_1,\ldots,A_n)$, are now fuzzy triangular numbers (see definition 2).

Such loss with the FWA could cause, in the context of banking risk management, relevant inaccuracy in estimating an expected loss, based on the aggregated expert estimates. This same argument, relevant loss of important information, holds also for evaluation of innovations—if information about the estimated minima and the maxima is lost, then ranking of evaluations, based on aggregated estimates, may be affected by the aggregation. Such effects on ranking may end up causing, for example, sub-optimal investment decisions with regards to the innovation alternatives.

The LFWA is based on the idea of adding up the "contribution" from each fuzzy number A_i that contributes to the LFWA, for each possible point x_j of the LFWA, in a way that individual contributing weights w_i are provided for each fuzzy number participating in the aggregation.

The procedure is described as follows (Luukka et al. 2018):

First, n fuzzy numbers, where n is the number of experts giving the aggregated estimates, are typically elicited.

Second, the experts evaluate credibility scores (CS) for each fuzzy number used in the creation of the LFWA. It is also possible to think that the credibility scores represent the credibility of the experts and via their credibility also the credibility of the fuzzy numbers generated through their estimations. These credibility scores are then transformed into individual weights for each fuzzy number by the following formula:

$$w_i = \frac{CS_i}{\sum_{i=1}^{n} CS_i} \qquad (3)$$

Third, with n fuzzy numbers and n weights, an average membership degree for each point $x \in X$ belonging to $A_1, A_2, \ldots, A_n$ is calculated. This is done by computing the weighted average $A_c(x)$ from n membership values of the contributing fuzzy numbers for each particular point $x \in X$ of the LFWA. This is formulated as:

$$A_c(x) = \sum_{i=1}^{n} w_i A_i(x) \qquad (4)$$

To set the universe or range for which this procedure is applied, an interval where non-zero averages are reached is defined. This is done by taking the union of the fuzzy numbers contributing to the LFWA and by computing the support area of this union:

$$X = cl\left(supp\left(A_1 \cup A_2 \cup \cdots \cup A_n\right)\right) = \left[x_1, x_2\right] \qquad (5)$$

where $supp(A) = \{x \in X | A(x) > 0\}$ used union is the standard max operator. With the closure used denoting that the support is bounded and leads to a closed interval $[x_1, x_2]$.

2.3   Scorecard Aggregation

The tool for data collection and data aggregation in this research is the scorecard, a simple management tool that rose to prominence in the world of business research in the wake of the successful "balanced score-card" (Kaplan and Norton 1992, 1993) in the 1990s. It has also been used in the management of research and development investments (Li and Dalton 2003; Bremser and Barsky 2004). In this research we are not interested in the balanced scorecard system but in using scorecards generally for the purpose of gathering (eliciting) information and aggre-gating it for the purposes of innovation management. From a formal point of view, a simple scorecard can be said to be a multiple-criteria decision-making system, where the weights of the criteria are uniform. We have previously studied the use of fuzzy scorecards, for example, in the contexts of research and development project management (Collan and Luukka 2016) and intellectual property management (Collan 2013), topics that are not dissimilar from management of innovations.

As already discussed, we are using scorecards in the collection of estimates from experts. The single-number "scenario" information collected is turned into triangular fuzzy numbers. What remains to be done in the process of creating a multi-expert evaluation of innovations is the aggregation of the experts' estimates with regards to each one of the criteria used in the scorecard and the aggregation with regards to the criteria used. These aggregations can be done in a number of ways and in different order. The different ways of aggregating fuzzy numbers has already been discussed and there are many relevant methods in existence—here we concentrate our study on and make use of fuzzy scorecard (FSC) aggregation (Collan 2013; Collan and Luukka 2016), FWA aggregation, and LFWA aggregation. The aggregation of the scorecard information can be done for each criterion first (aggregation of the experts' estimates for each criterion) and for all the criteria second, or for each expert's estimates for all criteria first and for all experts second, see Fig. 1.

The choice of in which order the aggregation is made may, at a first glance, seem trivial, however, it is not. The reason for the non-triviality is that the order may have an effect on the outcome of the aggregation. This is not an issue that is relevant for all aggregation operators, but an issue

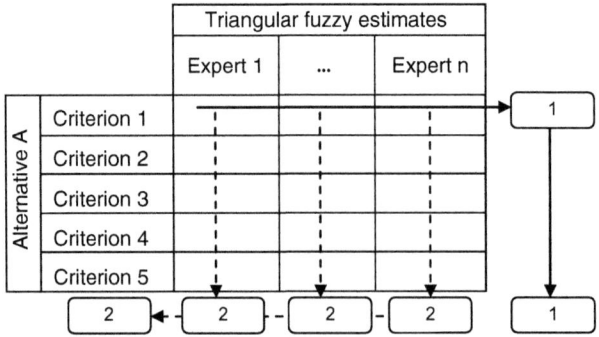

Fig. 1 Aggregating scorecard information: type 1. All estimates per criterion first, then all the criteria (solid line); type 2. All criteria per expert first, then all experts (dashed line)

that may be relevant for situations where aggregation operators are mixed, that is, when two different aggregation operators are used in the aggregation of the scorecard. In the context of managing innovations, it is important that there is clarity about what has affected the evaluation end result, and therefore it is of importance to understand the effects that the different ways of performing scorecard aggregation may have on the results. This observation that the order in which "a model is solved" may affect the end result is important and has to do with issues connected to behavioral operations research and path dependence (see e.g. Hämäläinen and Lahtinen 2016). One could even go as far as to call this a "parameter" of scorecard aggregation. These issues are interesting, but further analysis of them is left outside the focus of this chapter.

Now, using the previously presented methods, the next section of this chapter shows with a numerical illustration how innovation designs can be evaluated with scorecards, where the aggregation of information is done in a lossless way.

3   Numerical Case Illustration

In this numerical case we want to illustrate a number of things that are relevant to the multi-expert evaluation of innovation designs: the actual effect on the analysis end result of using a lossless aggregation versus not using one, and the difference that modeling choices, when using lossless aggregation, may cause.

The background case is the evaluation of three innovation designs. One innovation design is "cheap, but otherwise not so good," one is "good looking, but expensive," and one is "ugly, but cheap"—all are able to perform what is required of a minimum viable solution, but each one is different. Three experts have been asked to evaluate the future suitability for commercialization of the three innovation designs based on five criteria: "weight, durability, effect, cost, esthetics," on a scale from zero to ten (0–10). The experts were asked to rate the new designs in a way that they give not only the "best estimate" for each criterion but also the "minimum possible" estimate and a "maximum possible" estimate. The estimates given by the experts are shown in Table 1.

Table 1 Expert evaluations for the three innovation designs

		Expert 1			Expert 2			Expert 3		
		Minimum	Best estimate	Maximum	Minimum	Best estimate	Maximum	Minimum	Best estimate	Maximum
Design 1	Weight	3	4	5	4	4	6	4	6	7
	Durability	1	3	3	2	3	3	1	2	3
	Effect	4	5	5	4	5	6	4	5	5
	Cost	7	8	8	8	8	10	7	8	10
	Esthetics	4	4	5	3	4	5	3	4	4
Design 2	Weight	4	4	6	4	5	6	5	6	6
	Durability	4	5	6	5	5	6	3	4	6
	Effect	7	8	9	8	8	9	8	9	10
	Cost	2	3	4	3	3	4	1	3	4
	Esthetics	7	7	8	6	7	8	5	6	8
Design 3	Weight	3	4	5	4	4	6	4	6	7
	Durability	7	7	8	6	7	8	8	8	9
	Effect	9	10	10	8	9	10	9	10	10
	Cost	8	8	9	7	8	9	8	9	10
	Esthetics	1	2	3	2	3	3	1	2	4

As discussed, by asking for such three scenarios, one can capture the possible estimation imprecision about the future suitability of the three innovation designs—the range between the minimum and the maximum estimates is here understood (and assumed) to be a suitable yardstick to measure the imprecision, and the three scenario values are used to construct a triangular fuzzy number from each estimation. The triangular fuzzy numbers are then used in the aggregation of the scorecards. The different aggregations performed in this case are:

1. FSC + FWA: The scorecard is aggregated first by aggregating the estimates given by each expert with FSC and then aggregated expert estimates for all experts are aggregated with FWA (type 2 in Fig. 1)
2. FWA + FSC: The scorecard is aggregated first by aggregating the expert estimates for every single criterion with FWA and then these aggregated criterion evaluations are aggregated for all criteria with FSC (type 1 in Fig. 2)
3. FSC + LFWA: The scorecard is aggregated first by aggregating the estimates given by each expert with FSC and then aggregated expert estimates for all experts are aggregated with LFWA (see Fig. 2)
4. LFWA + SUM of COG: The scorecard is aggregated first by aggregating the expert estimates for every single criterion with LFWA, then the COG is calculated for each one of these aggregations and the resulting single numbers are summed up (type 1 in Fig. 1)
5. LFWA + LFWA: The scorecard is aggregated first by aggregating the expert estimates for every single criterion with LFWA and then these aggregated criterion evaluations are aggregated for all criteria with LFWA (see Fig. 2)

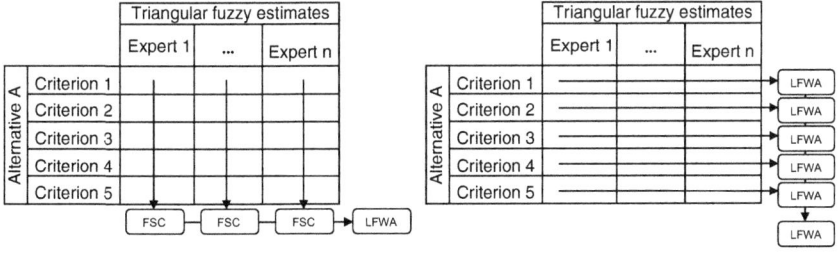

Fig. 2 The aggregation used in two of the four aggregations done for this illustration; FSC + LFWA and LFWA + LFWA

These five different aggregations include both types of aggregations described above (and in Fig. 1) and illustrate the joint use of different aggregation methods. Figure 2 shows the aggregation procedure of aggregations 3 and 5.

The results for the four first aggregations are shown in Table 2. What is reported for aggregations 1 and 2 are the resulting triangular fuzzy numbers (TFN), and what is reported for aggregation 3 are the minimum and maximum values, between which the resulting fuzzy set from the LFWA aggregation resides. It is clearly visible that the result from aggregations 1 and 2 are the same. This means that the order in which the aggregation is performed does not affect the result (Table 2).

Visual results for aggregations 1–3 are shown in Fig. 3. What can be clearly seen is that when the LFWA is used instead of FWA the results are quite different.

Figure 4 shows results from aggregation 5; the difference from aggregations 1–3 is very easy to see.

From the presented aggregation results, we calculate a single-number representation, or an expected value for each one of these results. In the multi-expert case, the expected value can also be interpreted as a consensus estimate of the expert evaluations. For the purposes of this research, we use the COG for deriving the expected value:

$$EX\left(A_c\right) = \frac{\int_{x_1}^{x_2} x A_c\left(x\right) dx}{\int_{x_1}^{x_2} A_c\left(x\right) dx}. \tag{6}$$

The resulting expected value (consensus estimate) is a representation that considers the original extremes of the expert estimates, which is an important issue in terms of the applicability of such representations in risk management (or for other purposes), where it is paramount not to exclude the extreme values. Here we look at the future suitability for commercialization of the alternatives—arguably, any loss of minimum, or especially of the maximum estimated values, is very negative from the point of view of decision-making.

Table 2 Results summary for the first four aggregations

Aggregation	Innovation 1	Innovation 2	Innovation 3
	For aggregations 1 and 2 the results are TFN		
Aggregation 1 – FSC + FWA	19.7, 24.4, 28.6	23.9, 27.75, 33.4	28.45, 32.6, 37.35
Aggregation 2 – FWA + FSC	19.7, 24.4, 28.6	23.9, 27.75, 33.4	28.45, 32.6, 37.35
	For aggregation 3 these are [min, max] values		
Aggregation 3 – FSC + LFWA	19, 30	22, 34	27, 40

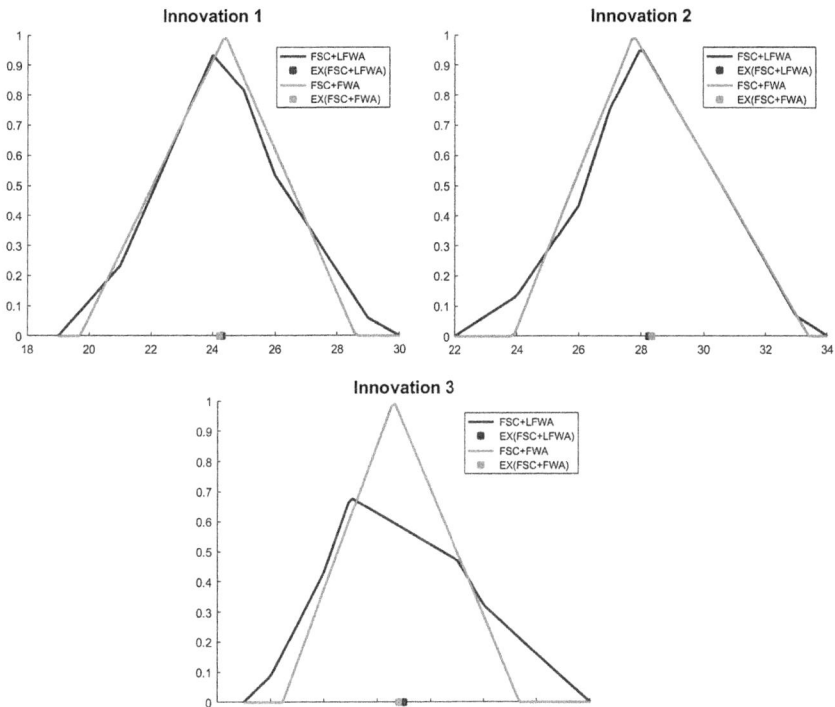

Fig. 3 Visualization of the results from aggregations 1–3 for the three innovations

In addition to the above results, we must calculate, for aggregation 4, a set of intermediary COG results for each criterion. These are added up to gain the end result from aggregation 4. The intermediate COG results are shown in Table 3.

In the context that we are looking at, we are interested in being able to put these innovation design alternatives into an order of "goodness," that

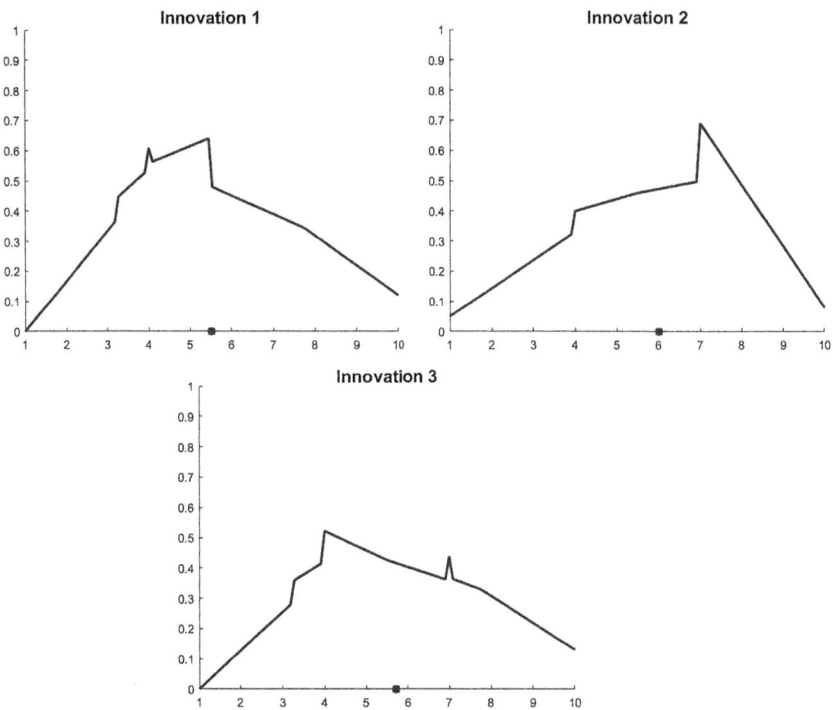

Fig. 4 Visualization of the results from aggregation 5 for the three innovation designs

Table 3 Intermediary COG results for each criterion of the three innovation designs, needed in aggregation 4

Innovation	C1	C2	C3	C4	C5
Innovation 1	5.02470	2.24800	4.83950	8.36130	3.96300
Innovation 2	5.0639	4.66830	8.55190	2.86240	6.66780
Innovation 3	5.0247	7.458900	9.327600	8.502500	2.312600

is to say, that we want to know which one is the best design for future commercialization. This means ranking the innovation designs—for the purposes of this illustration, we perform the ranking by ordering the innovation designs based on the COG of the aggregation results. By doing this, we get the orderings shown in Table 4.

Importantly, it can be noticed that the ordering is not the same for all five aggregations. In fact, in aggregation 5, where LFWA is used to first aggregate the expert opinions for each criterion and then used to aggregate the criteria for each innovation design, the order is different from the other aggregations. This illustrates the importance of the choice of aggregation operators used in aggregation. Arguably, using the combination of LFWA + LFWA is the way, where the least information is being lost in the process. One could say, or posit that the change in the ordering is caused by information loss—this is not trivial from the point of view of decision-making, because the end result that tells about the future suitability for commercialization of the alternatives, and possibly even the investment choice, is different.

4   Summary and Conclusions

This chapter has presented how multi-expert, multiple-criteria innovation evaluation can be performed in a way that captures estimation imprecision, but that is still straightforward enough to be easily done. The aggregation of the expert evaluations is the main focus of this chapter and we have shown that it is not indifferent to which aggregation methods are used and in which order aggregation is performed. The presented numerical illustration shows that rankings based on the resulting aggregations may change depending on the aggregations used; not a dramatic finding, but not a trivial finding either.

Table 4 COG results from the aggregations with ordering and scale

Aggregation	Ordering	COG results (1, 2, 3)	Scale
Aggregation 1 – FSC + FWA	3, 2, 1	32.80, 28.35, 24.23	0–50
Aggregation 2 – FWA + FSC	3, 2, 1	32.80, 28.35, 24.23	0–50
Aggregation 3 – FSC + LFWA	3, 2, 1	32.98, 28,26, 24.31	0–50
Aggregation 4 – LFWA + SUM	3, 2, 1	24.44, 27.81, 32.63	0–50
Aggregation 5 – LFWA + LFWA	2, 3, 1	5.52, 6.01, 5.70	0–10

The presented use of the LFWA method allows one to perform aggregation with minimum loss of information, which may be an important issue, when the potential and the downside of innovations are considered. This is the main reason why we believe that the presented methods are also relevant in the real-world evaluation of innovations. It is important to keep all the possible information present for as long as possible in computations, preferably until the last step.

Further research on this topic includes the study of the possible problems associated with the LFWA + LFWA approach, specifically cases where experts' estimates are far apart from each other, and the study of ranking of aggregations in such situations.

Acknowledgements This research would like to acknowledge the funding received from the Finnish Strategic Research Council, grant number 313396/MFG40 – Manufacturing 4.0.

References

Bojadziev, G., & Bojadziev, M. (2007). *Fuzzy logic for business, finance, and management* (Vol. 23). Washington, DC: World Scientific.

Bremser, W., & Barsky, N. (2004). Utilizing the balanced scorecard for R&D performance measurement. *Research Technology Management, 47*, 229–238.

Collan, M. (2013). Fuzzy or linguistic input scorecard for IPR evaluation. *Journal of Applied Operational Research, 5*, 22–29.

Collan, M., & Luukka, P. (2016). Strategic R & D project analysis: Keeping it simple and smart. In M. Collan, J. Kacprzyk, & M. Fedrizzi (Eds.), *Fuzzy technology* (Vol. 335, pp. 169–191). Heidelberg: Springer International Publishing.

Collan, M., Fedrizzi, M., & Luukka, P. (2013). A multi-expert system for ranking patents: An approach based on fuzzy pay-off distributions and a TOPSIS-AHP framework. *Expert Systems with Aplications, 40*, 4749–4759.

Collan, M., & Heikkilä, M. (2011). Enhancing patent valuation with the pay-off method. *Journal of Intellectual Property Rights, 16*(5), 377–384.

Hämäläinen, R., & Lahtinen, T. (2016). Path dependence in operational research – How the modeling process can influence the results. *Operations research perspectives, 3*, 14–20.

Kaplan, R. S., & Norton, D. P. (1992). The balanced scorecard – Measures that drive performance. *Harvard Business Review, 70,* 71–80.

Kaplan, R. S., & Norton, D. P. (1993). Putting the balanced scorecard to work. *Harvard Business Review, 71,* 134–147.

Kaufmann, M., & Gupta, M. (1985). *Introduction to fuzzy arithmetics: Theory and applications.* New York: Van Nostrand Reinhold.

Li, G., & Dalton, D. (2003). Balanced scorecard for R&D. *Pharmaceutical executive, 23*(10), 84–85.

Luukka, P., Collan, M., Tam, F., & Lawryshyn, Y. (2018). Estimating one-off operational risk events with the lossless fuzzy weighed average method. In M. Collan & J. Kacprzyk (Eds.), *Soft computing applications for group decision-making and consensus modeling* (Vol. 237, pp. 227–236). Heidelberg: Springer International Publishing AG.

Zadeh, L. A. (1965). Fuzzy sets. *Information and Control, 8,* 338–353.

Zadeh, L. A. (1978). Fuzzy sets as a basis for a theory of possibility. *Fuzzy Sets and Systems, 1,* 3–28.

Innovation Commercialisation: Processes, Tools and Implications

Mikko Pynnönen, Jukka Hallikas, and Mika Immonen

1   Introduction

Commercialisation-related activities have to simultaneously take into account the processes of generating innovations, developing products and services based on these aforementioned processes and then forming a business model around the product-service system. Therefore, it is necessary to take a multidisciplinary approach toward the commercialisation process by combining the views of technology and innovation management (Koen et al. 2001), new product development (Cho and Lee 2013), business model innovation (Foss and Saebi 2017), marketing (Aarikka-Stenroos and Lehtimäki 2014) and complex system approaches (Fleming and Sorenson 2001).

Innovation commercialisation is a process that aims to create and implement a feasible business model for an innovation-based product-service system in the surrounding business ecosystem. This view builds

M. Pynnönen (✉) • J. Hallikas • M. Immonen
School of Business and Management, Lappeenranta University of Technology, Lappeenranta, Finland
e-mail: mikko.pynnonen@lut.fi; jukka.hallikas@lut.fi; mika.immonen@lut.fi

© The Author(s) 2019   **341**
L. Chechurin, M. Collan (eds.), *Advances in Systematic Creativity*,
https://doi.org/10.1007/978-3-319-78075-7_19

on the business model innovation (BMI) framework, where the business model is seen as a value-creating subsystem in complex system of business models (Foss and Saebi 2017). BMI has four separate research streams: (1) conceptualising BMI, (2) BMI as a change process, (3) BMI as an outcome and (4) the consequences of BMI (Foss and Saebi 2017). In this chapter, we focus on the process view of creating new business models that are based on technological innovation.

There are many identified challenges in effective commercialisation. The lead time in the innovation commercialisation process has to be short enough to obtain a rapid response to the identified market gap. However, it is also essential that products, services and the entire business model be developed simultaneously. It is difficult to get different competence areas together for cooperation in the different stages of the commercialisation process. The stages of the process are seldom understood similarly between actors in the process, which makes communication difficult. Framing the stages of the commercialisation process is therefore needed to clarify the needed competences, collaboration and tools to finish different stages in the process. Most importantly, there is the need to concretely highlight different stages in an effective innovation commercialisation processes.

2   Generating Solutions and Business Models for Industrial Ecosystems

The key target of any high-technology commercialisation process is to guide the building of the initial innovation through a systematic process toward a business model based on customer needs while fitting it into the target ecosystem. To succeed in this, the process has to have several aspects. We highlight here the background that we build our process on.

It has been argued that creating customer value is the essence of a business' existence (Bowman and Ambrosini 2000). In this sense, customer value refers to what the customer wants, given certain limitations such as time and money resources. Customer value creation involves integrating the customer into the development activities at an early stage (Thomke and von Hippel 2002). To build a business model based on customer needs, it is essential to understand the customer's business goals, requirements, purchasing criteria, problems faced and so forth. In other words, it is essential to know the target customer's business so that the innovation team can help the customer better utilise the developed solution.

One key aspect of a commercialisation process is that it concentrates on integrating multiple value-adding products and services as solutions to customers' problems. Solutions are defined as systems of physical products, services and knowledge designed to fulfil customer-specific needs (Epp and Price 2011; Roehrich and Caldwell 2012). Solutions are jointly built through an evolving relationship between the providing company and the customer (Tuli et al. 2007). The development of a solution is based on the product–service system (PSS) methodology, according to which a company's product and service portfolio is mapped as a modular, interconnected system that supports the structure of an integrated solution (Dahmus et al. 1999; Geum et al. 2011; Ulrich 1995). The key principle of the system is its functional structure: the decomposition of its products and services into service functions that describe what the products and services do, not what they are (Ulrich 1995). One product or service may enable one or several service functions, and conversely, one service function could be enabled by several products and services (Geum et al. 2011). According to this logic, value is not delivered via the product or service itself, but via the solution that is based on these functions (Alonso-Rasgado et al. 2004).

To connect the solution's functional structure to a larger context, we need to extend the systemic analysis from the solution to business level. The *business model* is a particularly useful conceptual tool for this purpose in that it describes the architecture of the business system in terms of products, services and information streams (Timmers 2000) connecting the solution to the company's processes and stakeholders. From the company's perspective, *business-model innovation* requires the capability and processes to innovate and redesign the solutions (Chung et al. 2004). If the process is successful, it may even help the company redefine or refine the 'rules of the game' in the markets (Tidd and Bessant 2013). The creation of a business model also clarifies the earning logic of the company and helps to design the financial feasibility of the business.

Because solutions and business models are provider–customer-specific constructs, the industry-level approach must examine the holistic system of which the solutions are a part. To understand the complex solutions, the business network must be described on the levels of product-driven interdependencies, process-related ties and firm-to-firm interactions. In this relational view of business networks, the basic building blocks of the network are product components, production activities, resources and organisations that control activities. Because of the multilevel structure and ecosystem, the integrator of the complex solution must establish common rules, shared responsibilities and maintain fluid relationships among sequentially pooled and reciprocally connected stages of networked activities (Dubois et al. 2004; Karatzas et al. 2016).

The actors in the ecosystem can either share resource pools through the market mechanism or acquire specialised actors, depending on the costs of governing transactions where the goal is to stimulate growth or value creation potential (Holcomb and Hitt 2007). From the aspect of commercialisation, the diverse resource and capability requirements of the complex solutions are the drivers for change, which leads to business start-ups if the new actors are able to create self-reinforcing capability-building loops (M. G. Jacobides 2005). Later, however, the ecosystem tends to turn toward consolidation when commercialisation, market expansion and coordination of the platforms becomes relevant for the firms in an ecosystem (Andersson and Xiao 2016; Jacobides and Winter 2005).

Many innovation activities in the development and production stages require the capabilities and technologies of other companies. This has influenced the need to extend the traditional single organisation focus toward a value chain view of the innovation activities, which requires understanding the positioning of the company in the product or service value chain. The value chain (Porter 1985) emphasises joint processes, capabilities and efforts on finalising the customer solution. In other words, the value chain describes all the value-creating activities required for producing a customer solution in the most efficient way. This being said, it is necessary to design the flows of the activities, and it is also essential to establish collaborative relationships with the right partners and suppliers. According to Doz and Hamel (1998), alliance relationships differ from the traditional market transactions in that they can be characterised as strategic collaborative relationships with companies; they state that alliance relationships are essential in global competition because they allow the sharing of complementary resources between organisations.

In addition to the strategic partnership in the development stage of the product-service systems, the production design of the supply chain must be considered. Selection of the right suppliers for subcontracting may provide economies of scale and scope for the production. Supplier companies with high manufacturing volumes and capacity combined with flexible manufacturing systems make it unnecessary to heavily invest in production facilities at the early stage of the commercialisation process.

3   Commercialisation Experience in the Field of High-Technology Innovations

3.1   Case Description

Innovations based on high technology require hard work and a systematic approach, which if done right, still will take a long time to come to fruition. Often, the work is based on extensive research, for example, in universities. To boost the utilisation of promising new technological innovations, universities have been building different models. This is especially true of the technical universities around the world, which are more and more aiming to make an impact by attempting to commercialise their innovations. In Finland, this trend has been ongoing for a couple of years, and the models and processes are currently being built.

Universities typically have an administrative process for innovation commercialisation activities, for example, help in finding funding, patenting, legal issues and even a capital investment company to support the seed phase of spin-off companies. However, there are several challenges, including the idea itself and the innovation team needed to meet before they can approach the funders and markets related, for example, to the

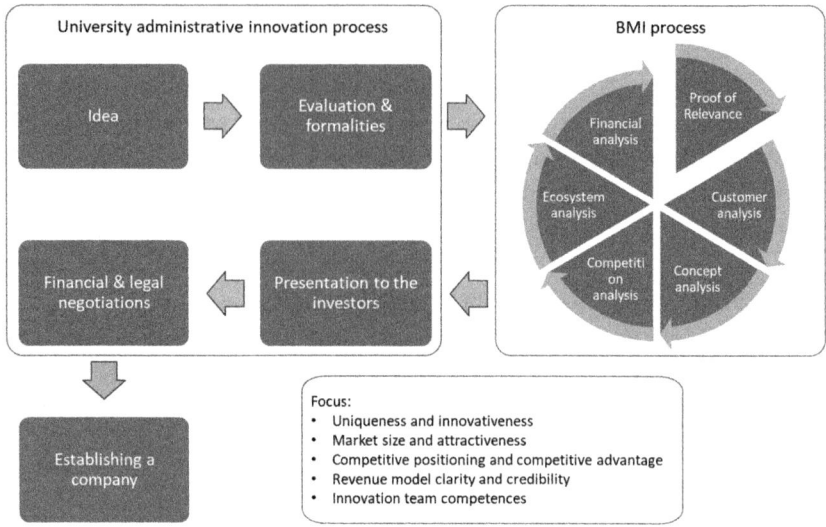

Fig. 1 The BMI process as part of the university innovation commercialisation process

feasible business model. The challenge in the process often is that the innovation team has a very strong technical background but not so much experience regarding the aspects of commercialisation.

Since 2012, we have been working in close collaboration with the university innovation process and the technical innovation teams attempting to provide a process and tools that can help teams strengthen the business aspects of their innovations and build a business model to commercialise the innovation (see Fig. 1).

During this time, we have been evaluating more than 40 concepts in the proof of relevance (PoR) phase and leading more than 10 full-scale business model development (BM development) processes for innovations that have passed the PoR phase. The duration of a PoR analysis is usually less than a week, and the full-scale BM development process runs from four to six months. The typical areas of the innovations and technologies are related to energy provision, Internet of Things, chemistry applications, industrial process improvement and others.

3.2   BMI Process Description

In the following, we describe a process that is used in our innovation commercialisation cases in the technical university context. The process has been developed and modified to serve the unique needs of innovation development groups. The process is iterative, and it has been divided into the following two parts: (1) proof of relevance and (2) BM development. The aim here is to build a business model for the idea and innovation. The general BMI process is illustrated in Fig. 1.

4   Challenges in the Process Stages and Lessons Learned

Here, we highlight some of the most common challenges related to the phases of our process. We also list the key issues that must be solved and highlight some examples of analysis tools and methods in the BMI process stages, including proof of relevance, customer analysis, concept analysis, competition analysis, ecosystem analysis and financial analysis.

4.1   Proof of Relevance

The BMI process usually starts and ends with the PoR phase. The target of the PoR phase is to find out if there is value potential for the innovation and where the potential would be the largest and most rapidly achieved. This phase often determines whether the process should continue. If there seems to be remarkable value potential within a reasonable time to market, the project can continue; otherwise, the decision should be to not continue the process. This PoR phase is often good to repeat after the BM development process to see if the chosen track is still relevant.

The major challenge in the very beginning of the commercialisation process of a technological innovation is that the technology itself can solve several problems, depending on the solution area. When seeking major business potential, the technology is usually not the solution, but rather, it is usually a piece of a solution. This means that one can plan a technology for a customer-specific problem, but outside this problem area, the options to utilise the technology are limited, and the business model created based on a single problem is usually not scalable. Bigger business possibilities lie in the solutions that the technology is a part of. By recognising the solution areas where the innovation can be a part of a solution and evaluating the value potential of the innovation in each solution area, as well as the potential market size and the time to market in those areas, the relevance of the business idea can be pointed out. For example, the innovation can be a new construct of a heat measurement sensor that provides quicker and more accurate measurements and that is smaller and more durable than the existing ones on the market. According to the innovation team, the heat sensor can be used inside marine diesel engines for measuring the energy losses of different buildings and also for measuring the energy consumption of an athlete. The solution, however, would be very different for these solution areas, and it would utilise very different features and attributes of the innovation concept.

The items to be solved are as follows:

- The problems, use cases, solution areas and so forth to which the innovation can contribute.
- What is the mechanism of the innovation's value provision in these areas?
- What is the value creation potential, market size and time to market in these areas?
- The design requirements in the innovation concept and solutions for different solution areas.
- The optimal (financially and technically) solution area and solution combination to continue with.

When mapping the contribution of the innovation, it is useful to have expertise from many different sectors. One way to obtain multiple perspectives on the possibilities of the new technology is a brainstorming workshop. These kinds of workshops are useful when organised with a structured and goal-oriented agenda that is supported with a group-decision support

(GDS) system. A GDS system could, for example, be an online platform that allows facilitated processes for simultaneous and anonymous input of ideas for a group of people (e.g., Meetingsphere, Thinktank). It also allows the grouping of the ideas and voting or rating the items generated. This kind of system enables a fast and efficient process for generating and selecting ideas for later analysis.

To illustrate the structure and results, we present a PoR workshop for a heat measurement concept that we organised with six specialists (experts in the technology itself, the heat measurement industry, applied electronics, control engineering, digital systems and innovation management). In this kind of workshop, the optimal size of the group is between five and 10 people. First, the participants identified interesting potential solution areas for the technology in a brainstorming session. Second, the participants were asked to identify how quickly the business could be established in these areas, assess how large the value potential of the innovation in a solution area would be and estimate the market size of the solution areas on a seven-point scale (with specific instructions on how to use the scale). Because of these rounds, 23 new solution areas for the technology were identified (see Table 1), three of which were selected for further study.

Table 1 Example of solution area identification

Nr	Item	Time to market Mean	SD	Value potential Mean	SD	Market size Mean	SD	Mean (row)
6	Internal combustion engine control	4.6	0.13	5	0.26	6	0.11	5.2
10	Combustion engine control systems	4.4	0.13	5.4	0.29	5.8	0.12	5.2
7	Emission minimisation control for ICE and power plants	4.6	0.17	6	0.11	5	0.35	5.2
2	Diagnostics, monitoring and control of cooling systems	5	0.15	5.2	0.19	5.2	0.16	5.13
1	Diagnostic and control of thermal machines and power plants	5	0.11	5.4	0.23	5	0.24	5.13

(continued)

Table 1 (continued)

Nr	Item	Time to market Mean	SD	Value potential Mean	SD	Market size Mean	SD	Mean (row)
11	Condition monitoring for heating systems in buildings	4.4	0.13	5.6	0.13	5.4	0.23	5.13
4	Heat exchangers	4.8	0.19	5.2	0.19	5	0.15	5
5	Boilers condition monitoring	4.8	0.27	6	0.18	4.2	0.32	5
8	Smart house energy system	4.6	0.17	5.4	0.17	4.6	0.31	4.87
14	Thermal behaviour and loss behaviour of power converters	4.4	0.29	5.6	0.2	4.6	0.34	4.87
3	Gas and steam turbines	4.8	0.19	5.2	0.12	4.4	0.27	4.8
17	Computer speed acceleration	3.4	0.17	4.8	0.32	6	0.18	4.73
12	Feedback signal in industrial chemical processes	4.4	0.17	4.6	0.17	5	0.35	4.67
9	Thermal management consulting	4.6	0.31	5.6	0.13	3.8	0.24	4.67
20	Electronics lifetime extension	3.2	0.12	4.8	0.27	5.6	0.25	4.53
13	Heat measurement in data centers	4.4	0.25	5	0.28	4	0.21	4.47
16	Measuring heat in consumer electronics	3.8	0.24	3.2	0.32	6	0.21	4.33
21	Heat management of GSM base station electronics	3.2	0.24	4.8	0.24	4.6	0.34	4.2
22	Thermal sensor for security systems	3.2	0.24	4.6	0.17	4.6	0.13	4.13
15	Precision heating equipment	3.8	0.22	4.8	0.29	3.2	0.31	3.93
23	Sports monitoring	3	0.15	3	0.28	4.6	0.31	3.53
19	Multiphase flow measurement	3.4	0.29	4.2	0.12	3	0.21	3.53
18	Robotic hand tactile/ thermo sensor	3.4	0.23	3.2	0.16	3.4	0.13	3.33
24	Body heat loss measurement	2.4	0.08	2.8	0.31	4.2	0.41	3.13
	Mean (column)	4.07	0.19	4.81	0.21	4.72	0.25	

Based on the results the further development was targeted towards the industrial applications in combustion engines and thermal control.

Third, a new round of brainstorming was conducted to identify potential customer companies and to recognise their special business areas. Usually, a requirement identification session is also implemented in the PoR phase.

4.2   Customer Analysis

The extensive BM development part of the BMI process after the PoR phase starts with a customer analysis. The target of the customer analysis phase is to find out, select and understand the potential customers for the selected solution areas. The essential part of the whole BMI process is that the innovation must be treated as a part of a solution to a problem. The more customer companies that have similar problems can be found, the bigger the market potential is. Also, to eventually create sales, it is essential to know the purchasing criteria of the customer group.

The typical challenge in this phase is that the innovation team knows that the solution can help the customers solve their problems, but they do not necessarily know how to articulate this to the customers. Also, customers usually can see the benefits in general, but it can be unclear how the solution actually works in their processes. By bringing a group of potential customer experts and the team into the same workshop, both parties can learn from each other. This helps the team understand the problems of the customers and adapt the innovation to become a solution for these problems.

The items to be solved are as follows:

- The innovation, problem and solution must be set into a certain business context.
- The most attractive customer or customer group with this problem must be identified and selected.
- The target customer's business model must be analysed to understand their business goals.

- The target customer's problems in pursuing their business goals must be analysed.
- The customer's business goals must be translated into the decision criteria they would use when rating their investment or purchasing decisions.

For example, a two-hour workshop we held to solve the problems related to this phase was done for an advanced automatic welding system that adopted neural networks to optimise the welding results. We used a three-phase GDS process and an online GDS system with a group of 18 experts from different potential customer companies. The first stage was to collect the problems the experts faced in the productivity in their current production process. The question was formulated as follows: 'What are the key questions and problems in achieving the productivity goals in production?' We obtained 15 ideas in five minutes. Then, by discussing the ideas and simultaneously categorising them, we ended up with nine key themes of problems affecting the productivity in the customers' production processes. The second stage aimed to spot the problems related to achieving quality in the production process. The question was formulated as follows: 'What are the key questions and problems in achieving the quality goals in production?' In this phase, we obtained 31 ideas in five minutes. In the discussion, we ended up with 18 key themes of problems affecting quality in the production processes. The third and final phase was to rate the problems to find the biggest obstacles the customers faced. We divided the production into three general phases to see what part of the production process the problems were affecting. The general phases were preparation, production and post-production check-up. The rating of the nine productivity-related problems and 18 quality-related problems was done with guiding statement: 'Assess the impact of a problem in the three stages of production'. The scale of the impact was from 0 (no impact) to 5 (huge impact).

To see whether the innovation concept could solve the problems of potential customer problems in production, we selected 10 major problems based on the previous phases. We also added two themes that the expert group suggested to bring into consideration at this stage of the process. The question was as follows: 'How well does the concept solve the current problems in the production?' The scale was from −2 (considerable negative impact) to 2 (considerable positive impact). This helped the team to focus to the real customer problems in their concept development work.

4.3   Concept Analysis

The target of the concept analysis phase is to construct a streamlined and efficient solution around the innovation. In the earlier phases, the innovation concept was treated as a general solution, but this phase analyses what the innovation concept actually does to solve the customer problem and how it does this. The problem we face again and again in our commercialisation cases is that the solution is defined as a bunch of technical attributes and features that are all equally important in the solution. The need for a customer, however, is how the innovation solves the problem on a detailed level, and maybe more importantly, what the innovation does not solve.

Of course, the technical attributes are important, but defining the solution as a set of functions helps the team see what is essential in the solution design and also the customers see how their problems are solved in detail. Using the sensor example, the customers would be interested in knowing how the sensor solution will solve their problem. The technical details would come into discussion in the later stages of negotiations. The functional structure in the consumer solution company case would probably lead to a result where the consumers need quicker and more accurate measurements fitted from a device that can fit into a wristband. Durability is not an issue for them. With this information, the team can proceed with the customer-relevant design and the innovation's technical attributes. The solution can be made more efficient for the purpose also by leaving out functions (e.g., extra durability).

The items to be solved are as follows:

- The functional structure of the solution must be mapped.
- The fit between the innovation concept functions and the target customer's decision criteria must be solved.

- The creation of a 'minimal functional design' of the solution based on the customer's requirements.
- A value proposition based on the functions of the solution.

To map the functional structure, we use the value flow mapping technique. The value flow map is an input–output map where the value streams are based on the products and services, describing what is being transferred, where the transaction originates from and to whom it goes (an example is shown in Fig. 2). The product or service stream often includes indirect and complementary value streams, which are revealed in the analyses of the actors and customers. The analysis also reveals new

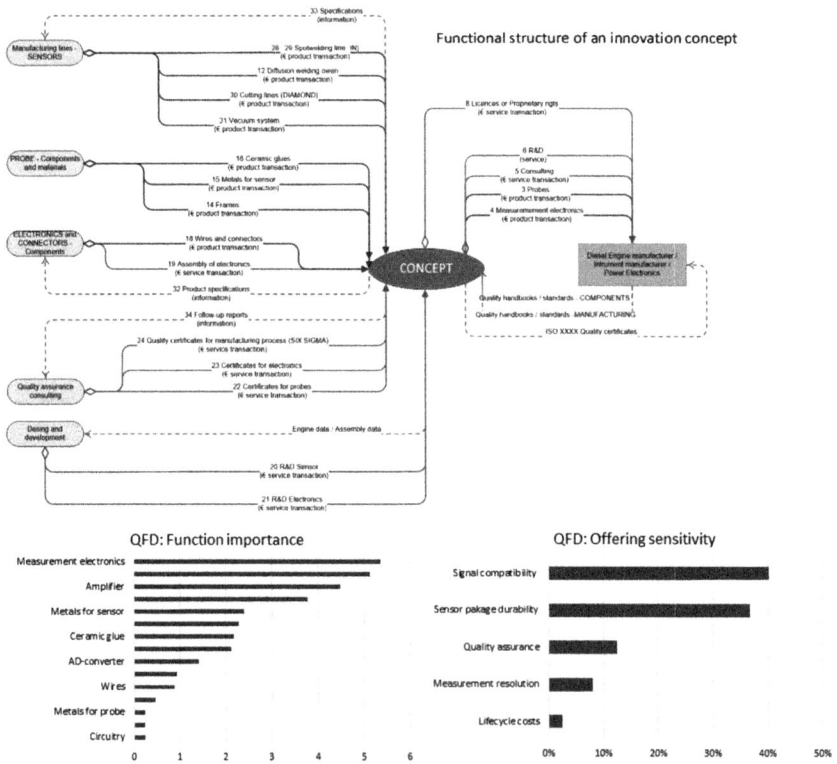

Fig. 2 Example of a functional structure and the QFD results of an innovation concept

actors, customers and value streams that are not visible until the process is underway. The solution should be designed so that it has relevant value streams that can deliver these benefits to the customer. We use the quality function deployment (QFD) matrix to assess the value of concept for customer segments. QFD is a method for converting customer demands into quality characteristics and for developing product designs by systematically deploying the relationships of customer demands and product characteristics. The prioritised customer value attributes from the customer analysis are connected to the functions of the concept by applying the QFD matrix. The analysis shows the importance order of solution elements in delivering the value for the customer. The QFD analysis also reveals the most sensitive customer requirements compared with the elements of the offering. The example of the results of a QFD analysis is presented in Fig. 2.

4.4 Competition Analysis

The target here is to find out the true competing solution for the one under development and to figure out the areas where the rival concept can be beat. The competition must be understood quite broadly in the commercialisation process. Also, the competition must be treated not only as competing concepts or companies, but also as being connected to the competitive advantage of the innovation concept.

The challenge in many cases is that although the innovation can be scientifically ground-breaking, the customer problem still might be already solved. For example, the patent analysis shows that there are no similar technologies yet patented, and initial analysis of competitors reveals no company that does exactly the same. The customer problem might have been solved using a totally different concept that did not show in the technology-based search for patenting. It might be that although the innovation in hand is technically outstanding, more common solutions are still ruling the market as the dominant design. A quite common case is that even though the customer has a problem and the solution is the dream fit to the problem, it is just not the time to invest. So the biggest competitor is not always a fancy competing technology but may just be the old technology in the form of an existing investment. For example, in the mining industry, a new filtration solution could significantly reduce energy consumption and related expenses of the sub-process. The companies, however, have low willingness to invest into process upgrades due to the very difficult situation in global economics.

The items to be solved are as follows:

- What are the real competing concepts for the solution under development?
- What is the dominant design in the target customer's industry?
- On what dimensions is the created concept better than the competing concepts, as reflected in the target customer's preferences?
- How should the competitive advantage be argued to the customer?
- How should the non-investment barriers in the target industry be overcome?

An example case of a competition analysis is based on a competitive advantage analysis of the solution. The aim here is to clarify how the developed solution is better than the competitive solutions in relation to the customer requirements. The analysis utilises the results from the customer analysis phase. The customer purchasing criteria are weighted using the analytic hierarchy process (AHP) model. The process is implemented in a workshop consisting of a group of technology and business experts in the field. In this case, first, we brainstormed the voice of the customer (VoC) model based on the purchasing criteria. The VoC consists of customer requirements that are grouped into categories. In this case the categories were: production-related requirements, product-related requirements and customer company's requirements. Each category consisted of three to four requirement attributes. After this, we first assessed the importance of the attributes from the customer's perspective by using a pairwise comparison technique to obtain a weight of the attributes of the priority tree. In the second phase, we assessed the performance of our concept and two competing solutions in the VoC attributes and in the categories, also by using a pairwise comparison technique. The analysis revealed the overall performance order of the alternative solutions and also how they perform in the unique attributes. This helped the teams to develop the solution further for the lower performing areas and to communicate the strengths better to the customers.

4.5   Ecosystem Analysis

The target of the ecosystem analysis is to position the business model to the target customer's value chain and the ecosystem around the solution. The aim is to clarify the industry's functional structure, business models, key players and financial and owner structures to position the solution into the existing structure with a business model that is designed to fit in the structure. Here, the fitting process can follow either an adaptive strategy or a disruptive strategy. The analysis focuses on the product architecture and the business ecosystems, including application areas, product concept and suppliers related to product. The aim is to recognise the opportunities for the potential start-up and the complementary technologies and services needed for the core concept.

The main challenge here is that the innovation is usually intertwined around the integrated, established technologies. For example, a high-speed generator technology that can be connected to any existing process utilising steam power (named HERGE in the Fig. 3) is in the development phase and facing this kind of a problem. The nature of the concept connects it as a part of smart grids, which have particular requirements regarding connectivity, control and demand forecasting. Therefore, the commercialisation of the innovation faces challenges where parts of the concepts are located in a diverse set of business clusters.

The items to be solved are as follows:

- The solution's position in value chain must be analysed.
- The business model's combinations must be solved.
- Identifying the supplementary business models needed to operate the main business model should be done.
- Differentiating partners and customers between business segments must be done.
- A business strategy must be built.

358   M. Pynnönen et al.

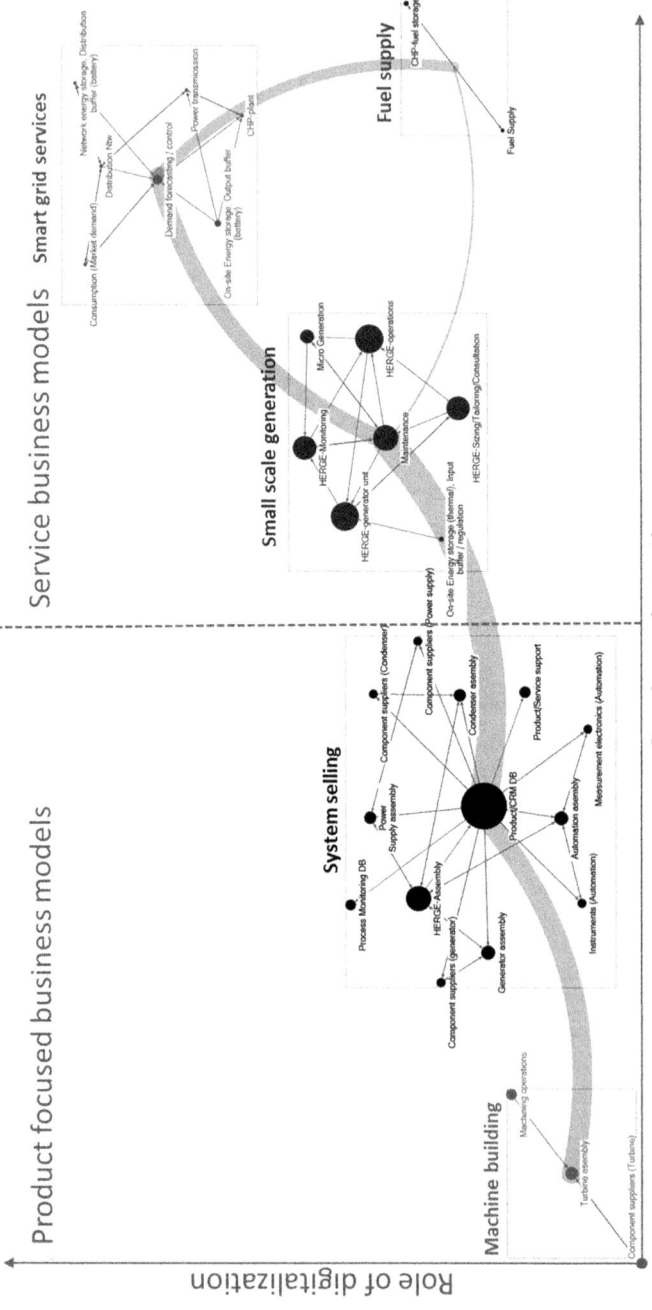

Fig. 3 Business cluster in the smart energy ecosystem

An approach to assess the ecosystem of a start-up is based on four phases of analysis: (1) activity mapping and specification, (2) interface specification between activities, (3) network modelling and (4) product strategy assessment. At first, the activity mapping and specification phase define the main structure of the activity network by the final offering and explain the prerequisites for the alternative business models. Second, the interface specification between activities handles the search operation, information, legal and co-operational linkages between recognised activities within the ecosystems. The linkages include both the direct and indirect strong relationships between activities, creating a framework for information exchange between actors and describing the phases of production activities. The third phase, network mapping analyses the activities of the network's position (by centrality metrics), clustering and visualisation of the network to deliver a comprehensive view of the ecosystem. The fourth phase, the strategy assessment, reveals alternative ways to renew structures through actor specialisation or by new business models and to manage the uncertainty related to component technology or markets within the ecosystem.

To illustrate the issues in ecosystem analysis, we present a case of a high-speed generator technology and the mapping of the ecosystem for this innovation.

The key proprietary rights and technology developed in the commercialisation project considered the core parts of the generator concept itself, which provide the premises for product-focused business models, either manufacturing or systems selling. The high-tech device manufacturing or assembly-based businesses are, however, complicated for a new entrant. Here incumbents have significantly better capabilities to invest in new products. Looking at the innovation in a broader scope there are, however, also other possibilities. The start-ups can build their businesses around digital products based on 'the energy efficiency as a service' mindset. These monitoring and production control services are located in the small-scale generation cluster of the ecosystem, which is expected to grow in the future, being driven by the megatrends in the energy industry. In this kind of a business, interfaces define the digital service platform that enable connecting the solution as part of smart grid services and provide a framework for sellers of systems to search new supplementary businesses (e.g., maintenance and product support).

Indeed, start-up companies tend to become too dependent on focal products because of the narrow portfolio. The system-selling business provides the start-up an alternative to get rid of manufacturing-specific investments, which increases flexibility for business planning and product variants. The new challenge here is a drastically increased need for effective supply management and the capability to manage product information. Compared to subsystem manufacturing, the system sellers are recognised as creating business value if they share information to supplementary service providers. The example of this analysis is highlighted in the Fig. 3.

4.6   Financial Analysis

The feasibility of the business model at the end depends on its potential to create revenue. The target of the financial analysis is to calculate and simulate the main assumptions about the income flows and costs associated with the business model. The challenge of the financial analysis is that very often, no exact historical data about the business model's financial flows is available for calculations. It is also sometimes considered challenging to include the risks and uncertainties associated with the input variables into financial calculations. The most obvious task of the financial analysis is to address the possible risks associated with the commercialisation project. The financial analysis may also trigger business model innovation. Sometimes, an original idea about the income flows of the business model changes during the commercialisation process. Therefore, it is essential to consider business innovation from the financial model point of view. In many cases, the financial analysis part has triggered novel ideas about possible revenue streams or cost-reduction opportunities. Therefore, whole teams should take part in the financial analysis process.

In a financial analysis, it is necessary to demonstrate and concretise the financial potential of the projects, for example, by providing concrete examples of the benefits of selected customer target groups (e.g., savings in EUR per year) and estimates of customer volumes. The financial analysis could also include a cash flow statement of revenue and expenditure of a commercialisation project (e.g., five years) and calculations of project present value (Net Present Value). Sometimes, it would also be useful to include product and production cost calculations for alternative production methods and material solutions.

The items to be solved are as follows:

- The revenue and cost streams of the business model must be identified.
- Best assumptions about the variable values must be judged and defined.
- Variables with uncertainty should be defined as distributions (e.g., normal, triangular, uniform)
- Calculate the financial outcome of the model by using appropriate software (e.g. @Risk, Crystall Ball).
- Interpret the result with the team and test different assumptions about the business model's financial streams.

Because exact historical data about the business model's financial flows are not often available, it is possible to use expert ratings and Monte Carlo–based simulation approaches for calculating the financial outcomes of the business model, as illustrated in Fig. 4. An advantage of the

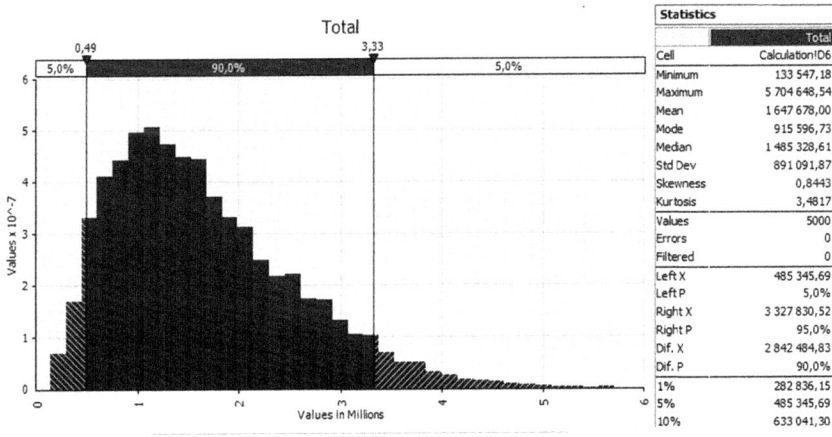

Fig. 4 Example of a Monte Carlo–based simulation for net present value of a business model

method is that input assumptions of the variables in the financial model can be defined as distributions (e.g., min, most likely max). This allows us to put inherent uncertainty into the financial model and analyse the probability of risks and opportunities. Here, an outcome of the financial potential of the model is shown as a probability distribution, as illustrated in Fig. 4. This allows us to identify the probability of the revenue generated by the business model and identify the probability of loss.

The potential for commercialisation of new business ideas needs to be evaluated at the early stages of research and development. This includes understanding the economic realities of the project, the funding required by the project, the competence of the core group and the risks of the project.

The commercialisation potential of product-service systems depends on the optimal design and management of the sales and costs factors. Sales can be connected to the market size, customer benefits, competitive positioning, pricing model and product market potential. Costs may be associated with the cost of the product structure, manufacturing cost, distribution cost, installation and maintenance cost and administrative costs.

5   Implications

We have presented a BMI process that aims to streamline the building of business models for commercialising high-technology innovations. In Table 2, we present the key challenges and key implications in BMI process phases.

Table 2   Key challenges and key implications in BMI process

BMI phase	The key challenges	The key implications
Proof of relevance	Technology itself can solve several problems, depending on the solution area, and is usually not the solution but rather is a piece of a solution	Find out the value potential for the innovation and where the potential would be the largest and most rapidly achieved
Customer analysis	The innovation team and key customers do not have a common view of the solution's possibilities	Find out, select and understand the potential customers' problems in the selected solution areas

(continued)

Table 2 (continued)

BMI phase	The key challenges	The key implications
Concept analysis	The solution is defined as a bunch of technical attributes and features that are all as important in the solution	Build a streamlined and efficient solution around the innovation that actually tells what the innovation does to solve the customer's problem and how it solves this problem
Competition analysis	The innovation can be scientifically ground-breaking and unique, but the customer's problem still might already be solved by a rival concept from another domain	Find out the true competing solution for the solution under development and figure out the areas where the rival concept can be beat
Ecosystem analysis	The innovation is usually intertwined with the integrated and established technologies and companies	Clarify the industry's functional structure, business models, key players and financial and owner structures to position the solution into the ecosystem
Financial analysis	Very often, there is no exact historical data about the business model's financial flows that are available for calculations	Calculate and simulate the main assumptions about the income flows and costs associated with the business model

6   Conclusion

This chapter highlighted the typical challenges we have noticed in the BMI process phases and the lessons learned when dealing with these challenges. We have tested the BMI process in a technical university context, putting it to the test with more than 40 ideas and 10 BMI projects. The presented process is a systematic and modularised toolset for building a business model based on an innovation, and it consists of the following six phases: proof of relevance, customer analysis, concept analysis, competition analysis, ecosystem analysis and financial analysis. We emphasised how the collaborative nature of the process can add to the reliability of the results.

The BMI process helps executives and researchers lead the commercialisation processes of high-technology innovations. We see that even though our case environment is based on university innovation commercialisation, the process is well suited for other areas of commercialising high-technology innovations. The process also helps universities streamline their commercialisation processes and make them more systematic and efficient.

References

Aarikka-Stenroos, L., & Lehtimäki, T. (2014). Commercializing a radical innovation: Probing the way to the market. *Industrial Marketing Management, 43*(8), 1372–1384. https://doi.org/10.1016/J.INDMARMAN.2014.08.004.

Alonso-Rasgado, T., Thompson, G., & Elfström, B.-O. (2004). The design of functional (total care) products. *Journal of Engineering Design, 15*(6), 515–540. https://doi.org/10.1080/09544820412331271176.

Andersson, M., & Xiao, J. (2016). Acquisitions of start-ups by incumbent businesses: A market selection process of "high-quality" entrants? *Research Policy, 45*(1), 272–290. https://doi.org/10.1016/J.RESPOL.2015.10.002.

Bowman, C., & Ambrosini, V. (2000). Value creation versus value capture: Towards a coherent definition of value in strategy. *British Journal of Management, 11*(1), 1–15. https://doi.org/10.1111/1467-8551.00147.

Cho, J., & Lee, J. (2013). Development of a new technology product evaluation model for assessing commercialization opportunities using Delphi method and fuzzy AHP approach. *Expert Systems with Applications, 40*(13), 5314–5330. Retrieved from https://www.sciencedirect.com/science/article/pii/S095741741300225X

Chung, W. W. C., Yam, A. Y. K., & Chan, M. F. S. (2004). Networked enterprise: A new business model for global sourcing. *International Journal of Production Economics, 87*(3), 267–280. https://doi.org/10.1016/S0925-5273(03)00222-6.

Dahmus, J. B., Gonzalez-Zugasti, J. P., & Otto, K. N. (1999). Modular product architecture. *Design Studies, 22*, 30–38. Retrieved from www.elsevier.com/locate/destud

Doz, Y. L., & Hamel, G. (1998). *Alliance advantage: The art of creating value through partnering*. Boston: Harvard Business School Press.

Dubois, A., Hulthén, K., & Pedersen, A.-C. (2004). Supply chains and interdependence: A theoretical analysis. *Journal of Purchasing and Supply Management, 10*(1), 3–9. https://doi.org/10.1016/J.PURSUP.2003.11.003.

Epp, A. M., & Price, L. L., (2011). Designing solutions around customer network identity goals. *Journal of Marketing, 75*(2), 36–54. https://doi.org/10.1509/jmkg.75.2.36.

Fleming, L., & Sorenson, O. (2001). Technology as a complex adaptive system: Evidence from patent data. *Research Policy, 30*(7), 1019–1039. https://doi.org/10.1016/S0048-7333(00)00135-9.

Foss, N. J., & Saebi, T. (2017). Business models and business model innovation: Between wicked and paradigmatic problems. *Long Range Planning*. https://doi.org/10.1016/j.lrp.2017.07.006.

Geum, Y., Lee, S., Kang, D., & Park, Y. (2011). Technology roadmapping for technology-based product–service integration: A case study. *Journal of Engineering and Technology Management, 28*(3), 128–146. https://doi.org/10.1016/J.JENGTECMAN.2011.03.002.

Holcomb, T. R., & Hitt, M. A. (2007). Toward a model of strategic outsourcing. *Journal of Operations Management, 25*(2), 464–481. https://doi.org/10.1016/J.JOM.2006.05.003.

Jacobides, M. G. (2005). Industry change through vertical disintegration: How and why markets emerged in mortgage banking. *Academy of Management Journal, 48*(3), 465–498. https://doi.org/10.5465/AMJ.2005.17407912.

Jacobides, M. G., & Winter, S. G. (2005). The co-evolution of capabilities and transaction costs: Explaining the institutional structure of production. *Strategic Management Journal, 26*(5), 395–413. https://doi.org/10.1002/smj.460.

Karatzas, A., Johnson, M., & Bastl, M. (2016). Relationship determinants of performance in service triads: A configurational approach. *Journal of Supply Chain Management, 52*(3), 28–47. https://doi.org/10.1111/jscm.12109.

Koen, P., Ajamian, G., Burkart, R., Clamen, A., Davidson, J., D'Amore, R., et al. (2001). Providing clarity and a common language to the "fuzzy front end". *Research-Technology Management, 44*(2), 46–55. https://doi.org/10.1080/08956308.2001.11671418.

Porter, M. E. (1985). *Competitive advantage: Creating and sustaining superior performance*. New York: Free Press.

Roehrich, J. K., & Caldwell, N. D. (2012). Delivering integrated solutions in the public sector: The unbundling paradox. *Industrial Marketing Management, 41*(6), 995–1007. https://doi.org/10.1016/J.INDMARMAN.2012.01.016.

Thomke, S., & von Hippel, E. (2002). Customers as innovators: A new way to create value. *Harvard Business Review, 80*(4), 74–81. Retrieved from http://search.ebscohost.com/login.aspx?direct=true&db=bth&AN=6413837&site=ehost-live

Tidd, J., & Bessant, J. R. (2013). *Managing innovation: Integrating technological, market and organizational change.* Chichester: Wiley.

Timmers, P. (2000). *Electronic commerce: Strategies and models for business to business trading.* Wiley. Retrieved from https://dl.acm.org/citation.cfm?id=555419

Tuli, K. R., Kohli, A. K., & Bharadwaj, S. G. (2007). Rethinking customer solutions: From product bundles to relational processes. *Journal of Marketing, 71*(3), 1–17. https://doi.org/10.1509/jmkg.71.3.1.

Ulrich, K. (1995). The role of product architecture in the manufacturing firm. *ELSEVIER Research Policy, 24*, 419–441. Retrieved from https://ac.els-cdn.com/0048733394007753/1-s2.0-0048733394007753-main.pdf?_tid=6fedb1d0-f9e2-11e7-91a6-00000aab0f6c&acdnat=1516013839_970f079eb9a4423c5bd0fa47d48c9e59

Index

© The Author(s) 2019
L. Chechurin, M. Collan (eds.), *Advances in Systematic Creativity*,
https://doi.org/10.1007/978-3-319-78075-7

367

Printed by Printforce, the Netherlands